# Medicinal and Environmental Chemistry: Experimental Advances and Simulations (Part II)

### Edited by

**Tahmeena Khan**
*Integral University*
*Department of Chemistry*
*India*

**Abdul Rahman Khan**
*Integral University*
*Department of Chemistry*
*India*

**Saman Raza**
*Isabella Thoburn College*
*Department of Chemistry*
*India*

**Iqbal Azad**
*Integral University*
*Department of Chemistry*
*India*

&

**Alfred J. Lawrence**
*Isabella Thoburn College*
*Department of Chemistry*
*India*

# Medicinal and Environmental Chemistry: Experimental Advances and Simulations *(Part II)*

Editors: Tahmeena Khan, Abdul Rahman Khan, Saman Raza, Iqbal Azad and Alfred J. Lawrence

ISBN (Online): 978-981-4998-30-7

ISBN (Print): 978-981-4998-31-4

ISBN (Paperback): 978-981-4998-32-1

©2021, Bentham Books imprint.

Published by Bentham Science Publishers Pte. Ltd. Singapore. All Rights Reserved.

# BENTHAM SCIENCE PUBLISHERS LTD.
## End User License Agreement (for non-institutional, personal use)

This is an agreement between you and Bentham Science Publishers Ltd. Please read this License Agreement carefully before using the book/echapter/ejournal (**"Work"**). Your use of the Work constitutes your agreement to the terms and conditions set forth in this License Agreement. If you do not agree to these terms and conditions then you should not use the Work.

Bentham Science Publishers agrees to grant you a non-exclusive, non-transferable limited license to use the Work subject to and in accordance with the following terms and conditions. This License Agreement is for non-library, personal use only. For a library / institutional / multi user license in respect of the Work, please contact: permission@benthamscience.net.

### Usage Rules:

1. All rights reserved: The Work is the subject of copyright and Bentham Science Publishers either owns the Work (and the copyright in it) or is licensed to distribute the Work. You shall not copy, reproduce, modify, remove, delete, augment, add to, publish, transmit, sell, resell, create derivative works from, or in any way exploit the Work or make the Work available for others to do any of the same, in any form or by any means, in whole or in part, in each case without the prior written permission of Bentham Science Publishers, unless stated otherwise in this License Agreement.
2. You may download a copy of the Work on one occasion to one personal computer (including tablet, laptop, desktop, or other such devices). You may make one back-up copy of the Work to avoid losing it.
3. The unauthorised use or distribution of copyrighted or other proprietary content is illegal and could subject you to liability for substantial money damages. You will be liable for any damage resulting from your misuse of the Work or any violation of this License Agreement, including any infringement by you of copyrights or proprietary rights.

### *Disclaimer:*

Bentham Science Publishers does not guarantee that the information in the Work is error-free, or warrant that it will meet your requirements or that access to the Work will be uninterrupted or error-free. The Work is provided "as is" without warranty of any kind, either express or implied or statutory, including, without limitation, implied warranties of merchantability and fitness for a particular purpose. The entire risk as to the results and performance of the Work is assumed by you. No responsibility is assumed by Bentham Science Publishers, its staff, editors and/or authors for any injury and/or damage to persons or property as a matter of products liability, negligence or otherwise, or from any use or operation of any methods, products instruction, advertisements or ideas contained in the Work.

### *Limitation of Liability:*

In no event will Bentham Science Publishers, its staff, editors and/or authors, be liable for any damages, including, without limitation, special, incidental and/or consequential damages and/or damages for lost data and/or profits arising out of (whether directly or indirectly) the use or inability to use the Work. The entire liability of Bentham Science Publishers shall be limited to the amount actually paid by you for the Work.

### General:

1. Any dispute or claim arising out of or in connection with this License Agreement or the Work (including non-contractual disputes or claims) will be governed by and construed in accordance with the laws of Singapore. Each party agrees that the courts of the state of Singapore shall have exclusive jurisdiction to settle any dispute or claim arising out of or in connection with this License Agreement or the Work (including non-contractual disputes or claims).
2. Your rights under this License Agreement will automatically terminate without notice and without the

need for a court order if at any point you breach any terms of this License Agreement. In no event will any delay or failure by Bentham Science Publishers in enforcing your compliance with this License Agreement constitute a waiver of any of its rights.

3. You acknowledge that you have read this License Agreement, and agree to be bound by its terms and conditions. To the extent that any other terms and conditions presented on any website of Bentham Science Publishers conflict with, or are inconsistent with, the terms and conditions set out in this License Agreement, you acknowledge that the terms and conditions set out in this License Agreement shall prevail.

**Bentham Science Publishers Pte. Ltd.**
80 Robinson Road #02-00
Singapore 068898
Singapore
Email: subscriptions@benthamscience.net

# CONTENTS

| | |
|---|---|
| FOREWORD | i |
| PREFACE | ii |
| LIST OF CONTRIBUTORS | iv |
| **CHAPTER 1  AIR POLLUTION AND ITS IMPACT ON RESPIRATORY HEALTH** | **1** |
| *Surya Kant* | |
| INTRODUCTION | 1 |
| MAJOR AIR POLLUTANTS | 3 |
| TYPES AND SOURCES OF AIR POLLUTION | 3 |
|     Primary Air Pollutants | 3 |
|     Secondary Air Pollutants | 4 |
| AIR POLLUTION CAN BE OF TWO TYPES | 4 |
| SECOND-HAND SMOKE | 4 |
| PARTICULATE MATTER (PM) | 5 |
| NATIONAL AIR QUALITY INDEX | 6 |
| OZONE ($O_3$) | 7 |
| CARBON MONOXIDE (CO) | 8 |
| NITROGEN OXIDES (NOX) | 8 |
| SULFUR DIOXIDE ($SO_2$) | 8 |
| LEAD | 8 |
| IMPACT OF AIR POLLUTION ON HEALTH | 9 |
| PREVENTION OF AIR POLLUTION | 12 |
| STRATEGIES AND SOLUTION | 13 |
| PRADHAN MANTRI UJJWALA YOJANA | 13 |
| PLANTATION | 13 |
| CONCLUDING REMARKS | 14 |
|     Suggestions of the Author to Combat Air Pollution | 14 |
| CONSENT FOR PUBLICATION | 14 |
| CONFLICT OF INTEREST | 14 |
| ACKNOWLEDGEMENT | 14 |
| REFERENCES | 14 |
| **CHAPTER 2  CYTOCHROME P450 AND HEALTH HAZARDS OF SMOG** | **18** |
| *Amber Rizvi* | |
| INTRODUCTION | 18 |
| TYPES OF SMOG | 19 |
| HEALTH HAZARDS OF SMOG | 20 |
| CYTOCHROME P450 | 20 |
| DISEASES CAUSED BY SMOG AND ITS EFFECT ON CYPS | 22 |
|     Myocardial Infarction (MI) | 22 |
|     Mechanism of Cardiotoxicity of Air Pollutants | 23 |
|     Role of CYPs in MIs | 23 |
|     Chronic Obstructive Pulmonary Disease (COPD) | 24 |
|     CYPs and COPD | 25 |
|     Atopic Dermatitis (Eczema) | 26 |
|     CYPs and Atopic Dermatitis | 27 |
|     Coughing | 27 |
|     CYPs and Coughing | 28 |
|     Experimental Work Done on CYP | 28 |

| | |
|---|---|
| CONCLUDING REMARKS | 29 |
| CONSENT FOR PUBLICATION | 30 |
| CONFLICT OF INTEREST | 30 |
| ACKNOWLEDGEMENT | 30 |
| REFERENCES | 30 |
| **CHAPTER 3 PHARMACEUTICAL AND MODELLING INTERVENTIONS FOR ENVIRONMENTAL POLLUTION RELATED CHRONIC OBSTRUCTIVE PULMONARY DISEASE** | **32** |
| *Tahmeena Khan, Alfred J. Lawrence, Iqal Azad, Shalini Dixit* and *Saman Raza* | |
| INTRODUCTION | 33 |
| INFLUENCING FACTORS FOR INITIATION AND MODULATION | 33 |
| COPD AND ITS ASSOCIATION WITH COMMON AIR POLLUTANTS: A WORLDWIDE PERSPECTIVE | 34 |
| COPD AND ITS STATUS IN INDIA | 35 |
| MECHANISTIC ACTION OF COPD- THE CHEMICAL AND BIOCHEMICAL APPROACH | 36 |
| OXIDATIVE STRESS | 36 |
| PROTEASE-ANTIPROTEASE IMBALANCE | 37 |
| ROLE OF MEDIATORS | 37 |
| DRUG AND PHARMACOLOGICAL ADVANCEMENT | 38 |
| Tissue and Systematic Inflammation | 38 |
| PHARMACEUTICAL INTERVENTIONS | 38 |
| Corticosteroids | 38 |
| Bronchodilators | 39 |
| SABA (Short-acting Inhaled Beta-agonists) | 39 |
| SABA & SAMA (Short-acting Muscarinic Antagonist) Combination Bronchodilators | 39 |
| Long-Acting Bronchodilators | 39 |
| Combination Therapies | 40 |
| Combination of Corticosteroids and Long-acting Bronchodilators | 40 |
| Triple Therapy | 41 |
| Combination of Inhaled Corticosteroids and Two Long-acting Bronchodilators (Triple Therapy) | 41 |
| Other Options | 41 |
| Drawbacks and Need for System Medicinal Approach | 41 |
| Simulated Medicinal Modeling and its Significance | 42 |
| COMPUTATIONAL MEDICINAL SIMULATION FOR COPD TREATMENT | 43 |
| Computational Lung Modelling | 43 |
| Multiscale Modelling | 44 |
| Mechanistic Models | 45 |
| Machine Learning Models | 45 |
| Statistical models | 46 |
| COPD and Advancement in Modelling | 46 |
| COPD Progression Modelling | 46 |
| Experimental vs. Computational medicinal modelling | 47 |
| LIMITATIONS AND NEED FOR FUTURE ADVANCEMENT | 48 |
| CONCLUDING REMARKS | 49 |
| CONSENT FOR PUBLICATION | 50 |
| CONFLICT OF INTEREST | 50 |
| ACKNOWLEDGEMENT | 50 |
| REFERENCES | 50 |

**CHAPTER 4 ARSENIC TOXICITY OF GROUNDWATER AND ITS REMEDIATION FOR DRINKING WATER** .................................................. 58
*Seema Joshi*
- INTRODUCTION .................................................. 58
- ESSENTIAL METALS .................................................. 59
- NON-ESSENTIAL METALS .................................................. 59
- CONCEPT OF TOXICITY .................................................. 59
  - 1. Solubility of the Metal Complexes .................................................. 60
  - 2. Oxidation State of the Metal .................................................. 60
  - 3. Ligand Attached to the Metal Atom .................................................. 60
- DETECTION OF METALS .................................................. 61
- TREATMENT FOR POISONING .................................................. 61
- ROLE OF METALS IN BIOLOGY .................................................. 61
- SOURCES OF ARSENIC .................................................. 63
- POTENTIAL FOR HUMAN EXPOSURE WITH SPECIAL EMPHASIS TO UTTAR PRADESH, INDIA .................................................. 63
  - Districts at High Risk .................................................. 64
  - District at Moderate Risk .................................................. 64
- MECHANISMS OF TOXICITY .................................................. 65
- CONSEQUENCES OF TOXICITY .................................................. 65
- REMEDIATION OF ARSENIC TOXICITY .................................................. 66
- CHELATING DRUGS USED IN THE TREATMENT OF ARSENIC POISONING .................................................. 68
  - 1. Dimercaprol (BAL) .................................................. 68
  - 2. Dimercaptosuccinic Acid (DMSA) .................................................. 69
  - 3. 2,3-Dimercapto-1-propanesulfonic Acid (DMPS) .................................................. 69
  - 4. Penicillamine .................................................. 70
- CHALLENGES OF CHELATION THERAPY .................................................. 70
- CASE STUDY FOR THE REMOVAL OF ARSENIC .................................................. 71
- CHEMICALS AND REAGENTS .................................................. 71
- PREPARATION OF FERRIC HYDROXIDE .................................................. 71
- REMOVAL OF ARSENIC FROM SPIKED TAP WATER USING FERRIC HYDROXIDE .................................................. 72
- PREPARATION OF IRON COATED CHARCOAL/SAND .................................................. 72
- EXPERIMENTAL DESIGN .................................................. 72
- CHARCOAL TREATMENT .................................................. 72
  - Iron Coated Charcoal Treatment .................................................. 73
  - Iron Coated Coarse Sand Treatment .................................................. 73
- DETERMINATION OF ARSENIC .................................................. 73
- FINDINGS OF THE STUDY .................................................. 73
- CONCLUDING REMARKS .................................................. 76
- CONSENT FOR PUBLICATION .................................................. 76
- CONFLICT OF INTEREST .................................................. 76
- ACKNOWLEDGEMENT .................................................. 76
- REFERENCES .................................................. 76

**CHAPTER 5 STUDIES ON POLYMERIC CERAMIC COMPOSITE MEMBRANES FOR WATER TREATMENT** .................................................. 82
*Fakhra Jabeen, Qazi Inamur Rahman* and *Miad Ali Siddiq*
- INTRODUCTION .................................................. 82
- WATER POLLUTION .................................................. 84
- PARAMETERS OF POLLUTION .................................................. 85
- MAIN SOURCES OF POLLUTANTS .................................................. 86

| MEMBRANES AND THEIR CLASSIFICATION | 88 |
|---|---|
| Synthetic Membrane | 88 |
| Biological Membrane | 89 |
| Organic Membranes | 89 |
| Inorganic Membrane | 89 |
| Metallic Membranes | 90 |
| Ceramic Membranes | 90 |
| Micro-Porous Memeberanes | 90 |
| Meso-Porous Membranes | 91 |
| Macro Porous Membranes | 91 |
| **APPLICATIONS OF CERAMIC MEMBRANES** | 97 |
| Chemical Industry | 97 |
| Metal Industry/Surface Engineering | 97 |
| Textiles/Pulp and Paper Industry | 98 |
| Food and Beverages | 98 |
| Recycling and the Environment | 98 |
| **POLYETHERSULFONE MEMBRANE CHARACTERISTICS AND ITS TYPES** | 100 |
| **DESALINATION FOR WATER TREATMENT** | 102 |
| **TYPES OF DESALINATION PROCESSES** | 104 |
| **ADVANTAGES AND DISADVANTAGES OF DIFFERENT DESALINATION PROCESSES** | 105 |
| **CONCLUDING REMARKS** | 110 |
| **CONSENT FOR PUBLICATION** | 111 |
| **CONFLICT OF INTEREST** | 111 |
| **ACKNOWLEDGEMENT** | 111 |
| **REFERENCES** | 111 |
| **CHAPTER 6 CHEMOSENSORS FOR ANIONS OF BIOLOGICAL AND ENVIRONMENTAL RELEVANCE** | 115 |
| *Shweta Agarwal* | |
| **INTRODUCTION** | 115 |
| Biological Significance of Anions | 116 |
| Important Techniques for Detection of Anions | 118 |
| Ion Chromatography (IC) | 118 |
| Capillary Electrophoresis (CE) | 118 |
| Chemosensors | 119 |
| Optical Chemosensors for Anions | 120 |
| *Challenges in Development of Chemosensors* | 120 |
| Sensing Mechanisms of Chemosensors | 122 |
| Binding Site-Signalling Subunit Approach | 122 |
| Displacement Approach | 123 |
| Chemodosimeter Approach | 123 |
| Optical (Colourimetric and Fluorescence) Chemosensors for Anions | 123 |
| Optical Anion Sensing by Discrete molecules | 124 |
| Hydrogen Bond Chemosensors | 124 |
| Halogen Bond Chemosensors | 126 |
| Boron Based Lewis Acid Chemosensors | 127 |
| Metal Complexes as Chemosensors | 128 |
| Anion-π Chemosensors | 130 |
| Chemosensors Based on Electrostatic Interactions | 131 |
| Chemodosimeters | 131 |

| | |
|---|---|
| Optical Sensing by Molecular Assemblies | 133 |
| **CONCLUDING REMARKS** | 134 |
| **LIST OF ABBREVIATIONS** | 134 |
| **CONSENT FOR PUBLICATION** | 135 |
| **CONFLICT OF INTEREST** | 135 |
| **ACKNOWLEDGEMENT** | 135 |
| **REFERENCES** | 135 |

## CHAPTER 7  ANTIBIOTIC POLLUTION: CHALLENGES AND STRATEGIES — 141
*Saman Raza* and *Tahmeena Khan*

| | |
|---|---|
| **INTRODUCTION** | 141 |
| **MECHANISM OF ACTION OF ANTIBIOTICS** | 144 |
| i. Inhibition of Bacterial Cell Wall Synthesis | 144 |
| ii. Inhibition of Bacterial Protein Synthesis | 145 |
| iii. Disruption of Cell Membranes | 145 |
| iv. Inhibition of Nucleic Acid Synthesis | 145 |
| v. Antimetabolite Activity | 145 |
| **USES OF ANTIBIOTICS** | 146 |
| **ANIMAL FARMING** | 146 |
| **AGRICULTURAL PURPOSES** | 146 |
| **AQUACULTURE** | 146 |
| **ANTIBIOTIC POLLUTION** | 147 |
| **EFFECTS OF ANTIBIOTIC POLLUTION** | 148 |
| **EFFECT OF ANTIBIOTIC POLLUTION ON HEALTH: ANTIBIOTIC RESISTANCE** | 149 |
| **EFFECTS OF ANTIBIOTIC POLLUTION ON THE ENVIRONMENT** | 150 |
| **STRATEGIES TO COUNTER ANTIBIOTIC POLLUTION AND RESISTANCE** | 151 |
| A. Methods for the Reduction of Antibiotic Pollution | 151 |
| *1. Removal of Antibiotic Residues from Water* | 151 |
| *2. Reduction in the Use of Antibiotics* | 152 |
| B. Methods to Counter Antibiotic Resistance | 154 |
| *1. Adjuvant Therapy* | 154 |
| *2. Development of New Antibiotics* | 158 |
| **CONCLUDING REMARKS** | 160 |
| **CONSENT FOR PUBLICATION** | 160 |
| **CONFLICT OF INTEREST** | 160 |
| **ACKNOWLEDGEMENT** | 161 |
| **REFERENCES** | 161 |

## CHAPTER 8  ANALYTICAL ADVANCEMENT FOR PHARMACEUTICALS QUANTIFICATION IN ENVIRONMENTAL MATRICES — 166
*Anushka Pandey, Manisha Bhateria* and *Sheelendra Pratap Singh*

| | |
|---|---|
| **INTRODUCTION** | 167 |
| Analytical Methods for the Determination of Pharmaceutical Residues in the Environment | 168 |
| Sample Preservation | 169 |
| *i. Filtration* | 169 |
| *ii. Non-acidic Preservative Agent* | 169 |
| *iii. Acidifying Agents* | 170 |
| Sample Preparation | 171 |
| *i. Liquid-Liquid Extraction (LLE)* | 171 |
| *ii. Dispersive Liquid-liquid Microextraction (DLLME)* | 174 |
| *iii. Solid – Phase Extraction (SPE)* | 176 |
| *iv. Solid-Phase Micro Extraction (SPME)* | 179 |

| | |
|---|---|
|     v. *Stir- bar Sorptive Extraction (SBSE)* | 181 |
|   Chromatographic Techniques for Pharmaceuticals Analysis | 183 |
|   Analysis of Pharmaceutical Compounds by Gas Chromatography (GC) | 183 |
|   Analysis of Pharmaceutical Compounds by Liquid Chromatography (LC) | 186 |
|   Methods Used for the Analysis of Pharmaceuticals in Different Environmental Analysis | 187 |
| **CONCLUDING REMARKS** | 193 |
| **CONSENT FOR PUBLICATION** | 193 |
| **CONFLICT OF INTEREST** | 193 |
| **ACKNOWLEDGEMENT** | 193 |
| **REFERENCES** | 193 |

**CHAPTER 9  USE OF BIOISOSTERIC FUNCTIONAL GROUP REPLACEMENTS OR MODIFICATIONS FOR IMPROVED ENVIRONMENTAL HEALTH** ........ 198
*Nidhi Singh* and *Jaya Pandey*

| | |
|---|---|
| **INTRODUCTION** | 198 |
| **BIOISOSTERISM - DIRECT EFFECT ON ENVIRONMENT** | 199 |
| **BIOISOSTERIC MODIFICATIONS FOR ANTHRANILIC DIAMIDES** | 200 |
| **BIOISOSTERIC MODIFICATIONS AT AROMATIC BRIDGED AMIDE FUNCTIONAL GROUP** | 200 |
| **BIOISOSTERIC MODIFICATIONS AT ALIPHATIC AMIDE FUNCTIONAL GROUPS** | 203 |
| **BIOISOSTERIC MODIFICATIONS FOR ORGANOCHLORINES** | 205 |
| **BIOISOSTERISM - INDIRECT EFFECT ON ENVIRONMENT** | 206 |
| **BIOISOSTERIC MODIFICATIONS FOR DIARYLPYRIMIDINE DERIVATIVES** | 207 |
| **BIOISOSTERIC MODIFICATIONS FOR CARBOHYDRATES** | 208 |
| **SOME IMPORTANT EXAMPLES OF BIOISOSTERIC FUNCTIONAL GROUP MODIFICATIONS FOR IMPROVED ENVIRONMENT** | 211 |
|   1. Ivacaftor | 211 |
|   2. Tetrabenazine | 211 |
|   3. JNJ-38877605 | 212 |
|   4. SCH-48461 | 212 |
|   5. Etofenprox | 213 |
|   6. Trifluoromethyl Ketone | 213 |
|   7. Pulegone | 214 |
|   8. Efavirenz | 215 |
|   9. Iloprost | 215 |
|   10. L-158809 | 216 |
| **CONCLUDING REMARKS** | 216 |
| **CONSENT FOR PUBLICATION** | 217 |
| **CONFLICT OF INTEREST** | 217 |
| **ACKNOWLEDGEMENT** | 217 |
| **REFERENCES** | 217 |

**CHAPTER 10  GOLD AND SILVER NANOPARTICLE SYNTHESIS BY PYRUS AND EURYA: ENVIRONMENT-FRIENDLY THERAPEUTIC AGENTS** ........ 220
*Dhara Shukla* and *Padma S. Vankar*

| | |
|---|---|
| **INTRODUCTION** | 221 |
| **MATERIAL AND METHODS** | 223 |
|   Material Collection | 223 |
|   Instrumentation | 223 |
|   Preparation of Bio-Extract | 224 |
| **RESULTS AND DISCUSSION** | 224 |

|  |  |
|---|---|
| Morphological Identification of Gold and Silver Nanoparticles Produced by Eurya acuminate Leaves | 226 |
| Effect of the Presence of Metal Ions on Nanoparticle formation | 228 |
| Biomedical or Therapeutic Applications Involving Gold and Silver NPs | 229 |
| **CONCLUDING REMARKS** | 231 |
| **LIST OF ABBREVIATIONS** | 232 |
| **CONSENT FOR PUBLICATION** | 233 |
| **CONFLICT OF INTEREST** | 233 |
| **ACKNOWLEDGEMENT** | 233 |
| **REFERENCES** | 233 |

## CHAPTER 11 NOVEL DRUG DEVELOPMENT STRATEGIES- A CASE STUDY WITH SARS-COV-2 ............ 238

*Iqbal Azad, Tahmeena Khan, Mohammad Irfan Azad* and *Abdul Rahman Khan*

|  |  |
|---|---|
| **INTRODUCTION** | 238 |
| **FACTORS AFFECTING THE SPREAD OF SARS-COV-2** | 239 |
| Environmental Factors | 239 |
| Food Materials, Handlers, and Packaging | 240 |
| Water and Wastewater | 240 |
| Air | 241 |
| Insects | 241 |
| Medicinal Intervention: The Scope of Virtual Screening | 242 |
| *Structure-based Virtual Screening (SBVS)* | 242 |
| *Ligand-based Virtual Screening (LBVS)* | 242 |
| In-silico Approaches | 243 |
| **LIGAND SELECTION CRITERION AS PHARMACEUTICAL LEADS** | 243 |
| **LIPINSKI'S RULE OF FIVE** | 243 |
| **GHOSE FILTER** | 244 |
| **VEBER'S RULES** | 244 |
| **MDDR-LIKE RULES** | 244 |
| **CMC LIKE RULES** | 245 |
| **WDI-LIKE RULES** | 245 |
| **BAYER FILTER** | 245 |
| **RULE OF THREE** | 245 |
| **WEIGHTED AND UNWEIGHTED QED** | 245 |
| **DRUG REPURPOSING** | 246 |
| **DRUG REPURPOSING ADVANTAGES** | 246 |
| **DRUG CANDIDATE SELECTION** | 247 |
| **DETECTION OF TARGETS FOR DRUGS AND THEIR MECHANISM OF ACTION** | 247 |
| **MOLECULAR DOCKING** | 247 |
| **TYPES OF MOLECULAR DOCKING** | 248 |
| **FLEXIBLE DOCKING** | 248 |
| **FLEXIBLE DOCKING: CHALLENGES AND REQUIREMENTS** | 248 |
| **RIGID DOCKING: CHALLENGES AND REQUIREMENTS** | 249 |
| **MOLECULAR DOCKING STUDIES OF PLANT-BASED ACTIVE CONSTITUENTS IN SEARCH OF A LEAD MOLECULE TO COMBAT SARS-COV-2** | 249 |
| Role of Immunity | 249 |
| **PLANT-BASED RESOURCES AS NATURAL IMMUNITY BOOSTERS** | 250 |
| **GINGER (ZINGIBER OFFICINALE)** | 250 |
| **GARLIC (ALLIUM SATIVUM L.)** | 252 |
| **GREEN TEA (CAMELLIA SINENSIS)** | 252 |

| | |
|---|---|
| **PURPLE CONEFLOWER (ECHINACEA)** | 253 |
| **BLACK CUMIN (NIGELLA SATIVA)** | 253 |
| **CITRUS FRUITS** | 253 |
| **MOLECULAR DOCKING STUDIES WITH SOME BIOACTIVE CONSTITUENTS OF CITRUS FRUITS** | 254 |
| **SOFTWARES USED** | 254 |
| **THE OPEN READING FRAME (ORF)** | 255 |
| **TARGET PROTEINS** | 255 |
|     Polyproteins (Proteases) | 255 |
| **SPIKE (S) PROTEIN** | 255 |
| **NUCLEOCAPSID (N) PROTEIN** | 255 |
| **ENVELOPE (E) PROTEIN** | 256 |
| **M-PROTEIN** | 256 |
| **SARS-COV HELICASE** | 256 |
| **PREPARATION OF THE RECEPTOR FOR DOCKING** | 257 |
| **PREPARATION OF LIGANDS FOR DOCKING** | 257 |
| **AUTODOCK VINA** | 258 |
| **IGEMDOCK** | 258 |
| **RESULTS AND DISCUSSION** | 259 |
| **CONCLUDING REMARKS** | 261 |
| **CONSENT FOR PUBLICATION** | 261 |
| **CONFLICT OF INTEREST** | 261 |
| **ACKNOWLEDGEMENT** | 261 |
| **REFERENCES** | 261 |
| **SUBJECT INDEX** | 268 |

# FOREWORD

In recent years, our environment has deteriorated at an alarming rate. Be it the air we breathe, the water we drink, or the food we eat—the hazards are hitting closer to home. Consequently, there has been a deluge of diseases and disorders associated with environmental pollution, industrialization, lifestyle changes, etc. From cardiovascular diseases and growth defects to neurological disorders and stress, these environmental diseases have been coupled with other environmental threats like pollution, climate change, food shortage, and novel infections and have made the study of environmental chemistry indispensable in present times. In the development of more effective and safer therapies that would cater to diseases both old and new, the study of medicinal chemistry is vital to determine accurate knowledge of drugs, their structure, synthesis, pharmacology, and pharmacokinetics.

Environmental diseases have brought about a close association between these two branches of chemistry as well as pharmaceutical chemistry. It gives me great pleasure that this book brings them together on one platform. This book aims to provide a better comprehension of environmental problems as well as remedial strategies to amend them and includes an assorted collection of topics presented by experts from academia, research, and development.

I think that the authors can be confident that readers will gain a broader perspective of the disciplines of environmental chemistry, medicinal chemistry, and pharmaceutical chemistry as a result of their efforts.

**Imran Ahmad**
Jina Pharmaceuticals Inc.
USA

# PREFACE

With the drastic disturbance in environmental harmony and balance, there has been a rise in global deaths and diseases, calling for the exploration of novel remediation strategies for innovative drug action mechanisms and target identification. The fine balance between human and ecological health is getting disturbed, leading to serious implications including the occurrence of new pathogens and diseases, including the novel corona virus SARS-CoV-2, being the most recent instance having gripped the entire globe.

Environmental diseases are non-communicable and are caused by chronic exposure to toxic pollutants. Other contributory causes of environmental diseases include radiation, pathogens, allergens, and psychological stress. Their increasing occurrence is due to industrialization, changes in farming protocols, and the increase in exposure to chemicals released into the environment. Lifestyle changes, including the increased use of tobacco and processed foods also greatly contribute to the environmental/lifestyle diseases burden.

Though medicinal chemistry and environmental chemistry have been widely explored separately, yet their close association and interdependence have been overlooked. By exploring the association between these two focal areas, the present book aims to provide solutions and curative strategies for the well-being of humans and the environment.

The twenty-one chapters included in the book are focused on diverse topics trying to blend the fields of environmental chemistry and medicinal chemistry and have been authored by expert scientists and academicians from renowned institutions. A wide range of topics has been explored in the book, to make it relevant to environmental chemists and students. The chapters have been designed to introduce environmental contaminants and techniques for their quantification and removal. Also, a medicinal perspective for remediation of environmental hazards, from therapeutic strategies available to the design of new and safer drugs, is introduced through experimental and simulation approaches.

Specialized chapters have been dedicated to persistent organic pollutants, heavy metals, antibiotics, and plastics, which have become a major source of pollution, along with their remediation. The biochemical aspect of Cytochrome $P_{450}$ and its association with mitigation strategy upon the exposure of smog on the human body, the effect of environmental xenoestrogens on human health, and the potential of natural curing agents to combat ecotoxicity have also been explored. Experimental techniques like the use of quantification methods for pharmaceuticals and persistent organic pollutants, chemosensors and polymeric ceramic composite membranes, and the concept of nanotechnology for the synthesis and use of gold and silver nanoparticles from plant-based sources have also been elaborated. To further elaborate on the importance of safe chemical practise, the concept of green chemistry has been introduced.

As we are aware that drug discovery for a particular disease is a time taking endeavour, therefore, a few chapters have also been dedicated to *in-silico* predictions like molecular docking and virtual models for biological properties, the software used and their utility to make futuristic and accurate predictions to make drug discovery efficient, quicker and cost-effective. Chapters summarizing the advances of biomolecular simulations for drug designing with respect to ecotoxicity, drug degradation, use of bioisosteric groups, and advances in pharmaceutical and modelling interventions for the treatment of COPD are also included. An interesting chapter has also explained the ligand identification for effective drug development through virtual screening by taking the example of COVID-19.

The book will prove beneficial for academicians, students of environmental chemistry and pharmacy, researchers, scientists, computational chemists, pharmacologists, environmentalists, policymakers, and postgraduate students. It would also provide researchers and medicinal chemists, information about the latest research done and the modern techniques used to develop more effective and safer drugs that would not be harmful to the environment. In this way, the proposed book would be highly beneficial to the audience it hopes to cater to.

**Tahmeena Khan**
Integral University
Department of Chemistry
India

**Abdul Rahman Khan**
Integral University
Department of Chemistry
India

**Saman Raza**
Isabella Thoburn College
Department of Chemistry
India

**Iqbal Azad**
Integral University
Department of Chemistry
India

&

**Alfred J. Lawrence**
Isabella Thoburn College
Department of Chemistry
India

# List of Contributors

| | |
|---|---|
| **Agarwal S.** | Isabella Thoburn College, Lucknow, India |
| **Ahmad I.** | Isabella Thoburn College, Lucknow, India |
| **Ahmad M.** | Zakir Husain College of Engineering and Technology, Aligarh Muslim University, Aligarh, India |
| **Alam Z.** | Shibli National PG College, Azamgarh, India |
| **Ali A.** | Zakir Husain College of Engineering and Technology, Aligarh Muslim University, Aligarh, India |
| **Ansari A.** | King George's Medical University, Lucknow, India<br>Shibli National PG College, Azamgarh, India |
| **Azad I.** | Integral University, Lucknow, India |
| **Azad M. I.** | Jamia Millia Islamia, New Delhi, India |
| **Bajpai S.** | Amity University, Lucknow, India |
| **Bhateria M.** | CSIR-Indian Institute of Toxicology Research (CSIR-IITR), Lucknow, India |
| **Bhatia S.** | Isabella Thoburn College, Lucknow, India |
| **Bhateria M.** | CSIR-Indian Institute of Toxicology Research (CSIR-IITR), Lucknow, India |
| **Bhatia S.** | Isabella Thoburn College, Lucknow, India |
| **Biswas K.** | Indian Institute of Technology Kanpur, Kanpur, India |
| **Dixit S.** | CSIR-Central Institute of Medicinal and Aromatic Plants (CSIR-CIMAP), Lucknow, India |
| **Gupta A.** | CSIR-Central Institute of Medicinal and Aromatic Plants (CSIR-CIMAP), Lucknow, India |
| **Gupta N.** | CSIR-Indian Institute of Toxicology Research, Lucknow, India |
| **Jabeen F.** | Jazan University, Jazan, Saudi Arabia |
| **Joshi S.** | Isabella Thoburn College, Lucknow, India |
| **Kant S.** | King George's Medical University, Lucknow, India |
| **Khan A. R.** | Integral University, Lucknow, India |
| **Khan M.A.** | K.K.L.K.M, Kathara, Kanpur, India |
| **Khan T.** | Integral University, Lucknow,, India |
| **Khare A.** | Indian Institute of Technology Kanpur, Kanpur, India |
| **Kumar S.** | CSIR-Indian Institute of Toxicology Research (CSIR-IITR), Lucknow, India |
| **Lawrence A. J.** | Isabella Thoburn College, Lucknow, India |
| **Mahdi A. A.** | King George's Medical University, Lucknow, India |
| **Mishra A.** | Indian Institute of Information Technology, Prayagraj, India |
| **Mishra N.** | Indian Institute of Information Technology, Prayagraj, India |
| **Mulpuru V.** | Indian Institute of Information Technology, Prayagraj, India |

| | |
|---|---|
| **Nagar P.K.** | Indian Institute of Technology Kanpur, Kanpur, India |
| **Nasibullah M.** | Integral University, Lucknow, India |
| **Pandey J.** | Amity University, Lucknow, India |
| **Patel D. K.** | CSIR-Indian Institute of Toxicology Research (CSIR-IITR), Lucknow, India |
| **Rahman Q. I.** | Integral University, Lucknow, India |
| **Raza S.** | Isabella Thoburn College, Lucknow, India |
| **Rizvi A.** | Previously at, CSIR- Central Drug Research Institute (CSIR-CDRI), Lucknow, India |
| **Sharma M.** | Indian Institute of Technology Kanpur, Kanpur, India |
| **Sharma P.** | Babasaheb Bhim Rao Ambedkar University, Lucknow, India |
| **Sharma V. P.** | CSIR-Indian Institute of Toxicology Research (CSIR-IITR), Lucknow, India |
| **Shukla D.** | S R Int, Knapur, India |
| **Siddiq M.A.,** | Jazan University, Jazan, Saudi Arabia |
| **Singh N.** | Amity University, Lucknow, India |
| **Singh S. P.** | CSIR-Indian Institute of Toxicology Research (CSIR-IITR), Lucknow, India |
| **Vankar P. S.** | Bombay Textile Research Association, Mumbai, India |
| **Verma J.** | CSIR-Indian Institute of Toxicology Research (CSIR-IITR), Lucknow, India |
| **Yadav A.** | Indian Institute of Technology Kanpur, Kanpur, India |

# CHAPTER 1

# Air Pollution and its Impact on Respiratory Health

Surya Kant[1,*]

[1] *King George's Medical University, Lucknow, India*

**Abstract:** Air pollution is a major environmental health threat due to the increasing rate of morbidity and mortality associated with it. The World Health Organization (WHO) classified particle pollution ($PM_{10}$ and $PM_{2.5}$), tropospheric ozone ($O_3$), carbon monoxide (CO), sulfur oxides ($SO_x$), nitrogen oxides ($NO_x$), and lead as six major air pollutants. Particulate matter (PM) can penetrate the respiratory system, causing respiratory and cardiovascular diseases. Stratospheric ozone plays a protective role against ultraviolet irradiation, but ozone is harmful when present in the troposphere, affecting the respiratory and cardiovascular systems. Nitrogen oxide, sulfur dioxide, carbon monoxide, and lead are harmful to humans causing respiratory problems, such as Chronic Obstructive Pulmonary Disease, asthma, bronchiolitis, lung cancer, and cardiovascular events. The only possible way to cope with this problem is through public awareness coupled with a multidisciplinary approach by scientific experts. The Government of India made the Pollution Prevention and Control Act, 1981, for the prevention of air pollution. Prime Minister Narendra Modi launched the Ujjwala scheme on 1$^{st}$ May 2016, from the Balia district in Uttar Pradesh. The scheme is aimed at replacing unclean cooking fuels. The Ministry of Environment, Forest, and Climate change has started the National Environment Health Profile (NHEP) study, involving 20 cities, to assess health effects associated with environmental exposure. The National Clean Air Programme (NCAP) has also been launched for pan-India implementation to tackle the increasing air pollution problem in the country (102 cities); the tentative national level target is 20%–30% reduction of particulate concentration by 2024.

**Keywords:** Air pollution, Asthma, Cardiovascular disease, Environment, Health, Particulate matter, Pollutants.

## INTRODUCTION

The interactions between humans and the surrounding environment have been extensively studied. The environment is an interplay of the biotic (living organisms) and the abiotic (hydrosphere, lithosphere, and atmosphere) components. Pollution is described as the addition of hazardous substances in the

---

[*] **Corresponding author Surya Kant:** King George's Medical University, Lucknow, U.P. India; E-mail: dr.kantskt@rediffmail.com

Tahmeena Khan, Abdul Rahman Khan, Saman Raza, Iqbal Azad and Alfred J. Lawrence (Eds.)
All rights reserved-© 2021 Bentham Science Publishers

environment that decreases the quality of the environment for living organisms. Human activities have the biggest adverse impact on the environment by polluting air, water, and soil. The industrial revolution has added a huge concentration of pollutants by emissions, which are harmful to human health. Globally, air pollution is considered as the major environmental health risk by the WHO [1]. Various studies have regularly revealed the detrimental effects of air pollution on human health. Air pollution leads to 7 million deaths globally due to its health hazards. In India also, 1.2 million deaths are attributed to it. The air we breathe consists of emissions from various sources like the industrial sector, automobiles, power industry, chemicals from factories, radioactive substances from nuclear power plants and household fuels along with tobacco smoke. Human lungs are the organs of respiration and are responsible for the delivery of oxygen to all the tissues. This oxygen that we breathe is given by plants and trees. Around 10,000 litres of air pass in and out through the lungs every 24 hours, and 10,000 litres of blood passes through the lungs every 24 hours; out of this 10,000 litre of air and 350 litres of oxygen is delivered every day to our body (Fig. **1**). We humans breathe 25,000 times a day. We can live without food for 3 weeks; we can live without water for 3 days, but we can live without air for only 3 minutes. That is why oxygen is called 'Pran-Vayu'.

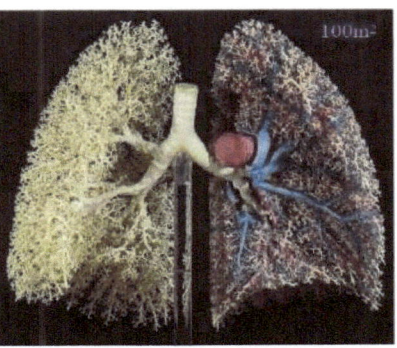

**Fig. (1).** Representation of the human lungs.

Air pollution has a huge impact on the normal morphology and functioning of the lungs. Air pollution's impacts on health have been extensively studied in recent years. Various studies show that air pollution is harmful to human health and predominantly for those who are already susceptible individuals, like children and the elderly or people having chronic health problems. The epidemiological studies suggest that harmful health effects are based on the concentrations of the pollutants and the time of exposure to them. The effect of long-term exposure is more hazardous than short-term exposure to air pollution [2, 3].

## MAJOR AIR POLLUTANTS

Air pollution has been defined as chemicals added in high concentrations to the atmosphere by natural events or human activities, enough to be harmful. Annually, various substances are released into the air from both natural sources and man-made (anthropogenic) activities. The use of fossil energy sources, growth of the manufacturing industry, and the use of chemicals result in growing air pollution [4]. Deforestation is also a major cause for the increase in air pollution; 50% of forests have been destroyed in the last 50 years in India which is leading to an imbalance in various environmental cycles and 6500 million trees are destroyed every year in our country. Smoking is also a significant contributor to air pollution.

## TYPES AND SOURCES OF AIR POLLUTION

There are two categories of air pollutants-

### Primary Air Pollutants

These are the harmful substances emitted directly into the atmosphere, for example- CO, $CO_2$, NO, $NO_2$, $SO_2$, most hydrocarbons, and most particulates.

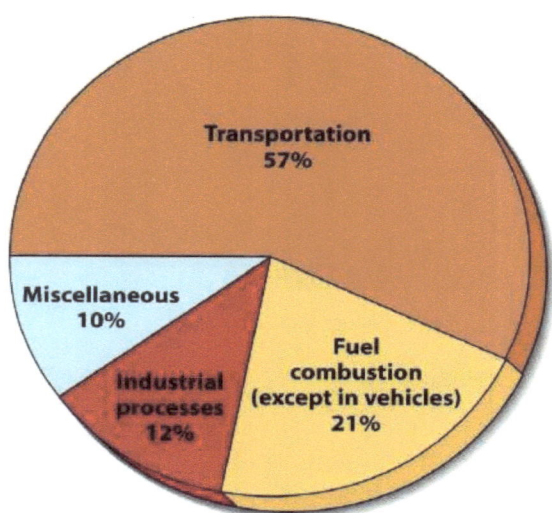

**Fig. (2).** Important sources of air pollution.

## Secondary Air Pollutants

These are the harmful substances formed in the atmosphere when a primary air pollutant reacts with substances normally found in the atmosphere or with other air pollutants. For example, $HNO_2$, $H_2SO_4$, $HNO_3$, $SO_3$, $H_2O_2$, $O_3$, most nitrates ($NO_3^-$), sulphates ($SO_4^{2-}$), and most PANs. The important sources of air pollution have been described in Fig. (2).

## AIR POLLUTION CAN BE OF TWO TYPES

-Indoor air pollution

-Outdoor air pollution

**Indoor air pollution** [5] is also a major contributor to total air pollution. There are many sources of indoor air pollution, the most important one being **biomass fuel** (Fig. 3). Around 3 billion people still do their cooking and heat their homes using solid fuels (*i.e.*, wood, crop wastes, charcoal, coal, and dung) in open fires and leaky stoves. Most are poor and live in low- and middle-income countries.

Other indoor air pollutants include radon, cigarette smoke, carbon monoxide, nitrogen dioxide, formaldehyde, pesticides, lead, cleaning solvents, ozone, and asbestos.

**Environmental tobacco smoking (ETS)** or **third-hand smoke** is a mix of chemicals that results from burning tobacco (bidi, cigarettes, cigars, and pipes) and exhaled smoke [6]. No level of exposure to ETS is safe and it can lead to serious health issues, like heart disease and lung cancer.

**Fig. (3).** (a) Active smoking, (b) Passive smoking, (c) Biomass fuel exposure.

## SECOND-HAND SMOKE

According to the Global Adult Tobacco Survey of India (2016-2017), 38.7% of adults are exposed to second-hand smoke at home. 30.2% of adults who work indoors are exposed to second-hand smoke at their workplace. 7.4% of adults are exposed to second-hand smoke at restaurants. The World Health Organization classified particle pollution ($PM_{10}$ and $PM_{2.5}$), ground-level ozone ($O_3$), carbon monoxide (CO), sulfur oxides ($SO_x$), nitrogen oxides ($NO_x$), and lead as six major air pollutants. Air pollution can harm the environment and poses a severe threat to living organisms. So, our interest is mainly to emphasize these pollutants as they are identified with wide-ranging and severe impacts on human health and the environment. Air pollution also has an important ecological impact on acid rain, global warming, the greenhouse gas effect, and climate change [1].

## PARTICULATE MATTER (PM)

Recent studies have indicated an association between PM and harmful health effects, focusing on either short-term (acute) or long-term (chronic) PM exposure. Chemical reactions between various types of pollutants in the atmosphere form particulate matter. The infiltration power of particles is mainly dependent on their size [6]. PM pollution includes particles with diameters of 10 micrometres (μm) or less, denoted as $PM_{10}$, and extremely fine particles with diameters of 2.5 micrometres (μm) and less, denoted as $PM_{2.5}$. Due to the small size, it can be inhaled and causes serious respiratory ill effects [7]. After inhalation, $PM_{10}$ can invade the lungs and even infiltrate the bloodstream. PM is divided into four main categories according to type and size (Table 1) [9].

Table 1. Types and sizes of particulate matter (PM).

| | Type | PM Diameter [μm] |
|---|---|---|
| Particulate contaminants | Smog | 0.01–1 |
| | Soot | 0.01–0.8 |
| | Tobacco smoke | 0.01–1 |
| | Fly ash | 1–100 |
| | Cement Dust | 8–100 |
| Biological Contaminants | Bacteria and bacterial spores | 0.7–10 |
| | Viruses | 0.01–1 |
| | Fungi and moulds | 2–12 |
| | Allergens (dog and cat hair, pollen, household dust) | 0.1–100 |
| Types of Dust | Atmospheric dust | 0.01–1 |
| | Heavy dust | 100–1000 |
| | Settling dust | 1–100 |
| Gases | Different gaseous Contaminants | 0.0001–0.01 |

The PM produces harmful effects related to its chemical and physical properties. In nature, $PM_{10}$ and $PM_{2.5}$ substances can be organic (polycyclic aromatic hydrocarbons, dioxins, benzene, 1-3 butadiene) or inorganic (carbon, chlorides, nitrates, sulfates, metals) [7].

Extremely fine particles, $PM_{2.5}$, pose a greater health risk due to greater penetration power as they can penetrate even the alveolar cells (Table **2**) [8].

Table 2. Penetrability of PMs according to particle size.

| Particle Size (µm) | Penetration Degree in the Human Respiratory System |
|---|---|
| >11 | Passage into nostrils and upper respiratory tract |
| 7–11 | Passage into the nasal cavity |
| 4.7–7 | Passage into larynx |
| 3.3–4.7 | Passage into the trachea-bronchial area |
| 2.1–3.3 | Secondary bronchial area passage |
| 1.1–2.1 | Terminal bronchial area passage |
| 0.65–1.1 | Bronchioles penetrability |
| 0.43–0.65 | Alveolar penetrability |

Half-lives of $PM_{10}$ and $PM_{2.5}$ particles in the environment are extended due to their smaller size; this allows their long-lasting suspension in the environment and even their transfer and coverage to distant destinations, where people and the environment may be exposed to the same level of pollution [6]. As stated, $PM_{2.5}$, because of its small size, causes more severe health effects. These above-mentioned fine particles are the main source of the 'haze' formation in various metropolitan cities [9].

## NATIONAL AIR QUALITY INDEX

AQI is an overall scheme that transforms individual air pollutant (*e.g.,* $SO_2$, CO, $PM_{10}$) levels into a single number, which is a simple and lucid description of air quality for the citizens. AQI relates to health impacts and helps citizens to avoid unnecessary exposure to air pollutants. AQI indicates compliance with National Air Quality Standards. It prompts local authorities to take quick actions to improve air quality (Fig. **4**). AQI guides policymakers to take broad decisions and encourages citizens to participate in air quality management.

Table 3. AQI range for different pollutants (Central Pollution Control Board).

| AQI Category (Range) | $PM_{10}$ (24 h) | $PM_{2.5}$ (24 h) | $NO_2$ (24 h) | $O_3$ (24 h) | CO (8 h) (mg/m³) | $SO_2$ (24 h) | $NH_3$ (24 h) | Pb (24 h) |
|---|---|---|---|---|---|---|---|---|
| Good (0-50) | 0-50 | 0-30 | 0-40 | 0-50 | 0-1.0 | 0-40 | 0-200 | 0-0.5 |
| Satisfactory (51-100) | 51-100 | 31-60 | 41-80 | 51-100 | 1.1-2.0 | 41-80 | 201-400 | 0.6-1.0 |
| Moderate (101-200) | 101-250 | 61-90 | 81-180 | 101-168 | 2.1-10 | 81-380 | 401-800 | 1.1-2.0 |
| Poor (201-300) | 251-350 | 91-120 | 181-280 | 169-208 | 10.1-17 | 381-800 | 801-1200 | 2.1-3.0 |
| Very poor (301-400) | 351-430 | 121-250 | 281-400 | 209-748 | 17.1-34 | 801-1600 | 1201-1800 | 3.1-3.5 |
| Severe (401-500) | 430+ | 250+ | 400+ | 748+ | 34+ | 1600+ | 1800+ | 3.5+ |

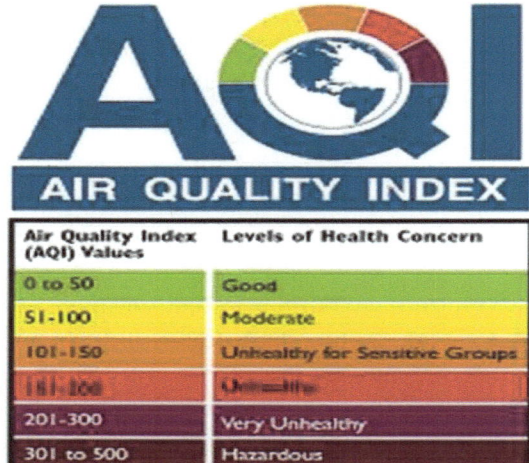

**Fig. (4).** AQI Indicators (Central Pollution Control Board).

# OZONE ($O_3$)

$O_3$ is a gaseous particle derived from oxygen, under the influence of high voltage electric discharge in the atmosphere [10]. Generally, it is formed in the stratosphere, but it could also form in the troposphere by chain reactions of photochemical smog [11]. Ozone moves with air and can travel to distant areas from the source [12]. Amazingly, ozone levels over cities are low in comparison to the increased amounts occurring in urban areas, and it could become detrimental for forests, and vegetation by reducing carbon assimilation [13, 14]. Ozone affects the growth and yield of crops [15, 16]. Ozone acts as a strong oxidizer and increases DNA damage in skin cells and leads to compromised cellular function [17]. Harmful effects of ozone are documented in urban areas all over the world, causing biochemical, morphological, functional, and immunological diseases [18].

## CARBON MONOXIDE (CO)

Carbon monoxide is formed by the incomplete combustion of fossil fuels. The inhalation of CO affects health, causing headache, dizziness, weakness, nausea, vomiting, and, finally, loss of consciousness. The binding affinity of CO to haemoglobin is 250 times greater than that of oxygen. When people are exposed to CO for a long time, it results in hypoxia, ischemia, and cardiovascular diseases due to the loss of oxygen by competitive inhibition of CO. Increasing level of CO affects the greenhouse gases equilibrium of the environment and is also responsible for global warming and climate change [19].

## NITROGEN OXIDES ($NO_x$)

Nitrogen oxides are transportation-related pollutants, mainly discharged from automobile combustion engines [20]. $NO_x$ acts as an irritant and penetrates deep in the lung, causing respiratory diseases, coughing, wheezing, dyspnea, bronchospasm, and even pulmonary oedema when inhaled at high concentration. Besides, $NO_x$ can damage the quality of fabrics by fading the colour [21].

## SULFUR DIOXIDE ($SO_2$)

$SO_2$ is a harmful gas, emitted mainly from the burning of fossil fuel or industrial activities. $SO_2$ emissions by industrial activities are responsible for respiratory inflammation, bronchitis, mucus production, and bronchospasm. Also, $SO_2$ damages the skin and eyes (lacrimation and corneal opacity) and mucous membranes by irritation. Susceptible people, such as those with lung disease, older people, and children, are more vulnerable to long-term exposure to $SO_2$ [21]. $SO_2$ also damages the environment by acidification of soil and causes acid rain [22].

## LEAD

The heavy metal, lead, is used in various industrial sectors and is emitted from petrol engines used in the automobile sector, batteries, radiators, waste incinerators, and wastewaters. Lead exposure can occur through inhalation, ingestion, and dermal absorption. After inhalation, it accumulates in the blood, soft tissue, liver, lungs, bones, and the cardiovascular, nervous, and reproductive systems [23]. It acts as a neurotoxin and causes mental disabilities, impedance of memory, and hyperactivity [24]. An increased concentration of lead in the atmosphere is harmful to plant and crop yield [25].

## IMPACT OF AIR POLLUTION ON HEALTH

Air pollution has a detrimental effect on the health of people leading to various disabilities and eventually to death. Many respiratory problems and other health problems are coming up these days due to increasing levels of air pollution. In India, 12.4 lakh people die every year due to air pollution [26].

People living in a polluted atmosphere face side effects of air pollution and experience disease symptoms. These effects are caused by both short-term and long-term exposure, affecting human health. It is necessary to make people aware of the harmful effects of air pollution, especially those who are vulnerable, such as older people, children, and people with diabetes and predisposing heart or lung disease, especially asthma.

Air Quality Index is also related to the health impacts on individuals. Cities with poor AQI will have serious health impacts as compared to cities having good AQI (Table **4-5**).

**Table 4. Health Impacts.**

| Indicator | Impact |
|---|---|
| Good | Minimal impact |
| Satisfactory | Minor breathing discomfort to sensitive people |
| Moderate | Breathing discomfort for the people with lung disease, such as asthma, and discomfort to people with heart disease, children, and older adults |
| Poor | Breathing discomfort to people on prolonged exposure and discomfort to people with heart disease, with short exposure |
| Very Poor | Respiratory illness to the people on prolonged exposure. The effect may be more pronounced in people with lung and heart disease |
| Severe | Respiratory effects even on healthy people and serious health impacts on people with lung/heart disease |

**Table 5. Various health problems due to air pollution.**

| S.No. | Symptom |
|---|---|
| 1 | Premature Death |
| 2 | Birth Defects |
| 3 | Respiratory Infections |
| 4 | Asthma |
| 5 | Emphysema |
| 6 | Lung Cancer |
| 7 | Hypertension |

(Table 5) cont.....

| S.No. | Symptom |
|---|---|
| 8 | Strokes |
| 9 | Chronic Headache |
| 10 | Allergic Disorders |
| 11 | Heart attacks |
| 12 | Impaired fertility |

Effects of short-term exposure to air pollutants are temporary and include uneasiness, irritation of the eyes, nose, skin, throat, wheezing, coughing and chest tightness, and breathing difficulties. Short-term exposure can be serious for patients having asthma, pneumonia, bronchitis, and lung and heart problems. It also causes headaches, nausea, and dizziness. Moreover, long-term exposure to air pollutants causes severe problems, being harmful to the neurological, reproductive, and respiratory systems, and causes cancer and even death. Air pollution's impact on health is also dependent on the geography of the nation, area, climatic conditions, and time. PMs, dust, benzene, and $O_3$ cause severe injury to the respiratory system [26]. Furthermore, there is an aggravated risk in case of a preexisting respiratory disease, like asthma [27].

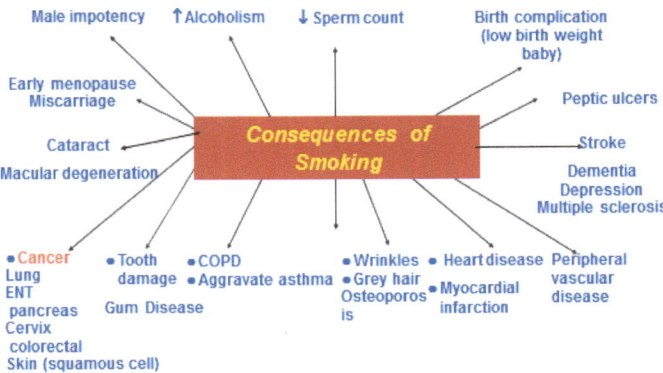

**Fig. (5).** Hazards of Smoking.

Various studies have found that exposure to ambient air pollutants is associated with decreased lung function and increased airway sensitivity in asthmatic children. The level of $O_3$ is directly related to the exacerbation and development of asthma [28, 29].

Chronic Obstructive Pulmonary Disease (COPD) is most commonly associated with tobacco smoking; however, various studies suggest that it may be induced by air pollution and increases the rate of morbidity and mortality in developing and developed countries. The prevalence of COPD in non-smokers ranges from 1.1 to 40% in different countries associated with traffic-related pollution, indoor air pollution by fossil fuel combustion, and secondary smoking [30, 31]. It is documented that improving indoor air quality in households decreases the frequency of COPD [32].

It is believed that tobacco smoking is the most important cause of lung cancer, however, recent studies demonstrate that tobacco smoking is not the only reason for lung cancer [33]; it may also occur due to long-term exposure to air pollutants [34, 35] (Fig. **6**). The International Agency for Research on Cancer announced outdoor air pollutants and related PM as class I human carcinogens, based on scientific data generated from humans, animal models, and mechanistic studies [36]. Children, the elderly and people with chronic diseases are most affected by respiratory infection through air pollution [37, 38].

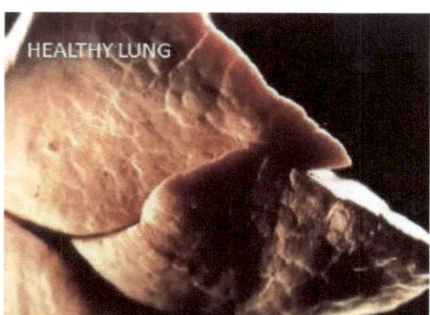 

**Fig. (6).** (a) Healthy and (b) Polluted Lung.

Exposure to air pollutants results in various cardiovascular effects. Long-term exposure to air pollutants modifies the blood cells and affects the cardiac system, causing defects like coronary arteriosclerosis [39]. Short-term exposure results in hypertension, stroke, myocardial infarction, and heart insufficiency [40]. Long-term exposure to air pollutants affects the nervous system of adults and children, causing psychological complications, autism, retinopathy, and low birth weight [22] (Fig. **7**). It is also seen that long-term exposure to air pollutants causes premature death, impaired fertility, and birth defects.

## PREVENTION OF AIR POLLUTION

In 2018, WHO organized a Global Conference on Air Pollution and Health and declared to achieve a goal of reducing two-thirds of deaths from air pollution by 2030. An effective solution could be envisaged for the regulation of increasing air pollution by the collaboration of authorities, environmental regulatory bodies, and doctors.

United Nations (UN) has time to time organized meetings and set a goal for the reduction of environmental pollution, to cope with global climate change, such as The Kyoto Protocol in 1997, Copenhagen summit of 2009, the Durban summit of 2011, and recently the Paris Agreement of 2015. Governments should focus on education, training, public awareness, and public participation for the awareness of climate change and environmental pollution [41].

The government of India (GOI) made Pollution Prevention and Control Act, 1981, for the prevention of air pollution, also called the 'Air Act'. According to the World Health Organization, 9 of the world's 10 most polluted cities are from India (Kanpur, Faridabad, Gaya, Varanasi, Patna, Delhi, Lucknow, Agra, and Gurgaon), among which Kanpur is the most polluted city ($PM_{2.5}$-173 mg/mm$^3$). GOI initiated a multicenter study titled 'National Environmental Health Profile study' in the 20 most polluted cities of India, in 2018; the main objective of the study is to find out the level of air pollution and its harmful effects on human health [42].

The National Clean Air Programme (NCAP), launched on 10$^{th}$ January 2019, by GOI, is a time-bound (five-year action plan) national strategy for pan India implementation, to tackle the increasing air pollution problem in the country (102 cities). Its tentative national level target is a 20%-30% reduction of $PM_{2.5}$ and $PM_{10}$ concentration by 2024 [43].

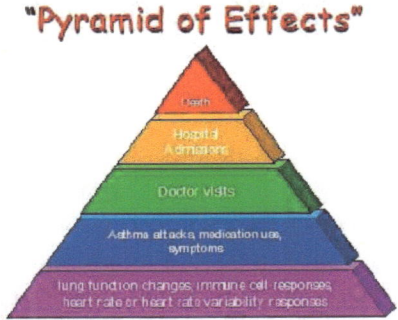

**Fig. (7).** Pyramid of Effects.

## STRATEGIES AND SOLUTION

The mainstay of tackling the menace of air pollution is the prevention, reduction, and management at the source of generation of the air pollutants. For example, the development of renewable energy in buildings, such as establishing solar lights, will help reduce air pollution due to biomass burning. Various awareness programs should be held to motivate the masses to reduce the emission of harmful air pollutants. Strict laws should be framed, and their proper enactment should be ensured. The health effects of air pollution, especially diseases like COPD, asthma, and Occupational Related Lung Disease must be treated accordingly [44].

## PRADHAN MANTRI UJJWALA YOJANA

Prime Minister Narendra Modi's Ujjwala scheme, which gives LPG connections at affordable rates to the women living below the poverty line, was launched on $1^{st}$ May 2016, from Balia in Uttar Pradesh. The scheme is aimed at replacing the unclean cooking fuels mostly used in rural India with the clean and more efficient LPG (Liquefied Petroleum Gas). This campaign (Fig. **8**) has benefitted 8 crores, poor families of India, to date [45].

**Fig. (8).** Pradhan Mantri Ujjwala Yojana.

## PLANTATION

Restrictions on deforestation and plantation of more and more plants, minimized urbanization processes, use of latest technology in industries, minimum use of vehicles and promotion of CNG vehicles, promoting solar energy technology, promoting LPG use in the rural areas, are some activities which can minimize air pollution. The Uttar Pradesh government planted 22 crore plants in 2019 to increase the forest cover of the state and it will also help to reduce the harmful effects of increasing air pollution [46].

## CONCLUDING REMARKS

The author has already started a campaign many years ago, of giving saplings in place of bouquets, on any auspicious occasion like birthday, anniversary, *etc*.

### Suggestions of the Author to Combat Air Pollution

1. Presenting plant saplings in the place of bouquets.
2. Making plantation a part of every ceremony, like birthday, wedding, anniversary, engagement, *etc*.
3. Increasing use of public transport.
4. Not smoking and becoming an advocate of the no-smoking campaign.
5. Taking the advantage of Ujjwala Yojana.
6. Using the mode of walking and cycling for transportation.
7. Taking responsibility for our actions by asking the question, 'Am I responsible for Air Pollution?' at the end of every day.

## CONSENT FOR PUBLICATION

Not Applicable.

## CONFLICT OF INTEREST

The author confirms that this chapter contents have no conflict of interest.

## ACKNOWLEDGEMENT

Declared none.

## REFERENCES

[1] Smith KR. Biofuels, air pollution, and health: a global review. Springer Science & Business Media 2013.

[2] Robinson DL. Air pollution in Australia: review of costs, sources and potential solutions. Health Promot J Austr 2005; 16(3): 213-20.
[http://dx.doi.org/10.1071/HE05213] [PMID: 16375037]

[3] Ghorani-Azam A, Riahi-Zanjani B, Balali-Mood M. Effects of air pollution on human health and practical measures for prevention in Iran. J Res Med Sci 2016; 21: 65.
[http://dx.doi.org/10.4103/1735-1995.189646] [PMID: 27904610]

[4] Domingo JL, Rovira J. Effects of air pollutants on the transmission and severity of respiratory viral infections. Environ Res 2020; 187109650
[http://dx.doi.org/10.1016/j.envres.2020.109650] [PMID: 32416357]

[5] Madhurmay; Suryakant; Kumar, H.; Kumar, S.; Prasad, R.; Verma, A.K.; Singh, A.K. Study of association between exposure to indoor air pollution and chronic obstructive pulmonary disease among non-smokers in a North Indian population – A case–control study. Indian J Respir Care 2019; 8: 71-5.

[6] Wilson WE, Suh HH. Fine particles and coarse particles: concentration relationships relevant to epidemiologic studies. J Air Waste Manag Assoc 1997; 47(12): 1238-49.

[http://dx.doi.org/10.1080/10473289.1997.10464074] [PMID: 9448515]

[7]  Cheung K, Daher N, Kam W, *et al.* Spatial and temporal variation of chemical composition and mass closure of ambient coarse particulate matter (PM10–2.5) in the Los Angeles area. Atmos Environ 2011; 45: 2651-62.
[http://dx.doi.org/10.1016/j.atmosenv.2011.02.066]

[8]  Zhang L, Yang Y, Li Y, *et al.* Short-term and long-term effects of $PM_{2.5}$ on acute nasopharyngitis in 10 communities of Guangdong, China. Sci Total Environ 2019; 688: 136-42.
[http://dx.doi.org/10.1016/j.scitotenv.2019.05.470] [PMID: 31229811]

[9]  Heal MR, Kumar P, Harrison RM. Particles, air quality, policy and health. Chem Soc Rev 2012; 41(19): 6606-30.
[http://dx.doi.org/10.1039/c2cs35076a] [PMID: 22660420]

[10] Bezirtzoglou E, Alexopoulos A. Ozone history and ecosystems: a goliath from impacts to advance industrial benefits and interests, to environmental and therapeutical strategies.Ozone Depletion, Chemistry and Impacts. New York: Nova Science Publishers 2009; pp. 135-45.

[11] Villányi V. Turk. B.; Franc, B.; Csintalan, Z. Ozone Pollution and its Bioindication.Air Pollution. London: Intech Open 2010.
[http://dx.doi.org/10.5772/10047]

[12] Bruce N, Perez-Padilla R, Albalak R. Indoor air pollution in developing countries: a major environmental and public health challenge. Bull World Health Organ 2000; 78(9): 1078-92.
[PMID: 11019457]

[13] Lorenzini G, Saitanis C. Ozone: A Novel Plant "Pathogen.".Abiotic Stresses in Plant. Springer Link 2003; pp. 205-29.
[http://dx.doi.org/10.1007/978-94-017-0255-3_8]

[14] Fares S, Vargas R, Detto M, *et al.* Tropospheric ozone reduces carbon assimilation in trees: estimates from analysis of continuous flux measurements. Glob Change Biol 2013; 19(8): 2427-43.
[http://dx.doi.org/10.1111/gcb.12222] [PMID: 23589473]

[15] Watson JT, Gayer M, Connolly MA. Epidemics after natural disasters. Emerg Infect Dis 2007; 13(1): 1-5.
[http://dx.doi.org/10.3201/eid1301.060779] [PMID: 17370508]

[16] McCarthy JT, Pelle E, Dong K, Brahmbhatt K, Yarosh D, Pernodet N. Effects of ozone in normal human epidermal keratinocytes. Exp Dermatol 2013; 22(5): 360-1.
[http://dx.doi.org/10.1111/exd.12125] [PMID: 23614745]

[17] Lippmann M. Health effects of ozone. A critical review. JAPCA 1989; 39(5): 672-95.
[http://dx.doi.org/10.1080/08940630.1989.10466554] [PMID: 2659744]

[18] Emberson LD, Pleijel H, Ainsworth EA, *et al.* Ozone effects on crops and consideration in crop models. Eur J Agron 2018; 100: 19-34.
[http://dx.doi.org/10.1016/j.eja.2018.06.002]

[19] Kant S, Srivastava K. Comparison of biochemical parameters among the confirmed TB patients using Biofuels with the healthy controls among north Indian population: A case-control study. J Int Acad Res Multidisc 2013; 1: 39-46.

[20] Chen T-M, Gokhale J, Shofer S, Kuschner WG. Outdoor air pollution: nitrogen dioxide, sulfur dioxide, and carbon monoxide health effects. Am J Med Sci 2007; 333(4): 249-56.
[http://dx.doi.org/10.1097/MAJ.0b013e31803b900f] [PMID: 17435420]

[21] Spengler JD, Ferris BG, Dockery DW, Speizer FE. Sulfur dioxide and nitrogen dioxide levels inside and outside homes and the implications on health effects research. Environ Sci Technol 1979; 13: 1276-80.
[http://dx.doi.org/10.1021/es60158a013]

[22] Farhat A, Mohammadzadeh A, Balali-Mood M, Aghajanpoor-Pasha M, Ravanshad Y. Correlation of blood lead level in mothers and exclusively breastfed infants: a study on infants aged less than six months. Asia Pac J Med Toxicol 2013; 2: 150-2.

[23] Assi MA, Hezmee MNM, Haron AW, Sabri MYM, Rajion MA. The detrimental effects of lead on human and animal health. Vet World 2016; 9(6): 660-71.
[http://dx.doi.org/10.14202/vetworld.2016.660-671] [PMID: 27397992]

[24] Balakrishnan K, Dey S, Gupta T, Dhaliwal RS, Kant R, Kant S. India State-Level Disease Burden Initiative Air Pollution Collaborators. The impact of air pollution on deaths, disease burden, and life expectancy across the states of India: the Global Burden of Disease Study 2017. Lancet Planet Health 2019; 3(1): e26-39.
[http://dx.doi.org/10.1016/S2542-5196(18)30261-4] [PMID: 30528905]

[25] Kurt OK, Zhang J, Pinkerton KE. Pulmonary health effects of air pollution. Curr Opin Pulm Med 2016; 22(2): 138-43.
[http://dx.doi.org/10.1097/MCP.0000000000000248] [PMID: 26761628]

[26] Guarnieri M, Balmes JR. Outdoor air pollution and asthma. Lancet 2014; 383(9928): 1581-92.
[http://dx.doi.org/10.1016/S0140-6736(14)60617-6] [PMID: 24792855]

[27] Ierodiakonou D, Zanobetti A, Coull BA, et al. Childhood Asthma Management Program Research Group. Ambient air pollution, lung function, and airway responsiveness in asthmatic children. J Allergy Clin Immunol 2016; 137(2): 390-9.
[http://dx.doi.org/10.1016/j.jaci.2015.05.028] [PMID: 26187234]

[28] McConnell R, Berhane K, Gilliland F, et al. Asthma in exercising children exposed to ozone: a cohort study. Lancet 2002; 359(9304): 386-91.
[http://dx.doi.org/10.1016/S0140-6736(02)07597-9] [PMID: 11844508]

[29] Jiang X-Q, Mei X-D, Feng D. Air pollution and chronic airway diseases: what should people know and do? J Thorac Dis 2016; 8(1): E31-40.
[PMID: 26904251]

[30] Bang KM. Chronic obstructive pulmonary disease in nonsmokers by occupation and exposure: a brief review. Curr Opin Pulm Med 2015; 21(2): 149-54.
[http://dx.doi.org/10.1097/MCP.0000000000000135] [PMID: 25590955]

[31] Shukla RK, Kant S, Bhattacharya S, Mittal B. Association of Clinical symptoms with smoking quantity of in northern Indian COPD patients at tertiary care hospital. Int J Bio Pharm Res 2012; 3: 545-9.

[32] Zhou Y, Zou Y, Li X, et al. Lung function and incidence of chronic obstructive pulmonary disease after improved cooking fuels and kitchen ventilation: a 9-year prospective cohort study. PLoS Med 2014; 11(3)e1001621
[http://dx.doi.org/10.1371/journal.pmed.1001621] [PMID: 24667834]

[33] Yu XJ, Yang MJ, Zhou B, et al. Characterization of somatic mutations in air pollution-related lung cancer. EBioMedicine 2015; 2(6): 583-90.
[http://dx.doi.org/10.1016/j.ebiom.2015.04.003] [PMID: 26288819]

[34] Dhar R, Singh S, Talwar D, et al. Bronchiectasis in India: results from the European Multicentre Bronchiectasis Audit and Research Collaboration (EMBARC) and Respiratory Research Network of India Registry. Lancet Glob Health 2019; 7(9): e1269-79.
[http://dx.doi.org/10.1016/S2214-109X(19)30327-4] [PMID: 31402007]

[35] Loomis D, Huang W, Chen G. The International Agency for Research on Cancer (IARC) evaluation of the carcinogenicity of outdoor air pollution: focus on China. Chin J Cancer 2014; 33(4): 189-96.
[http://dx.doi.org/10.5732/cjc.014.10028] [PMID: 24694836]

[36] Le TG, Ngo L, Mehta S, et al. HEI Collaborative Working Group on Air Pollution, Poverty, and Health in Ho Chi Minh City. Effects of short-term exposure to air pollution on hospital admissions of

young children for acute lower respiratory infections in Ho Chi Minh City, Vietnam. Res Rep Health Eff Inst 2012; 169(169): 5-72.
[PMID: 22849236]

[37] Darrow LA, Klein M, Flanders WD, Mulholland JA, Tolbert PE, Strickland MJ. Air pollution and acute respiratory infections among children 0-4 years of age: an 18-year time-series study. Am J Epidemiol 2014; 180(10): 968-77.
[http://dx.doi.org/10.1093/aje/kwu234] [PMID: 25324558]

[38] Hoffmann B, Moebus S, Möhlenkamp S, *et al.* Heinz Nixdorf Recall Study Investigative Group. Residential exposure to traffic is associated with coronary atherosclerosis. Circulation 2007; 116(5): 489-96.
[http://dx.doi.org/10.1161/CIRCULATIONAHA.107.693622] [PMID: 17638927]

[39] Leary PJ, Kaufman JD, Barr RG, *et al.* Traffic-related air pollution and the right ventricle. The multi-ethnic study of atherosclerosis. Am J Respir Crit Care Med 2014; 189(9): 1093-100.
[http://dx.doi.org/10.1164/rccm.201312-2298OC] [PMID: 24593877]

[40] Donaldson K, Stone V, Seaton A, MacNee W. Ambient particle inhalation and the cardiovascular system: potential mechanisms. Environ Health Perspect 2001; 109 (Suppl. 4): 523-7.
[PMID: 11544157]

[41] Conca K. Greening the United Nations: Environmental Organisations and the UN system. Third World Q 1995; 16: 441-58.
[http://dx.doi.org/10.1080/01436599550035997]

[42] Government launches National Clean Air Programme (NCAP), Ministry of Environment, Forest and Climate Change 2019.www.pib.gov.in/pressreleaseiframepage.aspx?prid=1559384

[42] The Ministry of Environment, Forest and Climate Change (MoEFCC), Government of India www.moef.gov.in

[44] Say C, Wood A. Sustainable rating systems around the world. Council on Tall Buildings and Urban Habitat Journal 2008; 2: 18-29. [CTBUH Review].

[45] Pradhan Mantri Ujjwala Yojna. www.pmuy.gov.in

[46] Environment, Forest and Climate Change Department, Government of Uttar Pradesh www.upforest.gov.in

# CHAPTER 2

# Cytochrome P$_{450}$ and Health Hazards of Smog

**Amber Rizvi**[1,*]

[1] *Previously at Central Drug Research Institute (CSIR-CDRI), Lucknow, India*

**Abstract:** The rising levels of smog, blanketing northern parts of India during October-January in recent years, have pushed pollution levels to an extremely hazardous point. The pollutants and the particulate matter (PM) generated by various activities have a very harmful effect on human health. This has resulted in an increase in human diseases, especially of the respiratory and cardiovascular organ systems. Combustion results in the formation of redox-active metals and aromatic hydrocarbons, which stay in the environment long after the activity has ceased. These moieties form air-stable, environmentally persistent free radicals on entrained particles that harm the pulmonary and cardiovascular systems. The protective mechanisms of the broncho-pulmonary tract are unable to stop the ultra-fine air pollutants from invading the body. The various by-products of smog enter the human body *via* several different routes, finally reaching the liver for detoxification by Cytochrome P$_{450}$ (also known as CYPs). Negative health effects of air pollutants have been shown on the cardiovascular system resulting in multiple respiratory diseases, including respiratory infections, asthma, chronic obstructive pulmonary disease (COPD), lung cancer, even in combination with stroke and heart diseases. The CYPs are endoplasmic reticulum resident enzyme systems that are involved in the metabolism of xenobiotics as well as drugs. The free radicals have a deleterious effect on these enzymes and have been found to inhibit six forms of P$_{450}$ in rat liver microsomes. These free radicals are thought to inhibit CYP2B4-mediated substrate metabolism by physically disrupting the CPR•P450 complex.

**Keywords:** Cardiovascular, COPD, Cytochrome P$_{450}$, Particulate matter, Pollution, Pulmonary, Smog.

## INTRODUCTION

Smog is a man-made haze, comprising air pollutants, that is seen over most parts of the world, especially with the onset of the winter months. The term is derived from the words smoke and fog and was used for the first time by H.A. Des Voeux,

---

[*] **Corresponding author Amber Rizvi:** Previously at, Biochemistry Division and Parasitology Division, Central Drug Research Institute (CDRI), Lucknow, India; E-mail: electronuag@gmail.com

Tahmeena Khan, Abdul Rahman Khan, Saman Raza, Iqbal Azad and Alfred J. Lawrence (Eds.)
All rights reserved-© 2021 Bentham Science Publishers

a Glasgow public health official in 1905, to describe the atmospheric conditions over many British towns. The Air Quality Index (AQI) world ranking puts India in the 4$^{th}$ spot, behind Mexico, Spain, and Romania, with an AQI of 367. This emphasizes the seriousness of the situation, even as the whole country has been in lockdown due to the novel coronavirus pandemic since March 2020.

The main components of smog are sulfur oxides, which are the by-products of the combustion of coal, residual particulate matter that wafts into the air when coal is burnt, volatile organic compounds (VOCs) from automobiles, nitrogen oxides, ozone, and peroxy-acyl nitrates (PANs) [1].

## TYPES OF SMOG

There are two types of smog formed in different weather conditions. They are known as sulfurous smog and photochemical smog (Fig. **1**) [2]. Sulfurous smog is also known as 'London smog,' as it was first observed in London. It has a high concentration of sulfur oxides and is formed due to the use of sulfur-bearing fossil fuels, especially coal. It is found in areas that have a heavy concentration of industries using coal as their major source of fuel. This type of smog is characterized by atmospheric dampness and an unusually high concentration of suspended particulate matter in the air.

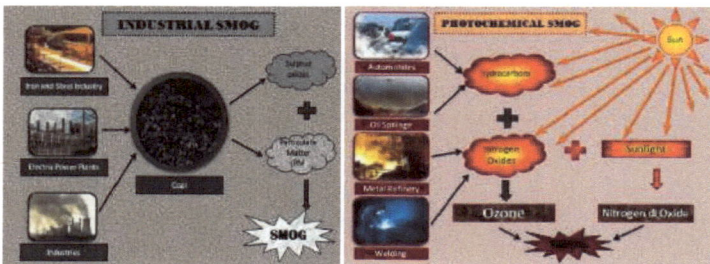

**Fig. (1).** Industrial and Photochemical Smog and their method of formation.

On the other hand, photochemical smog, which is also known as 'Los Angeles smog,' occurs in predominantly urban areas that have many vehicles. This type of smog has a high concentration of nitrogen oxides and hydrocarbon vapours emitted by automobiles and other sources, which then undergo photochemical reactions in the lower atmosphere. The highly toxic gas, ozone, is formed when nitrogen oxides react with hydrocarbon vapours in the presence of sunlight. Nitrogen dioxide is also produced from the reaction of nitrogen oxide with sunlight. The resulting smog has a light brown colour and it results in reduced visibility, plant damage, irritation of the eyes, and respiratory distress. Exposure to surface-level ozone concentrations is detrimental to human health, especially if the concentration is above 70 parts per billion, for eight hours or longer. These conditions are encountered in areas experiencing photochemical smog.

## HEALTH HAZARDS OF SMOG

Smog has been known to cause a range of diseases in humans, affecting many organ systems. Major among these are coughing, respiratory problems like COPD, asthma, respiratory tract infections like influenza, bronchopneumonia and purulent bronchitis, irritated eyes, a range of cardiovascular diseases like acute myocardial infarction, heart failure, cardiac arrhythmias, atherosclerosis, cardiac arrest, cerebrovascular diseases, skin diseases like atopic dermatitis and even a significant number of cases of cancer.

## CYTOCHROME $P_{450}$

Cytochrome $P_{450}$ enzymes (E.C.1.14.14.1) belong to a superfamily of monooxygenases that have a cysteine thiolate-ligated heme (Fig. 2). A hemoprotein is classified as belonging to the P-450 group depending on its absorption spectrum. The Fe (II)-CO complex gives a characteristic absorption maximum (Soret band) near 450nm due to axial ligation with a cysteine thiolate of the protein (with or without substrate protein). The important cysteine residue can be found in a relatively well-conserved region, ~80% into the protein from the N-terminus [3]. These enzymes catalyze the reaction in which an oxygen atom from molecular oxygen is transferred to several biological substrates to make them more water-soluble and to aid in their excretion from the human body. The second oxygen atom from the molecular oxygen is reduced by two electrons to a water molecule [4, 5]. These enzymes are involved in hydroxylation, epoxidation, heteroatom oxidation, and heteroatom de-alkylation reactions. The stoichiometry of the hydroxylation reaction catalyzed by CYPs can be written as equation **1**.

$$RH + O_2 + NADPH + H^+ \rightarrow ROH + H_2O + NADP^+ \quad (1)$$

Here RH refers to the substrate which binds to the CYP enzymes.

CYPs are found in almost all organisms, from unicellular yeasts and bacteria to multi-cellular birds, fish, plants, insects, and even mammalian tissues. These monooxygenase enzymes are involved in drug metabolism, biotransformation of naturally occurring molecules, the oxidative metabolism of xenobiotics, and synthesis of certain molecules like steroid hormones, cholesterol, some fatty acids, and bile acids. There are approximately 60 cytochrome P450 genes in humans.

The HUGO Gene Nomenclature Committee (HGNC) provides an index of gene groups and their member genes. The naming of Cytochrome P450 by HGNC follows the pattern where the first three alphabets denote the superfamily (*i.e.* CYP). This is followed by the group number, subgroup number and finally the gene number. For example- CYP3A4, where CYP indicates that this belongs to the Cytochrome P450, group 3, subgroup A and gene 4 [6].

These enzymes are mainly found in liver cells but are also present in other cells of the body. Inside the cells, cytochrome $P_{450}$ enzymes are in the endoplasmic reticulum and the mitochondria. The mitochondrial CYPs are generally involved in the synthesis and metabolism of internal substances, while endoplasmic reticulum CYPs metabolize xenobiotics, including medications and environmental pollutants [7].

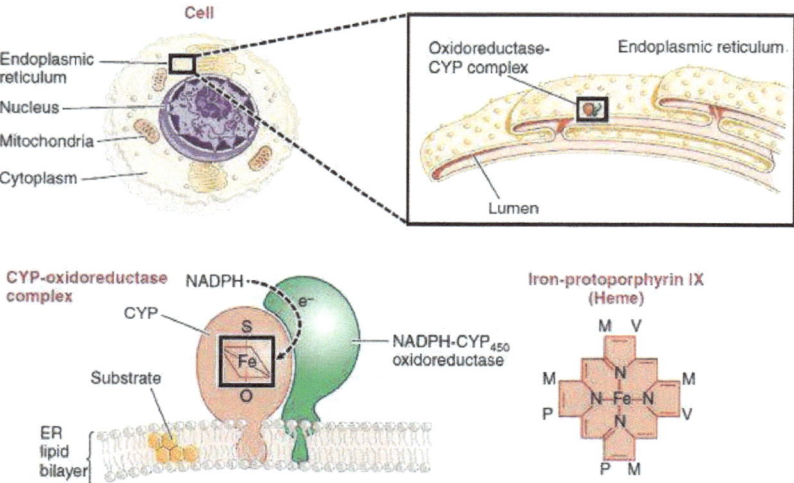

**Fig. (2).** Location of CYPs in the cell. The figure shows increasingly microscopic levels of detail, sequentially expanding the areas within the black boxes. CYPs are embedded in the phospholipid bilayer of the endoplasmic reticulum (ER). Most of the enzyme is located on the cytosolic surface of the ER. A second enzyme, NADPH-cytochrome $P_{450}$ oxidoreductase, transfers electrons to the CYP where it can in the presence of $O_2$, oxidizes xenobiotic substrates, many of which are hydrophobic and dissolved in the ER. A single NADPH-CYP oxidoreductase species transfers electrons to all CYP isoforms in the ER. Each CYP contains a molecule of iron-protoporphyrin IX that functions to bind and activate $O_2$. Substituents on the porphyrin ring are methyl (M), propionyl (P), and vinyl (V) groups. (Picture reproduced with permission from website Basicmedical Key) [8].

CYPs are involved in around 80% of drugs that undergo hepatic metabolism. Polymorphism is commonly encountered in CYP genes and this has been known to affect the functioning of the enzymes. Polymorphism in CYPs mainly affects the metabolism and subsequent drug dosage and effectiveness of the medication [7]. Depending on the gene and the polymorphism it exhibits, drugs can be metabolized quickly or slowly. If a particular CYP causes a drug to metabolize slowly, the drug stays active longer inside the cell and consequently, less dosage is required to get the desired effect. Thus, the effectiveness of the drug increases while its dosage decreases, which is advantageous to humans in the long run. On the other hand, a drug that is quickly metabolized is broken down quickly and a higher dose is needed to be effective. The polymorphism responsible for this outcome is disadvantageous to humans.

## DISEASES CAUSED BY SMOG AND ITS EFFECT ON CYPS

### Myocardial Infarction (MI)

Myocardial infarction is known as a heart attack in layman's terms. It occurs when one of the heart's coronary arteries is blocked suddenly or has extremely slow blood flow. The sudden blockage is due to the formation of a blood clot (thrombus). The blood clot usually forms in a coronary artery whose lumen has been narrowed by atherosclerosis (fatty deposits or plaques on the inside walls of blood vessels). Slow blood flow in a coronary artery occurs when the heart beats very fast or the person has low blood pressure. If the oxygen demand is greater than the supply, a heart attack can happen without the formation of a blood clot.

The function of the coronary arteries is to supply blood to the heart muscles. When a blockage occurs in these arteries, it causes pain and malfunction in the heart to which it supplies blood. The heart muscles work to pump blood from the heart to the entire body. When these muscles are deprived of oxygen and nutrients due to blockage, they cannot function smoothly and result in the heart's inability to pump blood efficiently. Some coronary arteries supply to areas of the heart that regulate heartbeat, so a blockage in that area results in potentially fatal abnormal heartbeats, called cardiac arrhythmias.

The risk factors for heart attack are abnormally high levels of blood cholesterol (hypercholesterolemia), very low level of HDL (high-density lipoprotein), high blood pressure (hypertension), diabetes, family history of coronary artery disease, cigarette smoking, obesity, and physical inactivity.

The major cause of MI is atherosclerosis. Other causes which contribute to this condition are congenital abnormalities of the coronary arteries, hypercoagulability (an abnormally increased tendency to form blood clots), a collagen vascular disease like rheumatoid arthritis or systemic lupus erythematosus (SLE, or lupus), cocaine abuse, a spasm of the coronary artery, or an embolus (small travelling blood clot), which floats into a coronary artery and lodges there.

Particulate matter and $NO_2$ increase the risk of severe MIs. In a study, elderly patients were found to have developed STEMI when exposed to $PM_{10}$ while those in the age group of 54 years and less showed STEMI on exposure to $NO_2$. These results were explained by Dr. Argacha: 'Considering that $NO_2$ is more related to vehicle emissions, one explanation for this finding could be that the younger population may be exposed to excess $NO_2$ from road traffic due to a higher level of social and professional activities. He further emphasized that the detrimental impact of $NO_2$ exceeds that of fine particles and raises new public health concerns [9].

## Mechanism of Cardiotoxicity of Air Pollutants

The various gases present in smog are responsible for several human diseases. The particulate matter present cannot be completely absolved of its part in harming human health. PM is the major cause of CVS mortality. PM is divided into two types- fine PM or $PM_{2.5}$ and coarse PM or $PM_{10}$. $PM_{2.5}$ is obtained due to combustion in vehicles and includes fine particles like nitrates and sulfates. On the other hand, $PM_{10}$ is obtained from natural sources like forest fires, bio-aeroso--endotoxins, fungal spores, pollen, windblown soil, and occupational exposure (grinding, smelting, *etc.*) [10]. PM affects the circulatory system both directly and indirectly. The direct effect is caused by the soluble components of $PM_{2.5}$ that can cross respiratory epithelium into the systemic bloodstream. The indirect effects that PM causes are pulmonary and systemic oxidative stress, resulting in inflammation [11, 12]. The circulatory inflammation, in turn, leads to alterations in blood rheology, increased fibrinogen-enhanced platelet aggregation, alteration in the cardiac autonomic system leading to rhythm disturbances, endothelial dysfunction leading to vascular spasms and plaque disturbance in short term and atherosclerosis in the long term. Increased $SO_2$ levels increase fibrinogen levels. Ozone is known to cause direct oxidation of both pulmonary and systemic vasculature, resulting in inflammation. It is also responsible for arterial vasoconstriction. Both carbon monoxide, which is a direct toxicant, and nitrogen oxides, impair ICD discharges [13].

## Role of CYPs in MIs

CYP isoforms play an important role in cardiovascular homeostasis and the pathophysiology of heart diseases. CYP2C and 2J subfamilies catalyze the conversion of AA (Arachidonic acid) into EETs (epoxyeicosatrienoic acids) in cardiovascular tissues. EETs are important for heart health as they have cardioprotective functions like vasodilatory, anti-inflammatory, anti-apoptotic, and anti-thrombotic actions. Patients with heart disease have a higher plasma concentration of EETs than healthy people. This can be because EETs have a vasodilatory effect in the presence of advanced atherosclerosis and help in maintaining the blood flow. EETs play an important role in protecting the heart from ischemia-reperfusion injury (IRI), exacerbation of myocardial damage induced by blood flow restoration, and tissue reperfusion after an infarction event. When EETs were administered directly to the coronary circulation, it caused a significant decrease in infarct size [14].

On the other hand, CYP2C9 and CYP2C8 are involved in the production of reactive oxygen species (ROS), which accelerates the damage in cardiac tissues [15]. ROS are produced during the reaction when electrons, which are needed for the reduction of the central heme iron, are transferred to the activated bound oxygen molecule. Endothelial cells, which have an increased expression of CYP2C8, showed decreased functional recovery and increased infarct size after ischemia-reperfusion. Thus, CYPs exert a protective, as well as deleterious effect on myocardial tissue, depending on the isoform present.

The constituents of smog, like particulate matter, $NO_2$, $SO_2$ and $O_3$, all affect heart health leading to MIs. These pollutants can cause inflammation, changes in the inner lining of blood vessels, and oxidative stress. These harmful changes are countered to a certain extent by CYPs present in the heart. The helpful CYPs are CYP2C and 2J subfamilies which, during the metabolic reactions they catalyze, result in the formation of EETs, which have a cardio-protective role. On the other hand, CYP2C9 and 2C8 produce ROS which further harm the cardiac tissues leading to a larger MI and unfavourable outcome for the patients.

## Chronic Obstructive Pulmonary Disease (COPD)

Chronic obstructive pulmonary disease (COPD) is a chronic inflammatory lung disease that results in gradual and progressive loss of lung function. Its symptoms are difficulty in breathing, coughing, wheezing, and mucus (sputum) production. The main causes resulting in COPD are exposure to irritating gases or particulate matter from cigarette smoke, biomass fuels from indoor cooking, dust, and oxidative stress.

The underlying pathological conditions that majorly contribute to COPD are emphysema and chronic bronchitis. These two conditions are almost always seen together in patients with COPD. Emphysema is the condition in which the alveoli (air sacs) present at the end of the smallest air passages (bronchioles) of the lungs are destroyed due to exposure to cigarette smoke, dust, smoke, fumes, and other air pollutants. Chronic bronchitis occurs when there is an inflammation of the lining of the bronchial tubes, which carry air to and from the alveoli. It results in cough and mucus formation. COPD symptoms appear only after there has been significant lung damage due to continuous exposure to smoking by-products.

The major symptoms of COPD are shortness of breath, especially during physical activities, wheezing, chest tightness, a chronic cough that may produce mucus (which can be clear, white, yellow, or greenish), frequent respiratory infections, lack of energy, unintended weight loss (in later stages), and swelling in ankles, feet, or legs.

## CYPs and COPD

Drug-metabolizing enzymes are present in the lungs, although their metabolic capacity is lower compared to the enzymes present in gastrointestinal and hepatic tissue systems. Certain enzymes do have a marked capacity in the lungs, like CYP 1A1 in smokers, and CYP 2E1.

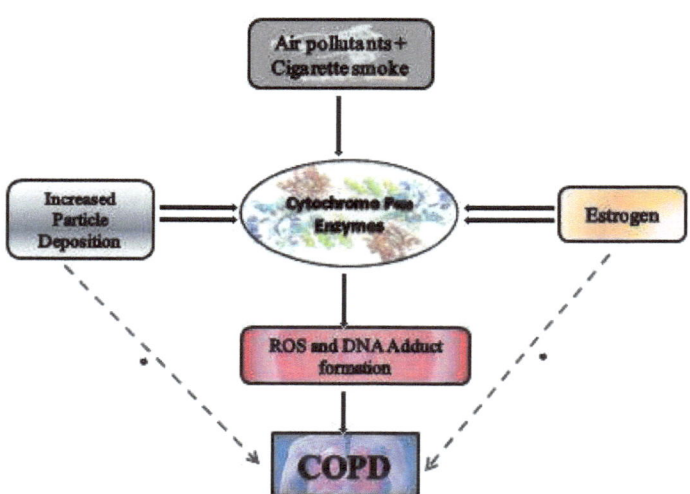

**Fig. (3).** A line diagram depicting why females are more susceptible to chronic obstructive pulmonary disease (COPD) and lung injury. Air pollutants and cigarette smoke cause up-regulation of CYP enzymes in the lungs. These enzymes metabolize the substrates leading to the formation of reactive oxygen species (ROS) and DNA adducts. ROS formed in the lungs leads to injury. Women may be particularly vulnerable to lung injury leading to COPD due to the hormone estrogen, which causes an increase in the CYP enzymes. Particle deposition deep inside the lungs increases as women have smaller airways and increased bronchial responsiveness. *Disjointed lines show that other pathways may be working to cause COPD as well.

Xenobiotic metabolism in humans is sex-dependent. Research has shown that female lungs have 2.4 times more mRNA for CYP enzymes than male lungs. Female hormones, especially estrogen, are implicated in the regulation of CYP expression. There are two major estrogen receptors (ERs) in the lungs: ERα and ERβ. These receptors are ligand-binding transcription factors and members of the nuclear receptor family. ERα is expressed more frequently in women than in men

(Fig. **3**), whereas ERβ is expressed with a similar frequency in men and women. Stimulation of the estrogen receptor in the lungs increases protein expression of CYP1A1 twofold [16, 17].

Females have an accelerated metabolism of xenobiotics present in the air pollutants through CYP mediated pathways. The metabolites related to N-nitroso derivatives and PAHs (like naphthalene), and formed through the CYP pathways, are potent oxidants and oxidizers. ERs do not affect the expression of phase II enzymes like glutathione S-transferase. These phase II enzymes are responsible for converting the active species formed by CYPs in phase I to harmless metabolites and finally excreting them from the body. Due to this reason, females have increased oxidative stress in the lungs, which may lead to extensive lung injury as compared to males [16].

**Atopic Dermatitis (Eczema)**

Atopic dermatitis (AD), also known as eczema, is a chronic relapsing inflammatory skin disease mostly found in children. It is caused due to defects in the skin barrier and immunologic dysregulation. The outermost layer of the epidermis, which is known as the stratum corneum, is made up of corneocyte layers. The corneocyte layers form a protective envelope in which numerous proteins like loricrin, involucrin, filaggrin, and small proline-rich proteins are present. Atopic dermatitis can develop due to primary defects in the skin barrier followed by penetration of allergens, or prior immunologic predisposition which leads to the formation of skin barrier defects [18].

The air pollutants responsible for producing lifetime eczema in children are $PM_{10}$, $NO_2$, $NO_x$, and CO. Research has shown that if benzene concentration increased by 1ppb, there was an increase of 27.38% in atopic dermatitis symptoms. Similarly, for a 1ppb increase in total VOC concentration, there was a corresponding increase of 25.86% in AD symptoms on the following day. $PM_{10}$ showed the smallest increase of 0.44% in AD symptoms on the following day for an increase in $PM_{10}$ concentration by 1 $mg/m^3$ [19].

Air pollutants lead to the production of reactive oxygen species (ROS) and reactive nitrogen species, which in turn cause damage to proteins, lipids, and DNA. The oxidative stress produced in the skin due to an imbalance between oxidants and anti-oxidants cause damage to the skin barrier. To determine that oxidative damage was the result of environmental pollution, a study focused on dinitrophenylhydrazine (DNP), a marker of oxidative protein damage. When immunohistochemical staining of DNP was done, it was found to be very intense in the top layers of the stratum corneum compared to the lower layers, indicating that oxidative damage had occurred. The oxidative protein damage leads to defects in the skin barrier, which in turn caused aggravation of AD [19].

## CYPs and Atopic Dermatitis

The skin is the outermost layer of the body and has the task of acting as a barrier against infectious agents and xenobiotics. It is the part where drugs are topically applied for a variety of reasons like burns, cosmetics, infections, *etc.* It is the function of the skin to metabolize these xenobiotics and drugs to eliminate them from the body. This function is performed with the help of CYP enzymes which are the major drug-metabolizing enzymes in the body. The main CYP enzymes which are present in keratinocytes are CYP1A1, CYP1B1, CYP2B6, CYP2E1, and CYP3A [20, 21].

CYP1A1 is among one of the few CYPs that is expressed in the skin. This CYP does not have any specific endogenous substrate. Among the pollutants, polycyclic aromatic hydrocarbons (PAHs) like BaP, 3-methylcholanthrene and 7, 12-dimethylbenz (a) anthracene, serve as substrates for CYP1A1. The reactions result in the formation of mutagenic and carcinogenic metabolites of the parent compounds. The formation of these compounds is extremely harmful, not only in the skin where these reactions occur but also throughout the body. Benz (a) anthracene, and b-naphthoflavone (b-NF) act as inducers of CYP1A1, causing an increase of aryl hydrocarbon hydroxylase (AHH) activity in rat epidermis as well as in human keratinocytes [22]. CYP1A1 expression is also dependent on the progression of cell differentiation. The study where b-NF was topically applied to mice was done by Stauber *et al.* in 1995. This resulted in an 87-fold increase in epidermal 7-ethoxyresorufin deethylase (EROD) activity per cell and a manifold increase in CYP1A1 expression in the epidermis. The superbasal differentiated cells of the epidermis had high CYP1A1 expression following induction by b-NF [21].

## Coughing

Coughing is a common reflex action that occurs to clear our throats of mucus, microbes, allergens, foreign particles, and irritants. During this process, the air is forced out of the lungs under high pressure to rid the nasal pathways of these invading and unwanted particles. There are different types of coughs categorized as acute, sub-acute and chronic, depending on their duration.

The causes of this simple action can be both temporary and permanent. The most common cause is to clear our throats. The airways might become clogged with mucus or foreign particles such as smoke or dust, making breathing difficult. To get more air in the body and to clear the particles, coughing occurs. Although this type of coughing occurs intermittently, its frequency will increase with an increase in exposure to air pollutants, such as dust, particulate matter, and chemicals. Smog causes health problems like coughing, and throat and chest irritation, due to its ozone content [23].

## CYPs and Coughing

Coughing caused by ozone present in smog changes some isozymes of CYPs while it does not have any effect on the other drug-metabolizing enzymes. Some enzymes which appeared to be resistant were CYP2E1, CYP2B and p-nitroanisole N-demethylase. On the other hand, lung benzo[a]pyrene hydroxylase and 7-ethoxycoumarin O-deethylase increased significantly, showing the protective nature of these xenobiotic-metabolizing enzymes. Proteomic studies showed that after 5 days of ozone exposure, several key hepatic protein levels were changed. Using 2D gel electrophoresis techniques and mass spectrometry, over 20 proteins were shown to have their levels altered, out of which 10 proteins were identified that were significantly altered by $O_3$ inhalation. These 10 proteins belonged to different groups, such as cytoskeletal, drug metabolism, energy metabolism, and ER stress-related proteins [24].

## Experimental Work Done on CYP

Since CYPs are such a huge family of enzymes which are involved in almost 75-80% of drugs metabolized in the body, it makes sense to study their mechanisms and pathways in almost all the diseases encountered by humans. These enzymes have been identified in all human tissues. Animals, plants, and micro-organisms contain these versatile enzymes as well.

During the study, these enzymes were estimated in white mice infected with rodent malarial parasites. The tissue examined was liver, both in healthy and infected mice, at different levels of parasitaemia. The enzyme estimation was done by the method of Omura and Sato [25]. It was found that the average amount of CYP present in healthy mice livers was 0.59 nmol/mg of protein. This was found to decrease when malaria infection took hold in mice and was inversely proportional to the level of parasitaemia (Fig. **4**). Such a decrease in CYP is observed in humans as well; however, the pattern of decrease is difficult to measure in humans as the priority is to clear the infection before any significant damage is done to the host.

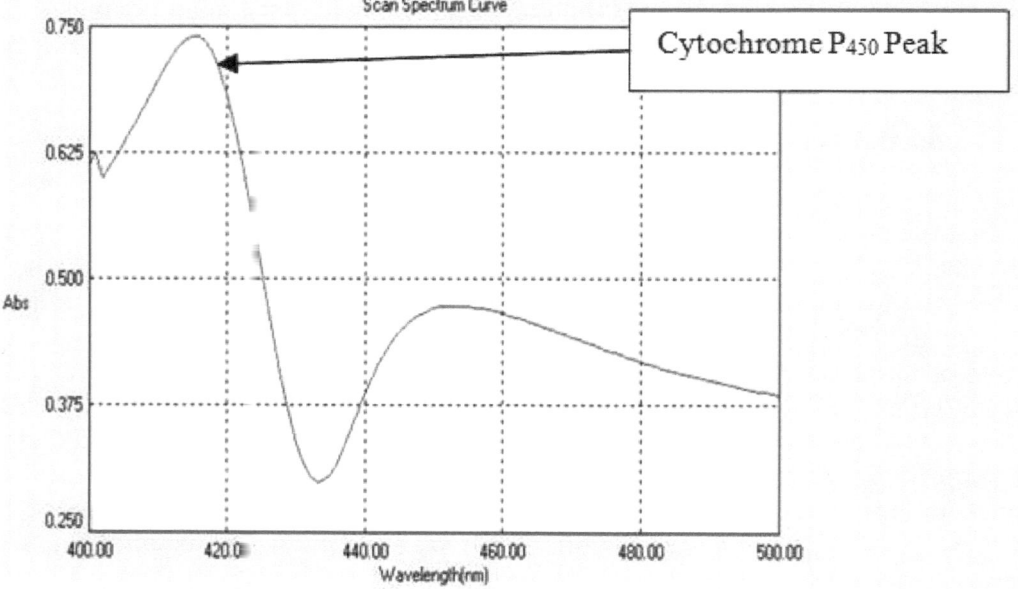

**Fig. (4).** A graph of Cytochrome $P_{450}$ estimated through the method of Omura and Sato, 1964 [25].

Drugs can either act as inhibitors or inducers of CYPs. If a particular drug inhibits CYPs, then the drug remains in the body for a longer period and hence brings about its effect with a lower dosage. On the other hand, if a drug causes induction of CYPs, then it is rapidly metabolized and eliminated from the body; so, it remains in the body for a shorter period and thus requires a higher dosage to bring about its required action.

The different diseases caused by smog are due to inhalation of the constituents of smog and their subsequent effects on various pathways. These components hamper CYP pathways too, which are vital for both removing these pollutants from the body and metabolizing the drugs that are given for the treatment of various diseases. It might be beneficial to find drugs that resolve the symptoms quickly, along with providing protective action to the host. CYPs that metabolize the oxides of sulphur and nitrogen, and ozone, can be used for eliminating these harmful constituents, thus providing a novel method of fighting smog.

## CONCLUDING REMARKS

The above discussion brings into sharp focus the fact that smog is extremely harmful to humans, not just for a short period but its long-term effects are equally disastrous. It causes a host of diseases that affect both the quality and duration of life. CYPs are the major xenobiotic-metabolizing enzymes and their numerous isoforms protect as well as harm the body, depending on their expression. Since these enzymes metabolize drugs, so they are potential targets during drug development for various diseases. The protective activities of various CYP isoforms can be harnessed for the benefit of humans.

## CONSENT FOR PUBLICATION

Not Applicable.

## CONFLICT OF INTEREST

The author confirms that this chapter contents have no conflict of interest.

## ACKNOWLEDGEMENT

Declared none.

## REFERENCES

[1] Balajee KL, Babu S, Suliankatchi RA, Meena S. Characteristics of the Ozone pollution and its Health Effects in India. Int J Med Public Health 2017; 7(1): 56-60.
[http://dx.doi.org/10.5530/ijmedph.2017.1.10]

[2] Cai H, Wang C. Surviving With Smog and Smoke: Precision Interventions? Chest 2017; 152(5): 925-9.
[http://dx.doi.org/10.1016/j.chest.2017.06.030] [PMID: 28694198]

[3] Guengerich FP. Reactions and significance of cytochrome P-450 enzymes. J Biol Chem 1991; 266(16): 10019-22.
[http://dx.doi.org/10.1016/S0021-9258(18)99177-5] [PMID: 2037557]

[4] Wislocki PG, Muva GT, Lu AYH. 1980.

[5] Nam W. Bio-coordination Chemistry Comprehensive Coordination Chemistry II 2003.

[6] Sim SC, Ingelman-Sundberg M. The Human Cytochrome P450 (CYP) Allele Nomenclature website: a peer-reviewed database of CYP variants and their associated effects. Hum Genomics 2010; 4(4): 278-81.
[http://dx.doi.org/10.1186/1479-7364-4-4-278] [PMID: 20511141]

[7] McDonnell AM, Dang CH. Basic review of the cytochrome p450 system. J Adv Pract Oncol 2013; 4(4): 263-8.
[PMID: 25032007]

[8] https://basicmedicalkey.com

[9] Argacha J. Air pollution and myocardial infarction
[http://dx.doi.org/10.1093/eurheartj/ehw622]

[10] Wu W, Jin Y, Carlsten C. Inflammatory health effects of indoor and outdoor particulate matter. J Allergy Clin Immunol 2018; 141(3): 833-44.
[http://dx.doi.org/10.1016/j.jaci.2017.12.981] [PMID: 29519450]

[11] Mishra S. Is smog innocuous? Air pollution and cardiovascular disease. Indian Heart J 2017; 69(4): 425-9.
[http://dx.doi.org/10.1016/j.ihj.2017.07.016] [PMID: 28822504]

[12] Tuan TS, Venâncio TS, Nascimento LFC. Effects of Air Pollutant Exposure on Acute Myocardial Infarction, According to Gender. Arq Bras Cardiol 2016; 107(3): 216-22.
[http://dx.doi.org/10.5935/abc.20160117] [PMID: 27533257]

[13] Ruidavets JB, Cournot M, Cassadou S, Giroux M, Meybeck M, Ferrières J. Ozone air pollution is associated with acute myocardial infarction. Circulation 2005; 111(5): 563-9.
[http://dx.doi.org/10.1161/01.CIR.0000154546.32135.6E] [PMID: 15699276]

[14] Chaudhary KR, Batchu SN, Seubert JM. Cytochrome P450 enzymes and the heart. IUBMB Life 2009; 61(10): 954-60.
[http://dx.doi.org/10.1002/iub.241] [PMID: 19787709]

[15] Ong CE, Pan Y, Mak JW. The roles of cytochromes P450 in vascular biology and cardiovascular homeostasis. Int J Clin Exp Med 2017; 10(1): 1624-36.

[16] Sin DD, Cohen SB, Day A, Coxson H, Paré PD. Understanding the biological differences in susceptibility to chronic obstructive pulmonary disease between men and women. Proc Am Thorac Soc 2007; 4(8): 671-4.
[http://dx.doi.org/10.1513/pats.200706-082SD] [PMID: 18073400]

[17] Takahashi Y, Miura T, Kubota K. *In vivo* effect of ozone inhalation on xenobiotic metabolism of lung and liver of rats. J Toxicol Environ Health 1985; 15(6): 855-64.
[http://dx.doi.org/10.1080/15287398509530711] [PMID: 3932670]

[18] Kim K. Influences of Environmental Chemicals on Atopic Dermatitis. Toxicol Res 2015; 31(2): 89-96.
[http://dx.doi.org/10.5487/TR.2015.31.2.089] [PMID: 26191377]

[19] Ahn K. The role of air pollutants in atopic dermatitis. J Allergy Clin Immunol 2014; 134(5): 993-9.
[http://dx.doi.org/10.1016/j.jaci.2014.09.023] [PMID: 25439225]

[20] Afaq F, Zaid MA, Pelle E, *et al.* Aryl hydrocarbon receptor is an ozone sensor in human skin. J Invest Dermatol 2009; 129(10): 2396-403.
[http://dx.doi.org/10.1038/jid.2009.85] [PMID: 19536146]

[21] Ahmad N, Mukhtar H. Cytochrome p450: a target for drug development for skin diseases. J Invest Dermatol 2004; 123(3): 417-25.
[http://dx.doi.org/10.1111/j.0022-202X.2004.23307.x] [PMID: 15304077]

[22] Zaid MA, Afaq F, Khan N, *et al.* Exposure of normal human epidermal keratinocytes to ozone results in increased expression of cytochrome 450.99th AACR Annual Meeting. San Diego, CA. 2008, Apr 12-16;

[23] Jo EJ, Song WJ. Environmental triggers for chronic cough. Asia Pac Allergy 2019; 9(2)e16
[http://dx.doi.org/10.5415/apallergy.2019.9.e16] [PMID: 31089458]

[24] Theis WS, Andringa KK, Millender-Swain T, Dickinson DA, Postlethwait EM, Bailey SM. Ozone inhalation modifies the rat liver proteome. Redox Biol 2014; 2: 52-60.
[http://dx.doi.org/10.1016/j.redox.2013.11.006] [PMID: 25544660]

[25] Omura T, Sato R. The carbon monoxide-binding pigment of liver microsomes I. Evidence for its hemoprotein nature. J Biol Chem 1964; 239: 2370-8.
[http://dx.doi.org/10.1016/S0021-9258(20)82244-3] [PMID: 14209971]

# CHAPTER 3

# Pharmaceutical and Modelling Interventions for Environmental Pollution Related Chronic Obstructive Pulmonary Disease

**Tahmeena Khan[1], Alfred J. Lawrence[2], Iqal Azad[1], Shalini Dixit[3,\*] and Saman Raza[2]**

[1] *Integral University, Lucknow, India*
[2] *Isabella Thoburn College, Lucknow, India*
[3] *CSIR-Central Institute of Medicinal and Aromatic Plants (CSIR-CIMAP), Lucknow, India*

**Abstract:** Chronic obstructive lung diseases, including asthma and chronic obstructive pulmonary disease (COPD), are causing an extreme burden on societal health, affecting above 500 million people worldwide and affecting lung physiology at a multi-biological level. The increasing burden of air pollution is a major contributing factor to the disease, other than smoking and living conditions. Over the years, several studies have been undertaken to understand lung function, airflow mechanisms, and impairment for better therapies and therapeutic interventions. Still, it is very unlikely to predict the morbidity and mortality associated with COPD due to limitations of early and timely prediction and progression which calls for personalized treatment interventions to avert exacerbation and refractory symptoms. This chapter presents an overview of the status of COPD worldwide with a special emphasis on Indian statistics, along with the drug and pharmacological advancement, and computational medicinal modelling, its applications, and limitations. Though experimental models may predict the prerequisites for the system medicine approach, they are unable to analyse the finer details, calling for more advanced molecular technologies. A computational model of system medicine mimics the functioning of a complex system and can predict future functioning as well. Working with large data sets, computational models may have greater benefits to minimize patient risk and assist in clinical decision-making.

**Keywords:** Chemistry, COPD, Environment, Lung, Modelling, Pharmaceutical, Pollution.

---

[\*] **Corresponding author Shalini Dixit:** Department of Analytical Chemistry, CSIR-Central Institute of Medicinal and Aromatic Plants, Lucknow, India; E-mail: shasddixit@gmail.com

Tahmeena Khan, Abdul Rahman Khan, Saman Raza, Iqbal Azad and Alfred J. Lawrence (Eds.)
All rights reserved-© 2021 Bentham Science Publishers

# INTRODUCTION

Exposure to air pollutants is rampant in cities because of the increasing population and emitting sources. The essential air quality assessment parameters are depicted in Fig. (**1**). The health effects of pollutants are numerous, ranging from short-term like cough, throat irritation, and asthma [1], to long-term, including chronic obstructive pulmonary disease (COPD) and cardiovascular health problems [2]. The particulate contamination which has a size smaller than 2.5 μm is especially detrimental for health because of its penetration in lungs and bloodstream and deposition in the brain and heart [3]. Chronic obstructive lung diseases, including asthma and COPD, are causing an extreme burden on societal health [4, 5]. Though a significant amount of work has been done on asthma, much must be explored when it comes to COPD [6, 7]. The existing challenge is to identify molecular interactions involved in the pathophysiology to work on treatment therapies. The exact contribution of air pollutants in the total COPD burden is very unclear at the present stage [8], though presently it is the fourth leading cause of death and by 2030, would become the third. Hence dealing with COPD and its consequences is a huge and momentous challenge for the scientific fraternity. Smoking is another important factor leading to COPD though it can also have other etiologies.

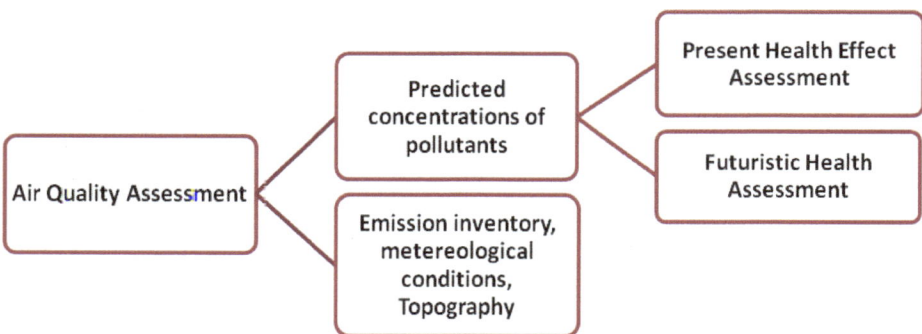

**Fig. (1).** Key parameters of air quality assessment.

# INFLUENCING FACTORS FOR INITIATION AND MODULATION

The overall assessment has shown that COPD is initiated by exposure to hazardous particles and gases. Passive exposure to tobacco, smoke from fuel burning, occupational hazards, and outdoor pollution also contributes significantly to COPD spread [9]. Respiratory toxicants containing complex mixtures of several thousands of compounds and their composition vary depending upon the source of origin [10]. There are also factors that do not directly contribute but influence the magnitude and progression of the disease. Women are more prone to the risk of COPD than men and have different comorbidities [11]. Diet and physical activity also modulate COPD to an appreciable extent. Food rich in vitamin C, vitamin E, and β-carotene is found to improve lung function and protects against COPD [12]. Proper and healthy functioning of the cardio-respiratory system is linked with lower occurrences and improved working in childhood and adolescence, leading to greater lung volume [13].

## COPD AND ITS ASSOCIATION WITH COMMON AIR POLLUTANTS: A WORLDWIDE PERSPECTIVE

COPD exacerbations have been linked with emergency hospital admissions and are associated with increased mortality and decreased quality of life. Patients have reportedly experienced one or more exacerbations [14]. A German study on the influence of Air Pollution on Lung, Inflammation, and Aging (SALISA) conducted for five years showed that the five-year average concentration of particulate matter with a diameter of 10 μm had a negative correlation with FVC and FEV1 and a positive relationship with the possibility of the occurrence of COPD [15]. In several time-series studies conducted in the USA, a nominal association between $PM_{10}$ and COPD related complaints and hospital visits has been found [16], though the studies did not consider delayed effects and temporal lags of the dependent variables in the progression. A good correlation between the $PM_{10}$ and $PM_{2.5}$ levels and frequent hospital visits has been established in several studies [17, 18]. Overall analysis showed a stronger correlation with $PM_{2.5}$ than with $PM_{10}$ because smaller particles penetrate deeper into the lungs. The influence of $PM_{2.5}$ was stronger in Asian countries with a great deal of heterogeneity involved. Doubly high concentrations of $PM_{2.5}$ than Europe have been reported in Asia, whereas these concentrations were four times higher than North America, where the $PM_{2.5}$ limit was lower than the WHO permissible limits. The meta-regression analysis hinted at the nonlinear relationship between COPD-related hospital visits and exposure to ambient air pollution. Though the association is quite clear, it still needs to be interpreted carefully because of the limited number of studies conducted to ascertain the close association. Regarding the effect of $PM_{10}$, very scarce information is available to assess seasonal variation and health outcomes, although a study conducted in a tropical climate has suggested a higher impact in the winter season [19]. A 2.7% rise in chances of COPD-related hospital admissions has been observed for a 10 μg/m³ increase in the concentration of $PM_{10}$ [20]. Not only particulate matter but gaseous pollutants have also contributed significantly to the rise of COPD. An association has been developed between CO and $SO_2$ levels and COPD exacerbations. The two pollutants have been found to have acute or short-term effects up to 2-3 days, respectively. The association was more prominent in Asian countries with an OR of 1.03; 95% CI, 1:00-1.06. Apart from geographical variation, a seasonal variation in $SO_2$ has also been observed. In two studies conducted in Taiwan, a significant association between $SO_2$ and COPD-related hospital variation has been observed in the winter season (temperature 28 °C) [21, 22]. Increased coal combustion has been identified as a potent factor in developing countries. There is not a very significant association between exposure to CO with COPD hospital admissions, except in North America and Europe. In Europe, a stronger association was observed than in North America because of higher concentrations of $SO_2$ in Europe (2.160 mg/m$^3$) than in North America (1.56 mg/m$^3$). Acute to lagged effects were reported which occurred up to 3 days [23, 24]. To assess the effect of $O_3$ on COPD, several studies have been conducted in different parts of North America, Europe, Asia, and Australia. Large heterogeneity (87%) was observed, and a positive effect was observed on COPD hospital admissions. The metaregression analysis showed a positive and strongest effect in Asia. A study conducted in Canada estimated twice as large complaints in the summer season [25], whereas a study conducted in a tropical climate showed that COPD complaints doubled in the winter season [26].

## COPD AND ITS STATUS IN INDIA

India has diverse demography and economic variation which affects respiratory health [27]. According to the 2017 India State-Level Disease Burden Initiative report, cases of COPD and their distribution and heterogeneity across the country have been reported from 1990-2016. The data was collected largely from household health surveys, monitoring data, tobacco eating patterns of youth and adults, and occupational surveys of the National Sample Survey Organization. The results revealed that chronic respiratory diseases caused 10.9% of the total deaths and 6·4% of the total DALYs in India in 2016, as compared with 9·6% and 4·5%, respectively, in 1990. Of the total global DALYs due to chronic respiratory diseases in 2016, 32% belonged to India [28, 29]. COPD contributed to 75·6% of the total DALYs due to chronic respiratory diseases in 2016 and asthma 20·0% [27]. The prevalence of COPD was highest in North Indian states falling under lower-middle to higher-middle state categories. Uttar Pradesh and Rajasthan were next in line, while in the northeast, Mizoram recorded the highest COPD cases. From the western and southern parts of the country, Maharashtra, Goa, Kerala, Karnataka, Andhra Pradesh, and Telangana reported the highest COPD complaints. As related to age, COPD cases increased after the age of 30 years, with a higher prevalence in men in the 80 years or older age group and among women in the 75–79 years age group. Of the COPD DALYs in India in 2016, 53·7% were due to air pollution, 25·4% due to tobacco consumption, and 16·5% due to occupational risks, making them leading causes of COPD. 33·6% of COPD DALYs might be accredited to ambient air pollution, 25·8% to household air pollution, and 21·0% to smoking. Household air pollution and smoke were found to be more dangerous to women, whereas smoking and occupational hazards were significant for men. Comparing to 1990, the prevalence of COPD increased up to 29% in 2016, and cases of asthma increased by 9%.

Studies have suggested that COPD is not usually reported in India, owing to a lack of awareness [30, 31]. The use of inhalers in advanced stages is associated with social stigma, especially in rural regions [32]. Women are more susceptible to delayed reporting of the diseases because they usually do not show early symptoms [33]. Other than smoking, exposure to air pollution, biomass burning, socioeconomic status, and nutritional malfunctioning are some other factors adding to the burden of COPD [34].

## MECHANISTIC ACTION OF COPD- THE CHEMICAL AND BIOCHEMICAL APPROACH

Several *in vivo* and *in vitro* studies have explained the possible biomedical mechanisms to disseminate the relationship between short-term air pollution exposure and COPD. The overall impact of COPD is depicted in Fig. (**2**).

## OXIDATIVE STRESS

An oxidative mediated mechanism is one such pathway leading to COPD [35]. Pollutants may absorb chemical substances from the environment leading to the reduction of reactive oxygen species [36]. Oxidative stress could cause damage to the epithelium of the airway, making the COPD sufferers more prone to damage and weakened immunity. Another probable mechanism is the inflammation in the lungs, reducing pulmonary function in COPD patients [37]. $O_3$ is a potential participant of photochemical reactions and shows seasonal fluctuations, affecting the exacerbation rate of COPD [38]. The formation of ozone in ambient air is mainly governed by solar radiation. Particulate matter consists of chemical components including elemental carbon, inorganic moieties like sulphates and nitrates, transition metal oxides, polycyclic aromatic hydrocarbons, and biological components, including pollen grains, bacteria, spores, *etc*.

Oxidants and pro-oxidants generate oxygen and nitrogen free radicals, inducing oxidative stress in the airways. The non-neutralized free radicals trigger inflammatory response and release of inflammatory cells and mediators, leading to subclinical inflammation [39]. The studied variations in the daily concentrations of $PM_{10}$, $SO_2$, CO, and $NO_2$ on the pulmonary performance of approximately 3000 students in Taipei showed a 69.8 mL reduction in FVC and a 73.7 mL decrease in FEV1 concerning CO and a 73.7 mL reduction in FEV1, with a 1-day lag effect. A 1 ppb rise in $SO_2$ concentration was linked to a 12.9 mL decrease in FVC and an 11.7 mL reduction in FEV1. Variations in $O_3$ and $PM_{10}$

showed a low but substantial negative relationship with FVC and FEV1 on the day of exposure [40]. The chronic inflammation of COPD leads to the assimilation of neutrophils, macrophages, B-cells, lymphoid, and CD8+ T-cells, especially in the small airways [41]. Oxidative stress and the redox status of the cells may regulate histone alterations, such as acetylation, methylation, and phosphorylation, causing higher stimulation of inflammatory mediators [42]. In AMs and bronchial epithelial cells, ROS increases the generation of inflammatory mediators and antioxidant enzymes (glutamate-cysteine ligase, Mn-SOD, and thioredoxin) [43] accounting for the enhanced expression of inflammatory mediators [44, 45]. $PM_{2.5}$ and ultrafine particles, particularly those obtained from diesel exhaust and residue oil fly ash (ROFA), have received significant attention; due to their larger surface area, their capacity to carry chemicals to the peripheral airways is more. ROFA is a complex mixture of sulphates, carbon, nitrogen-containing compounds, and metals like vanadium. Soluble metals and sulfate may cause airway hyper-reactivity and lung injury [46]. In respiratory epithelial cells, ROFA impacts the signalling events and cytokine/chemokine production. Polyaromatic hydrocarbons can also get chemisorbed on the surface of particulate matter through transition metal oxides. PMs derived by combustion are highly oxidizing and form free radicals, as their surface metals are involved in redox cycling or the depletion of antioxidant glutathione and protein-bound sulfhydryl groups [47]. The transition metals may form ROS through a Fenton-like chemical reaction, causing subsequent activation of cellular signalling, transcription and initiation of inflammatory mediator release, and airway hyper-responsiveness [47, 48]. Other components of air pollution that may contribute to oxidant generation include transition metals, such as chromium, iron, manganese, vanadium, and copper.

## PROTEASE-ANTIPROTEASE IMBALANCE

The imbalance between protease and antiprotease enzymes has been associated with the pathogenesis of emphysema [49]. In smokers suffering from COOD, low macrophage elastase activity was obtained due to decreased macrophage influx, showing a crucial role of neutrophil elastase and macrophage elastase in alveolar destruction in smokers [50].

## ROLE OF MEDIATORS

Several mediators have been identified to channelize COPD pathogenesis. Anti-inflammatory mediators, such as leukotrienes and opioids, have affected COPD pathogenesis. Histone acetylases and deacetylases are also key regulators of gene transcription and expression. The acetylation of core histones modifies the chromatin structure, affecting transcription, which is also regulated by histone deacetylase and histone acetyltransferase. The mechanism regulates cytokine production during COPD [51]. IL-13, a Th2 cytokine has also been involved in the progression of the disease. It was also overexpressed in asthma and the airway remodelling process [52].

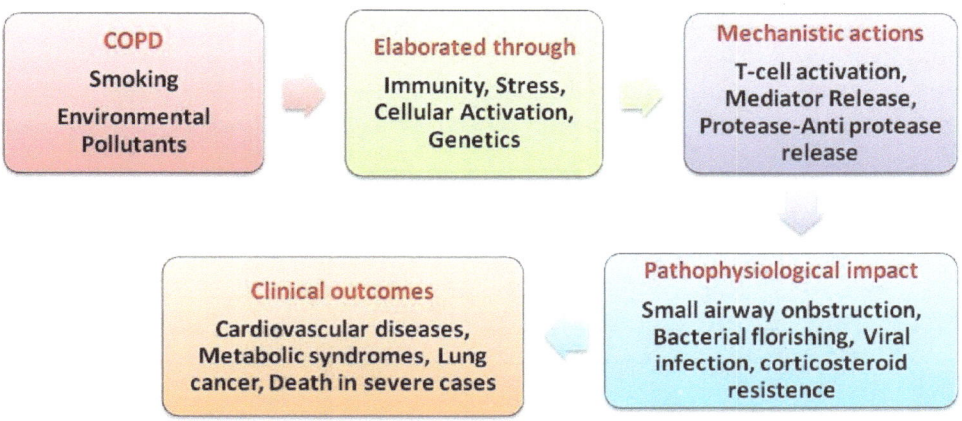

**Fig. (2).** COPD etiology, mechanisms, and impact.

# DRUG AND PHARMACOLOGICAL ADVANCEMENT

## Tissue and Systematic Inflammation

Tissue or local inflammation within the airways and lungs is more or less the primary condition to appear at the onset of the disease, which can be enhanced upon exacerbation, resulting in high neutrophil, lymphocyte, and eosinophil count [53]. Systematic or non-pulmonary artefacts in the form of muscle loss, cachexia, and cardiac failure, surface at the later stages of COPD. The systematic malfunctioning is mediated by inflammatory cells and cytokines.

# PHARMACEUTICAL INTERVENTIONS

## Corticosteroids

Inhaled corticosteroids (ICs) can be administered if symptoms persist, dependent on the lung function response. Corticosteroids are supposed to reduce inflammation, facilitating the airflow in the lungs. They may be inhaled, injected, or administered orally. The effective treatment plan must be tailor-made per the needs of patients. Factors such as age, gender, and lung conditions need to be considered before the treatment because so far, the predominant factors have been airflow obstruction, while eosinophil count has been the indicator. Other than these factors, bacterial growth, bronchiectasis, and asthmatic features must also be considered, including the search for biomarkers, which is of paramount importance.

***Fluticasone:*** Fluticasone or Flovent is available as an inhaler and has been used to treat COPD for a long time. Inhaled fluticasone or placebo was administered twice a day in the ISOLDE (Inhaled Steroids in Obstructive Lung Diseases in Europe) trial with 751 patients [54]. High dose fluticasone was also used for four months in another trial, where the administration of ICs was substituted with placebo and they showed enhanced exacerbations [55].

***Prednisolone*** has also been used in the preoperative regimen of coronary artery bypass surgery. A 20 mg dosage before 10 days of the surgery in patients, showed marked improvement in $FEV_1$. Patients having steroids needed lesser ventilation support, ICU retention, and hospital stay. The dosage of 2 mg nebulized budesonide and 30 mg prednisolone resulted in improved post-bronchodilator $FEV_1$ [56]. Prednisolone is available as a pill, liquid, or shot. It is also suggested in rescue treatment.

## Bronchodilators

Other than corticosteroids, bronchodilators are also used to open the airway which makes breathing easier. Bronchodilators and combinations of bronchodilators having different mechanisms are being given to the patients suffering from COPD. Bronchodilators may be short-acting and work within 15-20 minutes of administration or long-acting, which have to be taken once or twice daily by using inhalers or nebulizers. Long-acting bronchodilators can be either LABAs (long-acting beta2 agonists) or LAMAs (long-acting muscarinic antagonists) [57]. Tiotropium has a prolonged bronchodilator effect and has been shown to reduce exacerbations to an appreciable extent. Some common bronchodilators used for the purpose are as follows-

### SABA (Short-acting Inhaled Beta-agonists)

Proventil HFA®, ProAir®, Ventolin HFA® (albuterol)
Xopenex HFA®, Xopenex® (levalbuterol)

### SABA & SAMA (Short-acting Muscarinic Antagonist) Combination Bronchodilators

Combivent® (albuterol and ipratropium)
Duoneb® (albuterol and ipratropium)

### Long-Acting Bronchodilators

.1. **LAMAs:**
Incruse® (umeclidinium)
Seebri® (glycopyrrolate)
Spiriva® (tiotropium)
Tudorza® (aclidinium)
  1. LABAs:
Arcapta® (indacaterol)
Brovana® (arformoterol)
Perforomist® (formoterol)
Serevent® (salmeterol)
Stiverdi® (olodaterol)
Combinations of LAMA and LABA:
Anoro® (umeclidinium and vilanterol)
Stiolto® (olodaterol and tiotropium)
Utibron® (indacaterol and glycopyrrolate)
Bevespi® (glycopyrrolate and formoterol)

## Combination Therapies

Combination therapies of long-acting β-agonists (LABAs) and ICs have been tried [58]. Combining two drugs may improve the course of treatment. The synergistic activity of the ICs and LABAs has been demonstrated in several case studies [58]. Corticosteroids may translocate glucocorticoid from the cytoplasm to the nucleus which may be facilitated by a β-agonist leading to improved anti-inflammatory action. ICs act as activating agents to produce more β-receptors, which aid the bronchodilator activity of LABAs.

A 24-week trial of combined fluticasone and salmeterol showed marked improvement in lung functioning as compared to individual administration of placebo [59]. Pre-treatment FEV1 showed improvement as compared to placebo or with ICS or LABAs. A 17% significant decrease in risk was obtained when ICs were given in combination with LABAs, in a 44-week long study [60]. The TORCH (Towards a Revolution in COPD Health) trial involving nearly six thousand patients to fluticasone/salmeterol over three years, found a 17.5% risk reduction in death in combination therapy.

## Combination of Corticosteroids and Long-acting Bronchodilators

Budesonide/formoterol (Symbicort)
Fluticasone/salmeterol (Advair)
Fluticasone/vilanterol (Breo Ellipta)

## Triple Therapy

A combination of triple therapy with ICs, LABA, and LAMA may prove more beneficial for the improvement of lung function. A Single Inhaler Triple Therapy *vs* Inhaled Corticosteroid Plus Long-acting β2-Agonist Therapy (TRILOGY) study in patients with severe airflow defects, suggested that the triple therapy with beclomethasone dipropionate–formoterol fumarate–glycopyrronium bromide improved predose and postdose $FEV_1$ and SGRQ total score and reduced the exacerbation rate by 23% compared with beclomethasone dipropionate–formoterol fumarate [61]. A recent Lung Function and Quality of Life Assessment study showed that single-inhaler triple therapy was more beneficial than ICS-LABA [62]. Results of the COPD Treatment (IMPACT) study established that the triple combination of fluticasone furoate–umeclidinium–vilanterol lowered the rate of moderate or severe exacerbations more efficiently than the ICS-LABA (fluticasone furoate–vilanterol) and the LABA-LAMA (umeclidinium–vilanterol) combined dosing [63].

## Combination of Inhaled Corticosteroids and Two Long-acting Bronchodilators (Triple Therapy)

Fluticasone/vilanterol/umeclidinium (Trelegy Ellipta)

## Other Options

**Theophylline,** along with a bronchodilator, has been used in the initial stage of COPD. It works as an anti-inflammatory drug, relaxing the airway muscles. **Roflumilast** is another drug which is a phosphodiesterase-4 inhibitor, which eases the inflammation of the airway. Mucoactive drugs which have carbocysteine, erdosteine and N-acetylcysteine are used to relieve mucus.

## Drawbacks and Need for System Medicinal Approach

Current therapies and treatments for COPD are often insufficient to check its progression and are not very useful in completely eradicating the disease. Efficient and accurate tools are needed to predict personalized disease progression or response to treatment. Corticosteroids may result in [64]:

-renal disorders

-low bone density

-fracture risk

-increased risk of glaucoma

-skin bruising

-respiratory muscle weakness

The corticosteroid toxicity associated with the period of use and duration has been well documented [65]. Patients with stable COPD treated with systemic corticosteroids may have:

-glucose intolerance

-decreased serum levels of osteocalcin

-increased adrenal malfunction

-weight gain

-insomnia

-anxiety

-depression

Systems medicinal approach may improve and present a clear picture of lung health and disease progression, having the potential to facilitate the advancement of valuable, tailored and superlatively defensive treatment options. Targeted therapies with kinase inhibitors and therapeutic antibodies [64] can provide an explicit aid to control inter-and intracellular pathways. In conjugation, the two developments regulate the protease-antiprotease and oxidant-antioxidant balance, in a personalized manner [65].

## Simulated Medicinal Modeling and its Significance

Computational modelling simulation can prove handy to comprehend the human anatomical changes during disease development and to decode the functioning of living systems. Modelling simulations may provide deeper insight into the diagnostic advancements and therapeutic advancement, the stream popularly known as 'computational medicine'. Models not only mimic anatomical features but also physiological functions. Quantitative models can also disseminate the significant alterations and variations once the disease has set in [66]. Computational modelling has an edge over the conventional experimental approach, which involves molecule-to-molecule mapping to decode a biological jigsaw puzzle. High dimensional data obtained by virtual molecular medicinal modelling has the potential to frame molecular disease network, diagnosis, categorizing disease subtypes, prediction of clinical outcomes, and assessment of disease progression [67]. The main aim of computational simulations is to develop models mimicking biological systems, their physiology in the disease condition, any changes as the disease progresses, and ultimately the utilization of these calculations to produce improved and advanced therapeutic interventions. The modelling approach is crucial in having in-depth knowledge of the hierarchical nature of gene regulation and passing on of interactions among malignant phenotypes and other molecular agents within a biological network, thereof predicting disease progression, characterization, and clinical outcomes. Hence, modelling may elucidate the mechanism which is essential for drug design and computational learning paradigms for disease treatment [68].

## COMPUTATIONAL MEDICINAL SIMULATION FOR COPD TREATMENT

### Computational Lung Modelling

The diagnosis of chronic obstructive lung disease can be performed by assessing pulmonary lung function (PFT). But due to insufficient diagnostic methodologies, predictions of patient outcomes cannot be ascertained for individuals. Though the PFT measurements can be done to assess global lung functions, they do not measure the regional changes in the PFT, which is the peculiar characteristic of lung diseases. For accurate medical treatment and dose administration, patient-specific characteristics are required. Accurate measurements have been possible with the help of advanced imaging techniques, still, minutest of the morphological changes are not observed closely with either of them, like the small airways in the case of lungs. A combinatorial approach, taking into consideration both the imaging and patient-specific modelling predictions, can prove beneficial for superior healthcare [69]. Personalised airflow models have been used to check the response to therapy and ventilation levels in lungs affected with asthma [70]. Nonspecific representations were used in early lung models to represent airways and vessels, with the aid of imaging techniques; now, patient-specific models have been built to generate anatomical models. Functional outcomes are influenced by experimental and simulations. State-of-the-art models have been developed for 3D CFD, which includes high resolution computationally derived meshes of the central airway consisting of millions of computational elements. To have a clear overview of the full airway and vascular network, a one-dimensional strategy is required. The volume filling branching (VFB) algorithm has enabled to generate patient-specific network models, enabling the construction of airway branches with the help of high-resolution images [71]. To comprehend different aspects of the respiratory system, ventilation, perfusion, gas exchange, forced expiration, gas washout, impedance, particle transport, deposition, and tissue mechanics models have been developed. The main drawback of these modelling techniques is that they are more focused on healthy lung functioning and fewer studies have been focused on obstructive lung diseases. Fig. **3** represents the essential elements of a medicinal model.

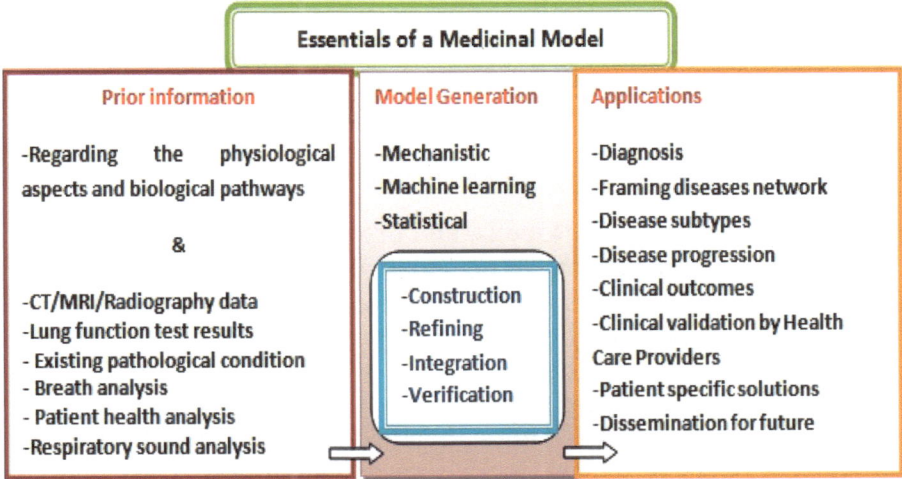

**Fig. (3).** Key parameters for the development of a medicinal model.

## Multiscale Modelling

Multiscale respiratory models are lesser in number and need more exploration. Donovan *et al.* have developed a multiscale model to study the impact of broncho construction on lung functions, linking the cellular and molecular level functions utilizing the central and distal airways [72]. Another potential model by Venegas *et al.* included airway network, distribution of airflow, and airway wall mechanics. The model observed ventilation in the functional images of asthma patients [73]. The behaviour could be simulated through the development of a fully integrated model to predict realistic outcomes, as shown in equation **1**.

$$P(t,n) = R(n)\dot{V}(t,n) + \frac{2^n}{C_L}\int_0^t \dot{V}(t',n)dt' + \frac{1}{C_{CW}}\int_0^t \dot{V}(t',0)dt' \quad (1)$$

$C_L$ and $C_W$ are the parameters representing lung parenchyma and chest wall. P, R, V, and t are pressure, resistance, flow, and time, respectively.

Another multiscale model to assess the airway edifice was developed considering the kinetics of agonist-receptor building and interactions with airway wall mechanisms, working on equations **2** and **3** [74].

$$\frac{dS_U}{dt} = -\alpha S_U + \beta S_\beta - \gamma S_U + S_{dep} \qquad (2)$$

$$\frac{dS_B}{dt} = \alpha S_U - \beta S_\beta \qquad (3)$$

$S_B$ and $S_U$ represent bound and unbound populations. α, β and γ represent circulation rates.

## Mechanistic Models

COPD affects several molecules, cells, and organs, thereby very few mechanistic models have been attempted for the personalized treatment, which is often based on ordinary differential equations (ODE) to monitor the change. These models are generally used to check the progression and exacerbation risk by exposure upon inhalation, inflammation, and the magnitude of airflow limitation to progression [75], thereby facilitating personalized estimation of ventilation/perfusion heterogeneity in the patients. Although none of these models has found any medical application, nevertheless, models based on computational fluid dynamics have found application in clinical optimization by the assessment of particle deposition in the airway tree [76] as extrapolative biomarkers. Mechanistic models may be further explored to make patient-specific predictions to assist clinical outcomes.

## Machine Learning Models

Machine learning algorithms make use of data to make improved and accurate predictions. Machine learning methods are divided into 1) Supervised ML algorithms, which are aided by annotated data like labelled emphysema in CT images, using them to identify new data samples. 2) Unsupervised ML algorithms that use unlabeled data to identify patterns like principal component analysis (PCA). Most of the COPD related machine-learning algorithms have been applied for a dataset, ranging from 16 to 300 patients [77], and no studies have been performed in a real-world clinical setting. To bridge this gap, a representative dataset is required, and the data must be based on an elaborate analysis of the patient to comprehend the multi-component background of the disease. So far, Machine Learning Mortality Prediction (MLMP-COPD) software has been used to predict all-cause mortality in COPD patients. The MLMP model was generated using the features of Cox regression. The MLMP model was also compared with the BMI, airflow obstruction, dyspnea, and exercise capacity index *etc* [78].

## Statistical models

The multilevel modelling approach works through a combination of classical statistical analysis and machine learning technique, for the personalized COPD risk assessment or course of treatment related to clinical and biomedical research data. Bayesian network algorithm is used to derive clinical variables of relevance for the prediction of exacerbation risk and recommend preventive action [79], along with survival risk factors by univariate analysis to develop probability distribution models to estimate mortality risk associated with COPD [80]. Many of these models work on the interface of personalized health behaviour and personal air exposure and exposure health association, like in the ODE model which combines exposure-health-relationship to a time-activity exposure [81]. The tailored prediction of disease risk remains the focal point of statistical models.

## COPD and Advancement in Modelling

A systematic medicinal approach is required to deal with the disease at a molecular level. Computational modelling may help in the personalized treatment of COPD through the integration of clinical and experimental factors. Recently, a 'System medicine-based clinical decision support for COPD patients' (SysMed-COPD) project was started internationally, considering the expertise of clinical scientists, computational researchers and bioinformatics engineers (Fig. **4**) [82]. There have been several studies focused on emphysema and fewer on COPD. The emphysema models have shown changes in parenchymal tissues and changes in the mechanical properties of tissues. The parenchyma tissues have been represented by a network of springs and individual rings depicting the lung alveolar tissues. Through modelling, the redistribution of forces in the tissues has been demonstrated in the progression of emphysema [83]. A 3D model was developed to study the correlation between the pattern of alveolar wall destruction and the mechanical properties of the tissues. The modelling predictions showed that the allocation of emphysematous lesions takes a significant part in organ functioning [84].

## COPD Progression Modelling

In approximately 70% of COPD patients, abnormalities, including emphysema, are first detected in small airways before changes occur in large airways, following a Tissue→Airway route. Only in 30% of the sufferers, the onset of the disease takes place in large airways, leading to small airway dysfunction and emphysema [85]. Early detection of COPD has remained a challenge for the medical fraternity. To understand the progression of COPD and observe the longitudinal changes in individuals, medicinal modelling has been used. An early stage of disease in healthy smokers is quite difficult, thereby preventing interventions in disease-modifying treatments. Modelling has been used to understand neurodegenerative diseases mimicking the long-term temporal progression [85]. With the application of SuStaIn, two subtypes were identified in healthy smokers.

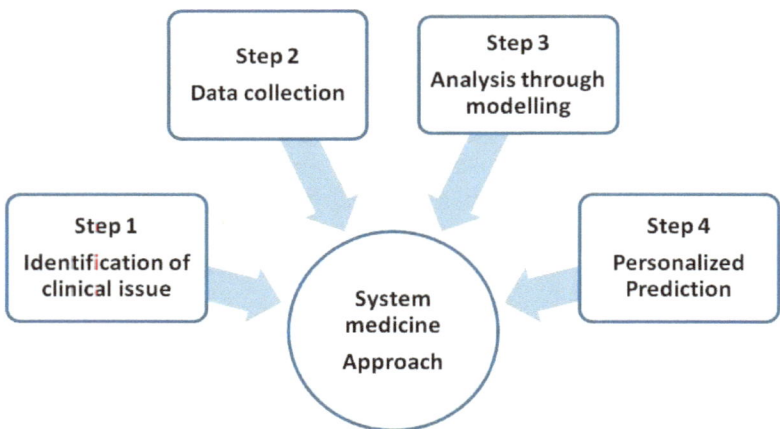

**Fig. (4).** Key elements of SysMed.

## Experimental *vs.* Computational medicinal modelling

Inbred mice models have been more frequently used for clinical studies [86]. However, the experimental models are not indispensable in every case, because of differences in pulmonary anatomy, immunological differences, and smoke-induced clinical phenotype. There are some changes, like *in vitro* cellular imaging of lung, early identification of COPD, and genetic manipulation, which cannot be analysed and performed in humans through experimental intervention [87]. *Ex vivo* cultures of surgically removed lungs have also been used as complementary models.

A computational model of system medicine mimics the functioning of a complex system and can predict future functioning as well. The model may have sub-classes as well. The mechanistic models are based upon the dynamic interactions of biological entities. The models sum up the existing knowledge about a biological system and predict the futuristic behaviour once a disease has occurred. Statistical and machine-learning models predict relationship rather than the underlying mode of action and offer accurate predictions, having clinically pertinent results. Some of the models from the beginning until the present, used to deal with different aspects of lung physiological studies and COPD, are summarized in Table **1**.

Table 1. Some simulated models dealing with different aspects of COPD.

| S.No. | Model | Function | Year | Reference |
|---|---|---|---|---|
| 1 | Symmetric Model | Airway structure simulation | 1963 | [88] |
| 2 | Asymmetric model | Morphometric information about the airway structure using resin casts | 1971 | [89] |
| 3 | Deterministic rule-based algorithm | 3D airway tree generation | 1999 | [90] |
| 4 | Space-filling algorithm | 3D airway tree generation defined by lung lobes. | 2004 | [91] |
| 5 | Static model | Asthma sensitivity and airway construction. | 2006 | [92] |
| 6 | Mathematical model | Prediction of topological distribution of inhaled air in the healthy lung. | 2012 | [93] |
| 7 | Multibranch airway lung model-Stochastic model | To predict the disparity in airway diameter. | 2012 | [94] |
| 8 | Machine learning framework | Estimation of lung tissue expansion and contraction in COPD diagnosis. | 2013 | [95] |
| 9 | CHRONIOUS system | Analysis and real-time evaluation of the patient's condition. | 2014 | [96] |
| 10 | Subject-specific and generic model | Quantitative comparison of lung hemodynamic response to vascular occlusion, and the FEV1. | 2015 | [97] |
| 11 | Patient-specific model | Construction of airway tree from high-resolution CT data. | 2015 | [98] |
| 12 | Neuro Fuzzy | Diagnosis of Asthma and COPD | 2018 | [99] |
| 13 | Expert Diagnostic System | Distinguish among patients with asthma, COPD or a normal lung function based on lung function. | 2018 | [100] |
| 14 | Deep Artificial Neural Networks including Feed-Forward Neural Networks (FFNN). | Prediction of exacerbation for classification of COPD patients and subsequent triage. | 2018 | [101] |
| 15 | Patient-specific model | Evaluation of COPD treatment options | 2019 | [102] |

# LIMITATIONS AND NEED FOR FUTURE ADVANCEMENT

Although the management of COPD over the years has become more personalized rather than general, yet there is still a great demand for P4 medicine for the management of the disease. The prediction of the effective or non-effective effect of treatment and medicinal intervention for specific individuals is still challenging, just as it is to understand the primary or secondary prevention of COPD. Underreporting and timely diagnosis is another persisting challenge. Patients with mild lung malfunctioning usually do not report for diagnosis [103].

There are several non-reported comorbidities, including cardiovascular comorbidities observed in patients where they are likely to receive a cardiac failure therapy, which could lead to more severe prospects [104] and the recommendation of beta-blockers in such case is not sufficient.

Effective medicinal COPD modelling must be accessible, interoperable, and reusable to bridge the gap between machine learning and mechanism-based modelling. Multi-scale, integrated models are needed to produce medicinally relevant outputs, answering specific medicinal queries. For the authentication of results, high-quality data is required to build association-based models.

Other than improving the methodology of model building, trained clinical modellers and translators are required to mediate between the clinician and the model. With the advancement of machine learning strategies, medical benefits would be greater, leading to minimal risk, and the development of a dynamic decision system in context to clinical treatment. Computational systems may assist clinical decision making, therefore there is an urgent need for user-friendly and accessible models for clinical researchers to obtain solutions to the problems through *in-silico* analysis.

## CONCLUDING REMARKS

COPD is categorized by airway and airflow impediments, posing a potent threat to human beings. The disease impacts the lung physiology at the tissue, cellular and sub-cellular scale. For effective treatment and diagnosis of the disease, it is of imperative importance to understand it at multiple biological scales. Experimental assessment of the complexity of the respiratory system is a difficult task to achieve; hence computational modelling offers a powerful approach to comprehend disease severity and disease sub-types which is essential for improved diagnosis. Simulating a realistic situation or anatomy is a challenging task. The modelling process needs many assumptions to work upon to mimic the real-life situation completely and accurately. Simplified assumptions should be considered to assess the outcomes of the model. To assess the progression of COPD, there is a greater need to come up with 'patient-specific models as only some parts of a model can be customized for specific patients, leaving many parameters like tissue properties, morphological changes in small airways, *etc*. Genetic changes related to the environment are also usually left undesired and can be explored through modelling, and the conceptualization of new hypotheses can be achieved by mathematical modelling. There is an emerging need for combined efforts of modellers and clinicians to develop patient-specific models, which may prove very useful for ascertaining the progression of the disease and the suitable course of medication. Validation of a model is also necessary from the accuracy, reliability, and precision point of view.

## CONSENT FOR PUBLICATION

Not Applicable.

## CONFLICT OF INTEREST

The author confirms that this chapter contents have no conflict of interest.

## ACKNOWLEDGEMENT

Declared none.

## REFERENCES

[1] Chen B, Kan H. Air pollution and population health: a global challenge. Environ Health Prev Med 2008; 13(2): 94-101.
[http://dx.doi.org/10.1007/s12199-007-0018-5] [PMID: 19568887]

[2] Gurjar BR, Butler TM, Lawrence MG, Lelieveld J. Evaluation of emissions and air quality in megacities. Atmos Environ 2008; 42: 1593-606.
[http://dx.doi.org/10.1016/j.atmosenv.2007.10.048]

[3] Grahame TJ, Klemm R, Schlesinger RB. Public health and components of particulate matter: the changing assessment of black carbon. J Air Waste Manag Assoc 2014; 64(6): 620-60.
[http://dx.doi.org/10.1080/10962247.2014.912692] [PMID: 25039199]

[4] Decramer M, Janssens W, Miravitlles M. Chronic obstructive pulmonary disease. Lancet 2012; 379(9823): 1341-51.
[http://dx.doi.org/10.1016/S0140-6736(11)60968-9] [PMID: 22314182]

[5] World Health Statistics Geneva. Switzerland 2008.

[6] Chung KF, Barnes PJ. Cytokines in asthma. Thorax 1999; 54(9): 825-57.
[http://dx.doi.org/10.1136/thx.54.9.825] [PMID: 10456976]

[7] Springer J, Geppetti P, Fischer A, Groneberg DA. Calcitonin gene-related peptide as inflammatory mediator. Pulm Pharmacol Ther 2003; 16(3): 121-30.
[http://dx.doi.org/10.1016/S1094-5539(03)00049-X] [PMID: 12749828]

[8] Ku¨nzli N, Perez L, Rapp R. Air quality and health. Lausanne: European Respiratory Society 2010.

[9] Adeloye D, Chua S, Lee C, *et al*. Global and regional estimates of COPD prevalence: Systematic review and meta-analysis. J Glob Health 2015; 5(2): 020415-5.
[http://dx.doi.org/10.7189/jogh.05.020415] [PMID: 26755942]

[10] Camp PG, Ramirez-Venegas A, Sansores RH, *et al*. COPD phenotypes in biomass smoke- *versus* tobacco smoke-exposed Mexican women. Eur Respir J 2014; 43(3): 725-34.
[http://dx.doi.org/10.1183/09031936.00206112] [PMID: 24114962]

[11] de Torres JP, Casanova C, Hernández C, *et al*. Gender and COPD in patients attending a pulmonary clinic. Chest 2005; 128(4): 2012-6.
[http://dx.doi.org/10.1378/chest.128.4.2012] [PMID: 16236849]

[12] Hanson C, Rutten EP, Wouters EF, Rennard S. Influence of diet and obesity on COPD development and outcomes. Int J Chron Obstruct Pulmon Dis 2014; 9: 723-33.

[http://dx.doi.org/10.2147/COPD.S50111] [PMID: 25125974]

[13] Hancox RJ, Rasmussen F. Does physical fitness enhance lung function in children and young adults? Eur Respir J 2018; 51(2)1701374
[http://dx.doi.org/10.1183/13993003.01374-2017] [PMID: 29386347]

[14] Hurst JR, Vestbo J, Anzueto A, et al. Susceptibility to exacerbation in chronic obstructive pulmonary disease. N Engl J Med 2010; 363(12): 1128-38.
[http://dx.doi.org/10.1056/NEJMoa0909883] [PMID: 20843247]

[15] Schikowski T, Sugiri D, Ranft U, et al. Long-term air pollution exposure and living close to busy roads are associated with COPD in women. Respir Res 2005; 6: 152.
[http://dx.doi.org/10.1186/1465-9921-6-152] [PMID: 16372913]

[16] Zanobetti A, Schwartz J, Dockery DW. Airborne particles are a risk factor for hospital admissions for heart and lung disease. Environ Health Perspect 2000; 108(11): 1071-7.
[http://dx.doi.org/10.1289/ehp.001081071] [PMID: 11102299]

[17] Stieb DM, Szyszkowicz M, Rowe BH, Leech JA. Air pollution and emergency department visits for cardiac and respiratory conditions: a multi-city time-series analysis. Environ Health 2009; 8: 25.
[http://dx.doi.org/10.1186/1476-069X-8-25] [PMID: 19515235]

[18] Belleudi V, Faustini A, Stafoggia M, et al. Impact of fine and ultrafine particles on emergency hospital admissions for cardiac and respiratory diseases. Epidemiology 2010; 21(3): 414-23.
[http://dx.doi.org/10.1097/EDE.0b013e3181d5c021] [PMID: 20386174]

[19] Tsai SS, Chiu HF, Liou SH, Yang CY. Short-term effects of fine particulate air pollution on hospital admissions for respiratory diseases: a case-crossover study in a tropical city. J Toxicol Environ Health A 2014; 77(18): 1091-101.
[http://dx.doi.org/10.1080/15287394.2014.922388] [PMID: 25072896]

[20] Atkinson RW, Mills IC, Walton HA, Anderson HR. Fine particle components and health--a systematic review and meta-analysis of epidemiological time series studies of daily mortality and hospital admissions. J Expo Sci Environ Epidemiol 2015; 25(2): 208-14.
[http://dx.doi.org/10.1038/jes.2014.63] [PMID: 25227730]

[21] Lee IM, Tsai SS, Chang CC, Ho CK, Yang CY. Air pollution and hospital admissions for chronic obstructive pulmonary disease in a tropical city: Kaohsiung, Taiwan. Inhal Toxicol 2007; 19(5): 393-8.
[http://dx.doi.org/10.1080/08958370601174818] [PMID: 17365044]

[22] Yang CY, Chen CJ. Air pollution and hospital admissions for chronic obstructive pulmonary disease in a subtropical city: Taipei, Taiwan. J Toxicol Environ Health A 2007; 70(14): 1214-9.
[http://dx.doi.org/10.1080/15287390701380880] [PMID: 17573635]

[23] Santus P, Russo A, Madonini E, et al. How air pollution influences clinical management of respiratory diseases. A case-crossover study in Milan. Respir Res 2012; 13: 95.
[http://dx.doi.org/10.1186/1465-9921-13-95] [PMID: 23078274]

[24] Slaughter JC, Kim E, Sheppard L, Sullivan JH, Larson TV, Claiborn C. Association between particulate matter and emergency room visits, hospital admissions and mortality in Spokane, Washington. J Expo Anal Environ Epidemiol 2005; 15(2): 153-9.
[http://dx.doi.org/10.1038/sj.jea.7500382] [PMID: 15187986]

[25] Mustafić H, Jabre P, Caussin C, et al. Main air pollutants and myocardial infarction: a systematic review and meta-analysis. JAMA 2012; 307(7): 713-21.
[http://dx.doi.org/10.1001/jama.2012.126] [PMID: 22337682]

[26] Liang WM, Liu WP, Kuo HW. Diurnal temperature range and emergency room admissions for chronic obstructive pulmonary disease in Taiwan. Int J Biometeorol 2009; 53(1): 17-23.
[http://dx.doi.org/10.1007/s00484-008-0187-y] [PMID: 18989710]

[27] Nations within a nation: variations in epidemiological transition across the states of India, 1990-2016 in the Global Burden of Disease Study. Lancet 2017; 390(10111): 2437-60.

[http://dx.doi.org/10.1016/S0140-6736(17)32804-0] [PMID: 29150201]

[28] 2017.https://mohfw.gov.in/documents/policy

[29] GBD compare data visualization https://vizhub.healthdata.org/gbd-compare/

[30] Salvi S, Agrawal A. India needs a national COPD prevention and control programme. J Assoc Physicians India 2012; 60 (Suppl.): 5-7.
[PMID: 23155805]

[31] Gupta D, Agarwal R, Aggarwal AN, *et al.* Guidelines for diagnosis and management of chronic obstructive pulmonary disease: Joint ICS/NCCP (I) recommendations. Lung India 2013; 30(3): 228-67.
[http://dx.doi.org/10.4103/0970-2113.116248] [PMID: 24049265]

[32] KalagoudaMahishale V, Angadi N, Metgudmath V, Lolly M, Eti A, Khan S. The prevalence of chronic obstructive pulmonary disease and the determinants of underdiagnosis in women exposed to biomass fuel in India—a cross section study. Chonnam Med J 2016; 52(2): 117-22.
[http://dx.doi.org/10.4068/cmj.2016.52.2.117] [PMID: 27231676]

[33] Salvi SS, Barnes PJ. Chronic obstructive pulmonary disease in non-smokers. Lancet 2009; 374(9691): 733-43.
[http://dx.doi.org/10.1016/S0140-6736(09)61303-9] [PMID: 19716966]

[34] Kodgule R, Salvi S. Exposure to biomass smoke as a cause for airway disease in women and children. Curr Opin Allergy Clin Immunol 2012; 12(1): 82-90.
[http://dx.doi.org/10.1097/ACI.0b013e32834ecb65] [PMID: 22157154]

[35] KalagoudaMahishale V, Angadi N, Metgudmath V, Lolly M, Eti A, Khan S. Angadi N, Metgudmath V, Lolly M, Eti A, Khan S. The prevalence of chronic obstructive pulmonary disease and the determinants of underdiagnosis in women exposed to biomass fuel in India—a cross section study. Chonnam Med J 2016; 52(2): 117-22.
[http://dx.doi.org/10.4068/cmj.2016.52.2.117] [PMID: 27231676]

[36] Salvi SS, Apte KK, Dhar R, *et al.* Asthma insights and management in India: lessons learnt from the Asia Pacific-Asthma Insights and Management (AP-AIM) study. J Assoc Physicians India 2015; 63(9): 36-43.
[PMID: 27608865]

[37] Vanjare N, Chhowala S, Madas S, Kodgule R, Gogtay J, Salvi S. Use of spirometry among chest physicians and primary care physicians in India. NPJ Prim Care Respir Med 2016; 26: 16036.
[http://dx.doi.org/10.1038/npjpcrm.2016.36] [PMID: 27385406]

[38] Global, regional, and national disability-adjusted life-years (DALYs) for 333 diseases and injuries and healthy life expectancy (HALE) for 195 countries and territories, 1990-2016: a systematic analysis for the Global Burden of Disease Study 2016. Lancet 2017; 390(10100): 1260-344.
[http://dx.doi.org/10.1016/S0140-6736(17)32130-X] [PMID: 28919118]

[39] Künzli N, Perez L, Rapp R. Air quality and health. Lausanne: European Respiratory Society 2010.

[40] Chang YK, Wu CC, Lee LT, Lin RS, Yu YH, Chen YC. The short-term effects of air pollution on adolescent lung function in Taiwan. Chemosphere 2012; 87(1): 26-30.
[http://dx.doi.org/10.1016/j.chemosphere.2011.11.048] [PMID: 22189374]

[41] Hogg JC. Pathophysiology of airflow limitation in chronic obstructive pulmonary disease. Lancet 2004; 364(9435): 709-21.
[http://dx.doi.org/10.1016/S0140-6736(04)16900-6] [PMID: 15325838]

[42] Mossman BT, Lounsbury KM, Reddy SP. Oxidants and signaling by mitogen-activated protein kinases in lung epithelium. Am J Respir Cell Mol Biol 2006; 34(6): 666-9.
[http://dx.doi.org/10.1165/rcmb.2006-0047SF] [PMID: 16484683]

[43] Haddad E-B, Salmon M, Koto H, Barnes PJ, Adcock I, Chung KF. Ozone induction of cytokine-

induced neutrophil chemoattractant (CINC) and nuclear factor-kappa b in rat lung: inhibition by corticosteroids. FEBS Lett 1996; 379(3): 265-8.
[http://dx.doi.org/10.1016/0014-5793(95)01524-8] [PMID: 8603703]

[44] Rahman I, Mulier B, Gilmour PS. T, Watchorn.; Donaldson, K.; Jeffery, P.K.; Nee, W.M. Oxidant-mediated lung epithelial cell tolerance: the role of intracellular glutathione and nuclear factor-kB. Biochem Pharmacol 2001; 62: 787-94.
[http://dx.doi.org/10.1016/S0006-2952(01)00702-X] [PMID: 11551525]

[45] Soler N, Ewig S, Torres A, Filella X, Gonzalez J, Zaubet A. Airway inflammation and bronchial microbial patterns in patients with stable chronic obstructive pulmonary disease. Eur Respir J 1999; 14(5): 1015-22.
[http://dx.doi.org/10.1183/09031936.99.14510159] [PMID: 10596683]

[46] Ghio AJ, Silbajoris R, Carson JL, Samet JM. Biologic effects of oil fly ash. Environ Health Perspect 2002; 110 (Suppl. 1): 89-94.
[http://dx.doi.org/10.1289/ehp.02110s1189] [PMID: 11834466]

[47] Valko M, Morris H, Cronin MT. Metals, toxicity and oxidative stress. Curr Med Chem 2005; 12(10): 1161-208.
[http://dx.doi.org/10.2174/0929867053764635] [PMID: 15892631]

[48] Paredi P, Kharitonov SA, Barnes PJ. Analysis of expired air for oxidation products. Am J Respir Crit Care Med 2002; 166(12 Pt 2): S31-7.
[http://dx.doi.org/10.1164/rccm.2206012] [PMID: 12471086]

[49] Shapiro SD. Proteolysis in the lung. Eur Respir J Suppl 2003; 44: 30s-2s.
[http://dx.doi.org/10.1183/09031936.03.00000903a] [PMID: 14582898]

[50] Hautamaki RD, Kobayashi DK, Senior RM, Shapiro SD. Requirement for macrophage elastase for cigarette smoke-induced emphysema in mice. Science 1997; 277(5334): 2002-4.
[http://dx.doi.org/10.1126/science.277.5334.2002] [PMID: 9302297]

[51] Groneberg DA, Witt H, Adcock IM, Hansen G, Springer J. Smads as intracellular mediators of airway inflammation. Exp Lung Res 2004; 30(3): 223-50.
[http://dx.doi.org/10.1080/01902140490276320] [PMID: 15195555]

[52] Wills-Karp M, Luyimbazi J, Xu X, *et al.* Interleukin-13: central mediator of allergic asthma. Science 1998; 282(5397): 2258-61.
[http://dx.doi.org/10.1126/science.282.5397.2258] [PMID: 9856949]

[53] Hurst JR, Perera WR, Wilkinson TMA, Donaldson GC, Wedzicha JA. Systemic and upper and lower airway inflammation at exacerbation of chronic obstructive pulmonary disease. Am J Respir Crit Care Med 2006; 173(1): 71-8.
[http://dx.doi.org/10.1164/rccm.200505-704OC] [PMID: 16179639]

[54] Burge PS, Calverley PM, Jones PW, Spencer S, Anderson JA, Maslen TK. Randomised, double blind, placebo controlled study of fluticasone propionate in patients with moderate to severe chronic obstructive pulmonary disease: the ISOLDE trial. BMJ 2000; 320(7245): 1297-303.
[http://dx.doi.org/10.1136/bmj.320.7245.1297] [PMID: 10807619]

[55] van der Valk P, Monninkhof E, van der Palen J, Zielhuis G, van Herwaarden C. Effect of discontinuation of inhaled corticosteroids in patients with chronic obstructive pulmonary disease: the COPE study. Am J Respir Crit Care Med 2002; 166(10): 1358-63.
[http://dx.doi.org/10.1164/rccm.200206-512OC] [PMID: 12406823]

[56] Maltais F, Ostinelli J, Bourbeau J, *et al.* Comparison of nebulized budesonide and oral prednisolone with placebo in the treatment of acute exacerbations of chronic obstructive pulmonary disease: a randomized controlled trial. Am J Respir Crit Care Med 2002; 165(5): 698-703.
[http://dx.doi.org/10.1164/ajrccm.165.5.2109093] [PMID: 11874817]

[57] Salpeter SR. Bronchodilators in COPD: impact of β-agonists and anticholinergics on severe

exacerbations and mortality. Int J Chron Obstruct Pulmon Dis 2007; 2(1): 11-8.
[http://dx.doi.org/10.2147/copd.2007.2.1.11] [PMID: 18044061]

[58] Szafranski W, Cukier A, Ramirez A, *et al.* Efficacy and safety of budesonide/formoterol in the management of chronic obstructive pulmonary disease. Eur Respir J 2003; 21(1): 74-81.
[http://dx.doi.org/10.1183/09031936.03.00031402] [PMID: 12570112]

[59] Mahler DA, Wire P, Horstman D, *et al.* Effectiveness of fluticasone propionate and salmeterol combination delivered *via* the Diskus device in the treatment of chronic obstructive pulmonary disease. Am J Respir Crit Care Med 2002; 166(8): 1084-91.
[http://dx.doi.org/10.1164/rccm.2112055] [PMID: 12379552]

[60] Kardos P, Wencker M, Glaab T, Vogelmeier C. Impact of salmeterol/fluticasone propionate *versus* salmeterol on exacerbations in severe chronic obstructive pulmonary disease. Am J Respir Crit Care Med 2007; 175(2): 144-9.
[http://dx.doi.org/10.1164/rccm.200602-244OC] [PMID: 17053207]

[61] Singh D, Papi A, Corradi M, *et al.* Single inhaler triple therapy *versus* inhaled corticosteroid plus long-acting β2-agonist therapy for chronic obstructive pulmonary disease (TRILOGY): a double-blind, parallel group, randomised controlled trial. Lancet 2016; 388(10048): 963-73.
[http://dx.doi.org/10.1016/S0140-6736(16)31354-X] [PMID: 27598678]

[62] Lipson DA, Barnacle H, Birk R, *et al.* Zhu, Chang-Qing.; Pascoe, S.J. FULFIL trial: once-daily triple therapy for patients with chronic obstructive pulmonary disease. Am J Respir Crit Care Med 2017; 196(4): 438-46.
[http://dx.doi.org/10.1164/rccm.201703-0449OC] [PMID: 28375647]

[63] Lipson DA, Barnhart F, Brealey N, *et al.* for the IMPACT Investigators Once-daily single-inhaler triple *versus* dual therapy in patients with COPD. N Engl J Med 2018; 378(18): 1671-80.
[http://dx.doi.org/10.1056/NEJMoa1713901] [PMID: 29668352]

[64] Tattersfield AE, Harrison TW, Hubbard RB, Mortimer K. Safety of inhaled corticosteroids. Proc Am Thorac Soc 2004; 1(3): 171-5.
[http://dx.doi.org/10.1513/pats.200402-016MS] [PMID: 16113431]

[65] Boumpas DT, Chrousos GP, Wilder RL, Cupps TR, Balow JE. Glucocorticoid therapy for immune-mediated diseases: basic and clinical correlates. Ann Intern Med 1993; 119(12): 1198-208.
[http://dx.doi.org/10.7326/0003-4819-119-12-199312150-00007] [PMID: 8239251]

[66] Hood L, Heath JR, Phelps ME, Lin B. Systems biology and new technologies enable predictive and preventative medicine. Science 2004; 306(5696): 640-3.
[http://dx.doi.org/10.1126/science.1104635] [PMID: 15499008]

[67] Schadt EE, Björkegren JL. NEW: network-enabled wisdom in biology, medicine, and health care. Sci Transl Med 2012; 4(115)115rv1
[http://dx.doi.org/10.1126/scitranslmed.3002132] [PMID: 22218693]

[68] Pinto-Plata VM, Cote C, Cabral H, Taylor J, Celli BR. The 6-min walk distance: change over time and value as a predictor of survival in severe COPD. Eur Respir J 2004; 23(1): 28-33.
[http://dx.doi.org/10.1183/09031936.03.00034603] [PMID: 14738227]

[69] Feng Yu, Xiaole C, Yang M, Dong KJ. Multiscale Computational Models for Respiratory Aerosol Dynamics with Medical Applications. Comput. Math Method M. 2019; p. 2.

[70] Pauwels RA, Buist AS, Calverley PM, Jenkins CR, Hurd SS. Global strategy for the diagnosis, management, and prevention of chronic obstructive pulmonary disease. NHLBI/WHO Global Initiative for Chronic Obstructive Lung Disease (GOLD) Workshop summary. Am J Respir Crit Care Med 2001; 163(5): 1256-76.
[http://dx.doi.org/10.1164/ajrccm.163.5.2101039] [PMID: 11316667]

[71] Cheng YK, Beroukhim R, Levine RL, Mellinghoff IK, Holland EC, Michor F. A mathematical methodology for determining the temporal order of pathway alterations arising during gliomagenesis.

PLOS Comput Biol 2012; 8(1)e1002337
[http://dx.doi.org/10.1371/journal.pcbi.1002337] [PMID: 22241976]

[72] Donovan GM, Sneyd J, Tawhai MH. The importance of synergy between deep inspirations and fluidization in reversing airway closure. PLoS One 2012; 7(11)e48552
[http://dx.doi.org/10.1371/journal.pone.0048552] [PMID: 23144901]

[73] Venegas JG, Winkler T, Musch G, *et al.* Self-organized patchiness in asthma as a prelude to catastrophic shifts. Nature 2005; 434(7034): 777-82.
[http://dx.doi.org/10.1038/nature03490] [PMID: 15772676]

[74] Amin SD, Majumdar A, Frey U, Suki B. Modeling the dynamics of airway constriction: effects of agonist transport and binding. J Appl Physiol (1985) 2010; 109(2): 553-63.
[http://dx.doi.org/10.1152/japplphysiol.01111.2009] [PMID: 20507971]

[75] Agustí A, Compte A, Faner R, *et al.* The EASI model: A first integrative computational approximation to the natural history of COPD. PLoS One 2017; 12(10)e0185502
[http://dx.doi.org/10.1371/journal.pone.0185502] [PMID: 29016620]

[76] Zhang B, Qi S, Yue Y, *et al.* Particle disposition in the realistic airway tree models of subjects with tracheal bronchus and COPD. BioMed Res Int 2018; 20187428609
[http://dx.doi.org/10.1155/2018/7428609] [PMID: 30155481]

[77] Gurbeta L, Badnjevic A, Maksimovic M, Omanovic-Miklicanin E, Sejdic E. A telehealth system for automated diagnosis of asthma and chronical obstructive pulmonary disease. J Am Med Inform Assoc 2018; 25(9): 1213-7.
[http://dx.doi.org/10.1093/jamia/ocy055] [PMID: 29788482]

[78] Moll M, Qiao D, Regan EA, *et al.* Machine Learning and Prediction of All-Cause Mortality in Chronic Obstructive Pulmonary Disease. Chest 2020.
[http://dx.doi.org/10.1016/j.chest.2020.02.079]

[79] Mets OM, Buckens CFM, Zanen P, *et al.* Identification of chronic obstructive pulmonary disease in lung cancer screening computed tomographic scans. JAMA 2011; 306(16): 1775-81.
[http://dx.doi.org/10.1001/jama.2011.1531] [PMID: 22028353]

[80] He H, Sun Y, Sun B, Zhan Q. Application of a parametric model in the mortality risk analysis of ICU patients with severe COPD. Clin Respir J 2018; 12(2): 491-8.
[http://dx.doi.org/10.1111/crj.12549] [PMID: 27606821]

[81] Moore E, Chatzidiakou L, Jones RL, Smeeth L, Beevers S, Kelly FJ. Linking e-healthrecords, patient-reported symptoms and environmental exposure data to characterise and model COPD exacerbations: protocol for the COPE study. BMJ Open 2016; 6011330
[http://dx.doi.org/10.1136/bmjopen-2016-011330]

[82] Agusti A, Sobradillo P, Celli B. Addressing the complexity of chronic obstructive pulmonary disease: from phenotypes and biomarkers to scale-free networks, systems biology, and P4 medicine. Am J Respir Crit Care Med 2011; 183(9): 1129-37.
[http://dx.doi.org/10.1164/rccm.201009-1414PP] [PMID: 21169466]

[83] Suki B, Lutchen KR, Ingenito EP. On the progressive nature of emphysema: roles of proteases, inflammation, and mechanical forces. Am J Respir Crit Care Med 2003; 168(5): 516-21.
[http://dx.doi.org/10.1164/rccm.200208-908PP] [PMID: 12941655]

[84] Parameswaran H, Majumdar A, Suki B. Linking microscopic spatial patterns of tissue destruction in emphysema to macroscopic decline in stiffness using a 3D computational model. PLOS Comput Biol 2011; 7(4)e1001125
[http://dx.doi.org/10.1371/journal.pcbi.1001125] [PMID: 21533072]

[85] Young AL, Oxtoby NP, Daga P, *et al.* A data-driven model of biomarker changes in sporadic Alzheimer's disease. Brain 2014; 137(Pt 9): 2564-77.
[http://dx.doi.org/10.1093/brain/awu176] [PMID: 25012224]

[86]  Seimetz M, Parajuli N, Pichl A, et al. Inducible NOS inhibition reverses tobacco-smoke-induced emphysema and pulmonary hypertension in mice. Cell 2011; 147(2): 293-305.
[http://dx.doi.org/10.1016/j.cell.2011.08.035] [PMID: 22000010]

[87]  Reyfman PA, Walter JM, Joshi N, et al. Single-cell transcriptomic analysis of human lung provides insights into the pathobiology of pulmonary fibrosis. Am J Respir Crit Care Med 2019; 199(12): 1517-36.
[http://dx.doi.org/10.1164/rccm.201712-2410OC] [PMID: 30554520]

[88]  Weibel E. Morphometry of the Human Lung 1963.http://link.springer.com/book/10
[http://dx.doi.org/10.1007/978-3-642-87553-3]

[89]  Horsfield K, Dart G, Olson DE, Filley GF, Cumming G. Models of the human bronchial tree. J Appl Physiol 1971; 31(2): 207-17.
[http://dx.doi.org/10.1152/jappl.1971.31.2.207] [PMID: 5558242]

[90]  Kitaoka H, Takaki R, Suki B. A three-dimensional model of the human airway tree. J Appl Physiol (1985) 1999; 87(6): 2207-17.
[http://dx.doi.org/10.1152/jappl.1999.87.6.2207] [PMID: 10601169]

[91]  Tawhai MH, Hunter P, Tschirren J, Reinhardt J, McLennan G, Hoffman EA. CT-based geometry analysis and finite element models of the human and ovine bronchial tree. J Appl Physiol (1985) 2004; 97(6): 2310-21.
[http://dx.doi.org/10.1152/japplphysiol.00520.2004] [PMID: 15322064]

[92]  Affonce DA, Lutchen KR. New perspectives on the mechanical basis for airway hyperreactivity and airway hypersensitivity in asthma. J Appl Physiol (1985) 2006; 101(6): 1710-9.
[http://dx.doi.org/10.1152/japplphysiol.00344.2006] [PMID: 16902064]

[93]  Swan AJ, Clark AR, Tawhai MH. A computational model of the topographic distribution of ventilation in healthy human lungs. J Theor Biol 2012; 300: 222-31.
[http://dx.doi.org/10.1016/j.jtbi.2012.01.042] [PMID: 22326472]

[94]  Leary D, Bhatawadekar SA, Parraga G, Maksym GN. Modeling stochastic and spatial heterogeneity in a human airway tree to determine variation in respiratory system resistance. J Appl Physiol (1985) 2012; 112(1): 167-75.
[http://dx.doi.org/10.1152/japplphysiol.00633.2011] [PMID: 21998266]

[95]  Bodduluri S, Newell JD Jr, Hoffman EA, Reinhardt JM. Registration-based lung mechanical analysis of chronic obstructive pulmonary disease (COPD) using a supervised machine learning framework. Acad Radiol 2013; 20(5): 527-36.
[http://dx.doi.org/10.1016/j.acra.2013.01.019] [PMID: 23570934]

[96]  Bellos CC, Papadopoulos A, Rosso R, Fotiadis DI. Identification of COPD patients' health status using an intelligent system in the CHRONIOUS wearable platform. IEEE J Biomed Health Inform 2014; 18(3): 731-8.
[http://dx.doi.org/10.1109/JBHI.2013.2293172] [PMID: 24808219]

[97]  Hedges KL, Clark AR, Tawhai MH. Comparison of generic and subject-specific models for simulation of pulmonary perfusion and forced expiration. Interface Focus 2015; 5(2)20140090
[http://dx.doi.org/10.1098/rsfs.2014.0090] [PMID: 25844154]

[98]  Bordas R, Lefevre C, Veeckmans B, et al. Development and analysis of patient-based complete conducting airways models. PLoS One 2015; 10(12)e0144105
[http://dx.doi.org/10.1371/journal.pone.0144105] [PMID: 26656288]

[99]  Rao A, Huynh E, Royston TJ, Kornblith A, Roy S. Acoustic Methods for Pulmonary Diagnosis. IEEE Rev Biomed Eng 2019; 12: 221-39.
[http://dx.doi.org/10.1109/RBME.2018.2874353] [PMID: 30371387]

[100]  Badnjevic A, Gurbeta L, Custovic E. An expert diagnostic system to automatically identify asthma and chronic obstructive pulmonary disease in clinical settings. Sci Rep 2018; 8(1): 11645.

[http://dx.doi.org/10.1038/s41598-018-30116-2] [PMID: 30076356]

[101] Nunavath V, Goodwin M, Fidje JT, Moe CE. Deep Neural Networks for Prediction of Exacerbations of Patients with Chronic Obstructive Pulmonary Disease. International Conference on Engineering Applications of Neural Networks. 217-28.
[http://dx.doi.org/10.1007/978-3-319-98204-5_18]

[102] Hoogendoorn M, Corro Ramos I, Baldwin M, Gonzalez-Rojas Guix N, Rutten-van Mölken MPMH. Broadening the perspective of cost-effectiveness modeling in chronic obstructive pulmonary disease: a new patient-level simulation model suitable to evaluate stratified medicine. Value Health 2019; 22(3): 313-21.
[http://dx.doi.org/10.1016/j.jval.2018.10.008] [PMID: 30832969]

[103] Lindberg A, Bjerg A, Rönmark E, Larsson LG, Lundbäck B. Prevalence and underdiagnosis of COPD by disease severity and the attributable fraction of smoking Report from the Obstructive Lung Disease in Northern Sweden Studies. Respir Med 2006; 100(2): 264-72.
[http://dx.doi.org/10.1016/j.rmed.2005.04.029] [PMID: 15975774]

[104] Fisher KA, Stefan MS, Darling C, Lessard D, Goldberg RJ. Impact of COPD on the mortality and treatment of patients hospitalized with acute decompensated heart failure: the Worcester Heart Failure Study. Chest 2015; 147(3): 637-45.
[http://dx.doi.org/10.1378/chest.14-0607] [PMID: 25188234]

# CHAPTER 4

# Arsenic Toxicity of Groundwater and Its Remediation for Drinking Water

## Seema Joshi[1,*]

[1] Isabella Thoburn College, Lucknow, India

**Abstract:** Due to the overall industrial development and human activities, the demand for clean water in India is continuously on the rise. There already exists a danger to the geochemical environment owing to the indiscriminate withdrawal of groundwater, resulting in the release of Arsenic (As). In some localized areas this level of As has already exceeded the World Health Organization's (WHO) permissible limits (10μg/L or 10ppb) for drinking water, leading to serious environmental and health consequences. Arsenic is predominantly present as inorganic species either as arsenate As (V) or arsenite As (III) in natural systems. In oxygen-rich environments where aerobic conditions persist, As (V) exists as mono-valent $(H_2AsO_4)^-$ or divalent $(HAsO_4)^{2-}$ anion, whereas, As (III) exists as an uncharged molecule $(H_3AsO_3)$ and anionic $(H_2AsO_3)^-$ species in moderately reducing atmosphere where anoxic conditions persist. The concentration of arsenic above its permissible level results in skin sclerosis. Arsenic gets deposited in the tissues of the vital organs and may cause cancer of the liver, lung, and urinary bladder. This study is an attempt to (a) review the arsenic problem in Uttar Pradesh, (b) to bring out the health issues due to arsenic, and (c) find sustainable solutions to address the issue.

**Keywords:** Arsenic, Cancer, Environment, Groundwater, Heavy metals, Inorganic, Remediation.

## INTRODUCTION

Inorganic substances are being mobilized and modified by human activity. These substances are not distinguished by the ecosystem as natural or anthropogenic, but rather as nutrients or at higher levels, as toxins [1]. Adverse effects of metal in their certain forms and specific doses cause metal toxicity or metal poisoning. The toxicity term generally refers to heavy metals but certain lighter metals like beryllium and lithium also become toxic under certain circumstances. Certain metalloids, like arsenic (As), are well known for their toxic effect. Even the trace elements, if present in abnormally high doses cause toxicity. Radioactive metals

---

* **Corresponding author Seema Joshi:** Department of Chemistry, Isabella Thoburn College, Lucknow, India; E-mail: seemjoshi1985@gmail.com

Tahmeena Khan, Abdul Rahman Khan, Saman Raza, Iqbal Azad and Alfred J. Lawrence (Eds.)
All rights reserved-© 2021 Bentham Science Publishers

have both radiological toxicity and chemical toxicity. All heavy metals are not necessarily toxic. Bismuth is mildly toxic and some metals, like iron, are also required essentially in the biological systems. For some metals, an oxidation state which is abnormal to the body may also cause toxicity. For example, chromium (III) is an essential trace element, but chromium (VI) is a carcinogen [2].

Some metals become toxic in the form of soluble compounds. Metals like lead, in any measurable amount, pose a negative impact on health. To summarize, metal toxicity or metal poisoning is the toxic effect of certain metals, in some specific forms and doses, on life. Therefore, for convenience, metals can be classified as essential and non-essential depending on their roles in biological systems.

## ESSENTIAL METALS

These metals play a crucial role in the biological system (Fig. 1). They are required essentially for various biochemical and physiological activities [3]. Their deficiency and excess both are detrimental to the biological system resulting in a variety of deficiency syndromes and metal poisoning. For some metals like chromium and copper, there is a very narrow difference between the beneficial and toxic concentrations [4 - 6].

## NON-ESSENTIAL METALS

The metals which do not play any biological role are termed non-essential metals (Fig. 1). Since no biological functions have been established for these metals, hence the term non-essential is used for them [7]. Such metals have a negative impact at all concentrations.

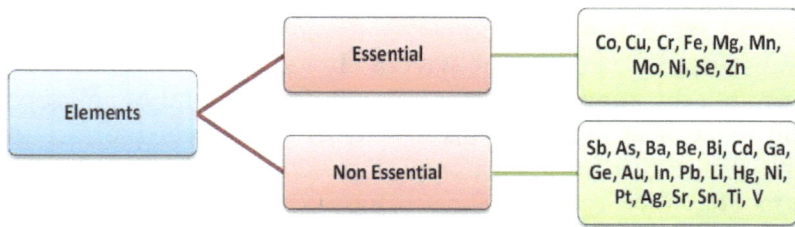

**Fig. (1).** An overview of essential and non-essential elements.

## CONCEPT OF TOXICITY

The toxicity of metals can be due to their accumulation in the vital cells or sometimes it is because of the imitation of the action of an essential element by some other metals in the body, interfering with the metabolic process and resulting in illness. Toxicity depends on the following factors:

## 1. Solubility of the Metal Complexes

Toxicity is a function of solubility. Insoluble compounds, as well as the metallic forms, often exhibit negligible toxicity.

## 2. Oxidation State of the Metal

Metals like chromium are non-toxic in lower oxidation state (III) but become carcinogenic at higher oxidation state (VI) [2]. Arsenic is less toxic in the oxidation state of +3 as compared to +5.

## 3. Ligand Attached to the Metal Atom

Toxicity of any metal decreases or increases depending on the ligand attached to it. For example, mercury (Hg) becomes more poisonous by coordinating to methyl group forming methyl mercury (Eq. **1**).

$$Hg + 4\,CH_4 \rightarrow Hg\,(CH_3)_4 + 2H_2 \tag{1}$$

The same is the case with lead (Pb), which becomes extremely toxic as tetraethyl lead (Eq. **2**).

$$Pb + 4C_2H_6 \rightarrow Pb(C_2H_5)_4 + 2H_2 \tag{2}$$

At the same time, the reverse is seen with cobalt (Co), the formation of its organometallic derivative, cobaltocenium cation (Eq. **3**) makes the metal less toxic.

$$Co + 2C_5H_5 \rightarrow [Co(C_5H_5)_2]^+ \tag{3}$$

Metals bioaccumulate in the body through the food chain and cause adverse and chronic effects. These accumulated metals start interfering with various biological activities. One such example is shown by the radioactive heavy metal radium, which imitates calcium and gets incorporated into human bones.

## DETECTION OF METALS

Detection of metals is not so easy. Atomic absorption spectral studies of the blood or tissue cells give the value of the respective metal in the sample. Comparison of these values with the standard values indicates the presence of metal ion. If the values are above the permissible level, then it is a case of metal poisoning. Following are some of the commonly used metal determination techniques:

- Atomic absorption spectrophotometer (AAS) by flame, graphite furnace (GFAAS), or cold vapour hydride generation.
- Atomic fluorescence
- X-ray fluorescence (XRF) and total reflection X-ray fluorescence (TXRF)
- Polarography
- Potentiometer (ion-selective electrodes, ISE)
- Neutron activation (NAA)
- Inductively coupled atomic emission plasma spectrometry (ICP-AES)
- Inductively coupled plasma spectrometry with mass spectrometry (ICP-MS) Mobilization.

## TREATMENT FOR POISONING

Generally, increased exposure to heavy metals in the environment increases the risk of developing cancer [8]. As toxic metals are elements, so these cannot be destroyedeasily. Their toxicity management is quite different from those of organic toxins. Metallic elements are first converted to an insoluble form which then is excreted through the body. An option for the treatment of metal poisoning may be chelation therapy, a medical procedure to remove metals as insoluble complex from the body, which is achieved either by the administration of chelating agents or through bioremediation. Chelating agents can form stable coordination complexes with metal ions. Due to complexation, metal ions become inactive to react with other molecules in the body and are eliminated through urine. Chelation drugs can have significant side effects, so their administration must be done under careful medical supervision. Apart from the removal of accumulated toxic metals from the body, chelation also removes essential elements. So, appropriate measures are required to prevent their loss. Toxicity is a function of concentration, so dilution techniques are also in use.

## ROLE OF METALS IN BIOLOGY

Recent studies have shown that essential heavy metals affect the functioning of both animals and plants [5, 9]. Being important constituents of their enzymatic system, they affect their biological redox reactions [9 - 12]. Various metalloenzymes catalyze biological activities like haemoglobin formation, carbohydrate metabolism, catecholamine biosynthesis, and cross-linking of collagen, elastin, and hair keratin, *etc*. There is a continuous interchange of the oxidation state of metals present in the metalloenzymes, resulting in important enzymatic redox reactions. These oxidation change cycles, sometimes also generate superoxide and hydroxyl radicals which makes the metal potentially toxic [13 - 16]. The mechanism can be explained by Fenton's reaction in which hydrogen peroxide disproportionates in two different radical species. In the first step, metal is oxidized by hydrogen peroxide generating hydroxyl radical and hydroxide (Eq. **4**). In the second step, the oxidized metal is reduced back by reacting with another molecule of hydrogen peroxide, producing hydroperoxyl radical and a proton (Eq. **5**), which can be responsible for the generation of these free radicals (Eq. **6**).

$$M^{2+} + H_2O_2 \rightarrow M^{3+} + HO^\bullet + OH^- \tag{4}$$

$$M^{3+} + H_2O_2 \rightarrow M^{2+} + HOO^\bullet + H^+ \tag{5}$$

$$2H_2O_2 \rightarrow HO^\bullet + HOO^\bullet + H_2 \tag{6}$$

These free radicals carry out secondary reactions, like the oxidation of organic compounds, and convert the contaminants primarily to carbon dioxide and water. Hydroxyl is a selective powerful oxidant. Sometimes, the radicals produced during such reactions get involved in the oxidation of heavy metal contaminants, if present in the biological system, and becomes detrimental by disturbing the essential redox cycles. The kinetics of these reactions is affected by the pH. The reactions are very fast in an acidic medium and slow down under alkaline conditions. The interaction of the heavy metals with nucleic acid like DNA and nuclear proteins *etc*. is well reported [17]. It has been found that such interactions cause DNA damage and conformational changes resulting in carcinogenesis. Several studies have demonstrated that reactive oxygen species (ROS) production and oxidative stress play a key role in the toxicity and carcinogenicity of heavy metals [18 - 21]. Out of all metals, due to their high degree of toxicity, arsenic [22 - 24] cadmium [25], chromium [26, 27], lead [28], and mercury [29] are known as human carcinogens.

There are various mechanisms to explain heavy metal-induced toxicity and carcinogenicity [30, 31]. Each metal has its unique features and therefore exhibits different physicochemical properties, resulting in its specific toxicological mechanisms of action. This chapter provides an analysis of the environmental occurrence of arsenic in Uttar Pradesh, its potential for human exposure, molecular mechanisms of toxicity, genotoxicity, carcinogenicity, and sustainable method for its remediation.

## SOURCES OF ARSENIC

Arsenic in its pure form is a harmless element, naturally present on the earth's surface, but when it combines with elements such as oxygen, it turns highly toxic. Most arsenic enters water supplies either from natural deposits in the earth or from industrial and agricultural pollution. Arsenic is present in low concentrations in virtually all environmental matrices [32]. Volcanic eruptions, soil erosion, and anthropogenic activities pollute the environment in terms of arsenic [33]. Apart from these natural phenomena, industrial production and consumption of arsenic-containing agricultural compounds and medicine [34, 35] *etc.* also increase its concentration in the environment. There are various medicines in use which contain arsenic and recently, Food and Drug Administration (FDA) has also approved arsenic trioxide for the treatment of acute promyelocytic leukaemia [36 - 38].

## POTENTIAL FOR HUMAN EXPOSURE WITH SPECIAL EMPHASIS TO UTTAR PRADESH, INDIA

Over thousands of years, arsenic sediments have been washing down from the Himalayas with the water of the river Ganges and leaching into the grounds [39]. Several research papers have reported the chronic exposure to large populations in several countries, including Argentina, Taiwan, Vietnam, United States of America (USA), Mexico, China, Bangladesh, India, Chile, Uruguay, and Mexico, *etc* [40 - 48].

For the last two decades in some places like Ballia district in Uttar Pradesh, the concentration of arsenic in drinking water has been found to be above the permissible limits. The inhabitants of rural areas are particularly affected by As toxicity due to poor socio-economic conditions, malnutrition, food habits, illiteracy, and meagre resources to have an access to drinking water. In these places, the concentration of arsenic in the groundwater is quite high [49]. The story of arsenic in various states of India began late in the 20$^{th}$ century when due to the sewage system, poor quality of water was available, resulting in severe water-born calamities [50]. The government on the advisory of UNICEF and the World Bank constructed millions of wells to get deeper groundwater.

Many hand pumps were constructed throughout Bihar and Uttar Pradesh (U.P.) to provide clean and easy-to-access water to the villagers [51, 52]. As the hand pumps were comfortable to use, the villagers neglected existing open wells. Initially, these hand pumps and tube wells helped in checking water-borne diseases. But in later years, excessive use of water in industrial development, human activities at the domestic level, and farming created a danger to the geochemical environment. Indiscriminate withdrawal of groundwater resulted in the lowering of the water table. The presence of high arsenic contents in sands is at around 25 m depth of the ground. Various studies have confirmed that the groundwater is highly contaminated with arsenic in tube wells with a depth of about 30 m, whereas shallow tube wells with a depth of about 10 m scarcely have As [52]. The first case of arsenic contamination of groundwater in Uttar Pradesh goes back to 2003, and recent studies have shown that around 2.34 crore people of Uttar Pradesh [53] from forty districts are exposed to high levels of arsenic in groundwater. Most of the districts are at the banks of the major rivers flowing in the state.

## Districts at High Risk

Ballia, Barabanki, Gorakhpur, Ghazipur, Gonda, Faizabad, and Lakhimpur Kheri.

## District at Moderate Risk

Shahjahanpur, Unnao, Chandauli, Varanasi, Pratapgarh, Kushinagar, Mau, Balrampur, Deoria, and Siddharthnagar.

A total of 78% of the state's population lives in villages and uses groundwater for irrigation, drinking, cooking, and other domestic requirements. Results of testing of groundwater samples from hand pumps and wells have confirmed the presence of a high concentration of arsenic in these identified regions in Ballia, Varanasi, Ghazipur, Gorakhpur, Faizabad, and Deoria [54 - 57]. In these rural areas, natural contamination of groundwater with arsenic is posing a health threat as people rely on hand pumps or tube wells for drinking water. These places are experiencing a public health crisis due to arsenic exposure. These are extremely fertile regions with rice being the primary crop. Through the crops, arsenic can also enter the food chain. In Ballia district alone, 310 villages are exposed to a high level of arsenic According to the studies, in these places As levels are between 150-500 µg/L, exceeding the WHO (10µg/L or 10ppb) and even Indian standard (50µg/L or 50ppb) [58, 59]. The case studies from these places have shown that arsenic poisoning causes skin lesions, cancer of the skin, bladder, lungs, and cardiovascular diseases, as well as reduced intellectual function in children, making future generations highly vulnerable [11, 14, 15].

Exposure to arsenic may occur through ingestion, inhalation, and dermal contact, *etc* [21 - 24]. The natural concentration of arsenic in air, water, and soil is generally quite below the permissible levels but pesticide application or waste disposal can produce much higher values [34]. Daily exposure to arsenic from all these different sources remains low but, in the areas, having groundwater contamination, there is strong evidence of long-term arsenic exposure, resulting in the promotion of carcinogenesis. Clinical manifestations of long-term arsenic exposure through water include hyperkeratosis and hyper-pigmentation patterns in the sole as well as arsenical dermatitis [60 - 62]. It can even result in cancer of the skin, liver, lung, and urinary bladder [14, 15, 29, 30, 40, 63].

## MECHANISMS OF TOXICITY

The toxic effect of arsenic depends on its oxidation state, solubility, and many other factors [37]. The inorganic form of As (III) is several times more toxic than As (V). Due to the affinity of As to sulphur (Hard Soft Acid Base principle) (Fig. 2), it can bind effectively to the thiol groups in the proteins. Such associations hinder various biological enzymatic reactions and are responsible for arsenic's widespread effects on different organ systems. As (V) also has an affinity for phosphorus and can replace phosphate, involved in various biochemical reactions. Methylation in human beings involves inorganic arsenic. During metabolism, methylate arsenic trioxide is first converted to monomethyl arsonic acid (MMA), which is further methylated enzymatically to dimethyl arsenic acid (DMA) (Fig. 3) which is then excreted in the urine [64]. Earlier, methylation was considered to be a pathway for arsenic detoxification, but recent studies have shown that if methylated metabolites are left with As (III), then they become more toxic than As (III) itself. Weak acid salts of As (III) are soluble and studies have shown that contamination of groundwater occurs much more with As (III) as compared to As (V). In these studies, arsenic concentration was observed to be in negative correlation with the oxidation-reduction potential, suggesting the reducing condition in groundwater [52]. Arsenic is leached out into groundwater under reduced conditions [65]. The concentrations of As, $Fe^{2+}$, and N are found to be correlated with each other [52]. It has been reported that oxidative stress plays a key role in arsenic-induced cytotoxicity.

## CONSEQUENCES OF TOXICITY

Arsenic toxicity or poisoning can lead to several health conditions. Arsenic, being a soft acid, binds effectively with the thiol groups in metabolic enzymes, which increases the concentration of arsenic in the tissues of vital body organs like lungs, skin, kidneys, and liver. This accumulation of arsenic also increases the production of free radicals causing oxidative stress and compromised immune system, causing both short- and long-term health consequences on the neurological, respiratory, hematologic, cardiovascular, gastrointestinal, and other systems. There are various social factors like nutritional status, inter-individual and population differences, *etc.* which affect the extent of metal toxicity.

Swallowing a large quantity of arsenic results in gastrointestinal disorders, vomiting, and diarrhoea, whereas a lower level of exposure for a longer duration leads to skin ailments, cancer, and adverse effects on the muscular system.

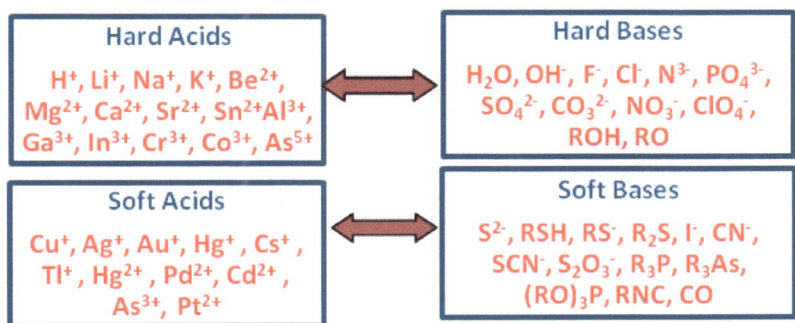

**Fig. (2).** Illustration of the HSAB concept.

$$\text{HO-As-OH} \atop \text{O}\ \ (\text{Methylarsonic acid (MMA)})\qquad \text{H}_3\text{C-As-OH} \atop \text{CH}_3\ \ (\text{Dimethylarsinic acid (DMA)})$$

**Fig. (3).** Structures of MMA and DMA.

The chelate complexation between arsenic and enzymes inhibits the enzymatic activity resulting in metal toxicity. Acute exposure to arsenic results in carcinogenesis in multiple organ systems whereas chronic arsenic poisoning (arsenicosis) causes gastrointestinal and hepatic effects. The diagnosis of arsenic poisoning is generally made by measuring the amount of arsenic excreted in the urine. Chelation therapy is in use for the treatment of arsenic poisoning. The chelating agents play a vital role in the detoxification of harmful metals from the human body. Chelating agents form bonds between organic molecules and accumulated metal in the blood, major organs, and blood vessels. Chelators used to treat arsenic poisoning include dimercaprol, succimer, and penicillamine.

## REMEDIATION OF ARSENIC TOXICITY

Much work has been reported during the past decade on the removal of As from drinking water. The Environmental Protection Agency (EPA) has also reported various technologies to remove As from drinking water. In these technologies, effective treatment is done in arsenate form. As (III) can be oxidized to As (V) by using various oxidants but sometimes this peroxidation may also create an undesirable concentration of by-products.

Japan International Cooperation Agency (JICA) and the University of Miyazaki, under their collaborative work on arsenic mitigation project in Bahraich, have focused on spreading awareness among villagers, installed alternative water supply units, and have taken care of arsenicosis patients at 2 villages [66 - 68]. Few reports have shown that the medium basin of the groundwater of the Ganges is contaminated with arsenic [69]. These studies suggest that deeper uncontaminated aquifers could be a source of arsenic-free water and digging deep tube wells or deep hand pumps in rural areas can be a solution to this problem [70]. But the huge costs and the long-term safety of these have placed a question mark on the viability of this solution. Over-consumption or improper installation would lead to pollution, eventually. Project Sankhiya is a model for arsenic mitigation. Under this project, UPJN & SWSM have revived dug wells in the entire district of Ballia. The construction work required reviving and the bacteriological cleaning done by the Panchayati Raj Institutions (PRI) in their villages. As the population has suffered a lot due to arsenic exposure, so villagers are also coming forward to contribute their bit for arsenic mitigation. Now old, disused, and discarded wells are being cleaned and revived. The wells that were once an integral part of the village culture, are slowly regaining their lost value and importance. The government has started piped water schemes to provide safe water. These projects require a long time to complete and a lot of money to be invested. The water treatment in such plants requires energy which makes their sustainability difficult. The community filters in many arsenic-affected areas are not in working conditions due to lack of proper maintenance. Arsenic removal plants and overhead tanks installed by Uttar Pradesh Jal Nigam (UPJN) are also not giving desired results due to lack of power supply and their poor maintenance.

A sustainable alternative to provide arsenic-free water is the use of a simple device at the domestic level for arsenic decontamination in rural areas, cost-effectively. The technologies recommended for the removal of arsenic from water include ion exchange, activated ammonia coagulation, coagulation with ferric chloride, and pressurized iron particles, iron oxide-doped alginate, manganese dioxide-coated sand, polymeric ligand exchange, zero-valent iron, filtration, and lime softening. The use of iron salt is an effective method in the case of large-scale fixed-bed treatment (adsorption), which is getting increasingly popular for arsenic removal in the small-scale treatment system. Due to their simplicity, iron oxide and oxyhydroxide play an important role in a variety of industrial operations. The major problems with granulated iron hydroxide include colouration, along with some quantity of iron being released during the process of decontamination.

The conventional methods are based on the use of synthetic coagulants for the purification of water [71, 72]. Biosorption technology has also gained importance during recent years. In this context, the seed powder of plant *Moringa oleifera* Lamarck (Miracle Tree) has been used for water treatment, which showed decontamination of both As (III) (60%) and As (V) (85%) in a batch study.

In India, the Jal-TARA Slow Sand Filter is a common water purification system comprising chlorination, coagulation, and sand filtration techniques. Of all these techniques, sand filtration offers a chemical-free, reliable, and economical treatment. Jal-TARA Slow Sand Filter is designed to treat drinking water using slow sand filtration technique, as per the specifications of WHO. Jal-TARA filter is a biological filter merged with the advanced technique of fabric protection to improve and simplify the traditional process of slow sand filtration. The two filters act together to improve the quality of water. The gravel filter removes turbidity, and the sand bed filters remove pathogenic bacteria from raw water.

A household arsenic filter has been developed under the Nepal water project which not only removes carcinogenic chemicals but also pathogens. The project is being implemented jointly by the Massachusetts Institute of Technology (MIT). Field research by MIT and ENPHO (Environment and Public Health Organization) shows arsenic removal by using such filters is in the range of 85-95%.

## CHELATING DRUGS USED IN THE TREATMENT OF ARSENIC POISONING

### 1. Dimercaprol (BAL)

During World War II, arsenic-based warfare was in use resulting in arsenic poisoning. The drug Dimercaprol, commonly known as British anti-Lewisite (BAL) (Fig. **4**) was developed by British biochemists as an antidote for lewisite. This drug forms stable chelates *in vivo* with arsenic and is used in chelation therapy for arsenic poisoning. Dimercaprol has a strong affinity for arsenic and binds to the metal, preferably to the thiol groups. This prevents the accumulation of metal in the body. Arsenic-Dimercaprol chelate complex being hydrophilic gets excreted from the body through urine.

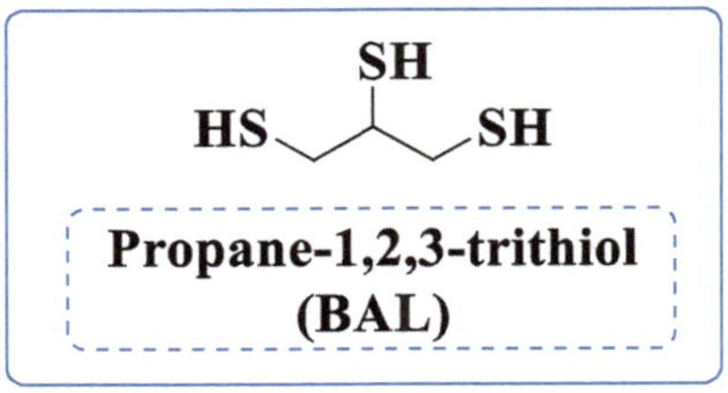

```
        SH
HS ∕∖ SH
Propane-1,2,3-trithiol
      (BAL)
```

**Fig. (4).** Structure of BAL.

## 2. Dimercaptosuccinic Acid (DMSA)

DMSA, commonly known as succimer, is an essential class of medicine on the WHO's list of medicines. This drug has been used medically since the 1950s to treat arsenic poisoning. This drug is administered orally for 19 days in a cycle and successive cycles are at intervals of two weeks. DMSA binds to arsenic and forms a chelated complex, which is excreted from the body along with urine. DMSA (Fig. 5) is available in capsule or suppository form. Researchers have found DMSA to be safer and three times less toxic than DMPS.

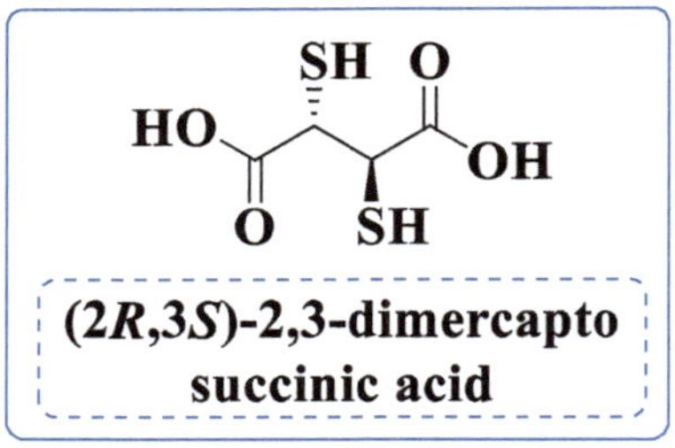

```
      SH  O
HO          OH
   O   SH
(2R,3S)-2,3-dimercapto
     succinic acid
```

**Fig. (5).** Structure of DMSA.

## 3. 2,3-Dimercapto-1-propanesulfonic Acid (DMPS)

DMPS was first synthesized in 1956. DMPS (Fig. 6) and its sodium salt unithiol form chelated complexes with heavy metals. It is used as medicine in severe acute arsenic poisoning. DMPS shows a protective effect enhancing the survival time. The thiol group of DMPS gets chelated to arsenic which prevents complexation between arsenic and enzymes of the body system. Thus, enzymatic reactions of the body retain their normal function and vital organs also get protected from metal accumulation. DMPS can be taken orally, intravenously, or as a suppository but is injected mostly intravenous. This drug is not approved by the FDA and therefore is used for metal detoxification only as a last alternative medication.

(S)-2-amino-3-mercapto-3-methylbutanoic acid

**Fig. (6).** Structure of DMPS.

## 4. Penicillamine

Penicillamine (Fig. 7) is commonly sold by the name of Cuprimine and is used as a second-line treatment for arsenic poisoning. The medicine is used orally for arsenic poisoning. Penicillamine carries an amine, a thiol, and a carboxylic acid functional group. Penicillamine binds to arsenic. The resulting penicillamine–arsenic complexes are then removed from the body in the urine. Its L-penicillamine enantiomer shows no antibiotic properties and inhibits the action of Vit- B-6.

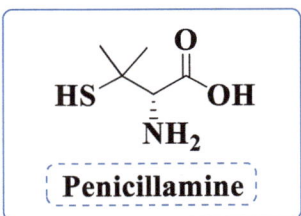

Penicillamine

**Fig. (7).** Structure of Penicillamine.

## CHALLENGES OF CHELATION THERAPY

Although chelation therapy is used for metal toxicity, yet it has several challenges. For effective medication, high doses of the drugs are to be used which results in various side effects also. Some of the major challenges are:

1. The common side effects of all the drugs used in chelation therapy are high blood pressure, pain at the site of the injection, vomiting, and fever.
2. These chelators must be administered in high doses. Therefore, during therapy, they also bind to the essential minerals/metals like zinc, magnesium, manganese, molybdenum, selenium calcium, iron, and vital nutrients like vitamin C and E, which also get excreted along with the toxic metal. The deficiency of minerals influences the metabolic reaction of the body, thus

adversely affecting the immune system.
3. Another big challenge is that these chelators cannot cross the blood-brain barrier, hence are not effective for the arsenic accumulated in the brain tissues.
4. With DMSA medication, it has been reported that arsenic, instead of getting excreted, sometimes gets reabsorbed in different parts of the body, thus affecting the immune system.
5. Several studies have shown that DMPS quite rapidly excretes the accumulated arsenic from the body organs into the bloodstream and then to the kidneys. Thus, at once, a large concentration of the metal gets dumped into the bloodstream and causes overloading on the kidney and liver. The toxic metal gets reabsorbed by the vital organs and affects the biological systems.
6. Dimercaprol, which is used widely for arsenic poisoning, is toxic itself. It tends to concentrate arsenic in the liver and kidney, causing nephrotoxicity and hypertension.
7. The common side effects of penicillamine chelator are bone marrow suppression, dysgeusia, anorexia, vomiting, diarrhoea, and unwanted breast growth.

## CASE STUDY FOR THE REMOVAL OF ARSENIC

In the present study, iron hydroxide prepared in the laboratory was used for the removal of arsenic present at different concentrations in the water. The capacity for arsenic removal was also examined after one month of storage. Later, because of the difficulties in the handling of the material for online use, the material was adsorbed on charcoal and sand separately and then iron coated charcoal or sand was used for arsenic decontamination for its eventual use in online system or columns.

## CHEMICALS AND REAGENTS

Sodium arsenite ($NaAs_2O_3$) and sodium arsenate ($NaAs_2O_5$) were procured from Sigma Aldrich, India, ferric chloride ($FeCl_3$) from Ranbaxy India, sodium borohydride and sodium hydroxide from Merck India, charcoal, and hydrochloric acid AR from Qualigens India.

## PREPARATION OF FERRIC HYDROXIDE

A solution of ferric chloride was mixed with the solution of sodium hydroxide in the molar ratio of 1:10. A brown precipitate of ferric hydroxide ($Fe(OH)_3$) was obtained, which was allowed to settle and separated from the supernatant. It was washed with distilled water twice and suspended in 100 ml distilled water.

## REMOVAL OF ARSENIC FROM SPIKED TAP WATER USING FERRIC HYDROXIDE

Tap water in triplicate was mixed with As (III) or As (V) stock solution and 50 ml arsenic solution of 50 ppb concentration was prepared. The prepared Arsenic solution of 50 ppb was mixed with 10 ml of ferric hydroxide suspension prepared above. The pH of the solution was checked and adjusted to 7.0. A control without ferric hydroxide was also run with tap water. The mixture could interact at room temperature for 2 h, with intermittent shaking. The efficiency of ferric hydroxide upon storage up to one month was also examined, using 200 ml of tap water spiked with 200 and 400 ppb arsenic concentration, as described above.

## PREPARATION OF IRON COATED CHARCOAL/SAND

Charcoal (brand 1) was used in the experiment directly, while river sand procured from the local market was sieved to get a homogeneous particle size, washed with deionized water, and dried at 100 $^0$C for 3 days, 8 h a day before iron coating. The suspension of freshly prepared ferric hydroxide was added to charcoal, or sand and the material was mixed after gentle agitation. The mixture was heated at 110 $^0$C for 4 h, followed by heating at 350 $^0$C for 3 consecutive days 8 h a day. After cooling, the iron-coated charcoal or sand material was washed with deionized water to remove unabsorbed ferric hydroxide. The material was dried for hours and was again subjected to a second treatment with ferric hydroxide suspension, dried, and washed in the same manner as described above. After complete drying of the material for several hours, the material was mechanically homogenized by breaking the grain in the case of charcoal, which was not required for sand material. The iron-coated charcoal or sand was stored in a stoppered glass bottle for use in the experiment.

## EXPERIMENTAL DESIGN

Batch studies were conducted with charcoal and iron-coated sand at room temperature for decontamination of arsenic from spiked tap water.

## CHARCOAL TREATMENT

Tap water spiked with 50 ppb arsenic was separately taken in beakers and experiments were carried out in four sets along with a control. The mixture could interact with intermittent shaking using a glass rod every 15 minutes. The charcoal added samples were filtered using Whatman filter paper at 1, 2, 3, or 4 h respectively. The filtrate from charcoal treated samples or aliquot from control was used for the determination of arsenic.

## Iron Coated Charcoal Treatment

Tap water spiked with 50 ppb arsenic was separately taken in beakers experiments were carried out in four sets along with a control. Iron-coated charcoal was added to all the beakers except one, which served as control. The mixture could interact with intermittent shaking and the samples were drawn out and filtered at 1, 2, 3, or 4 h, as described above. Filtrate or aliquot from control was used for the determination of arsenic.

## Iron Coated Coarse Sand Treatment

Tap water spiked with 50 ppb arsenic was taken in beakers and experiments were carried out in four sets along with a control. Iron-coated sand material was added in all the beakers except one, which served as control. The mixture could interact as described above and samples were filtered at 1, 2, 3, or 4 h respectively. The filtrate from iron-coated sand treated samples or aliquot from control was used for the determination of arsenic.

## DETERMINATION OF ARSENIC

Determination of As was carried out using Atomic Absorption Spectrometer (AAS) Varian 250+ equipped with vapour generation assembly (VGA 77) with the lamp current of 10 mA. Wavelength 193.3 nm and a low flame were used with other conditions, as described in the user's manual. The acid channel of VGA was supplied with 0.6% sodium borohydride in 0.5% sodium hydroxide. All the samples were prepared in 6 M HCl, so were the standards used for calibration, and these conditions were found to give optimum sensitivity in the condition used in our experiments.

## FINDINGS OF THE STUDY

The results of the study showed that iron hydroxide was successful in removing arsenic from spiked tap water. It was observed that the efficiency of ferric hydroxide was more for the removal of As (V) (98%) while for As (III), it was relatively less. As (III) could not be removed beyond 93% at 50 ppb (Table **1** & **2**). When the material was stored for up to 1 month and reused for the removal of arsenic, it was observed that removal efficiency was maintained. The efficiency of iron-coated coarse sand was found in the range of 75% to 89% at variable concentration, pH, and temperature conditions (Table **3**). The results suggested that pH 7 is quite suitable for the removal of arsenic from water.

Experiments were carried out to find the optimum concentration of arsenic. The results of the study suggested that the arsenic decontamination efficiency of the adsorbent was concentration-dependent. The data reveal that the removal capacity of sand was 85% at 30 ppb level, 76% at 50ppb level, and 70% at 100ppb level spiked water of arsenic was 85% for 200 ppb, and 92% for 400 ppb arsenic spiked samples (Table **4**).

A comparative study using charcoal, iron-coated charcoal, and iron-coated sand to examine their efficiency for removal of arsenic spiked 50 ppb tap water sample was done. The results are presented in Table **5**.

The results indicate that in the case of only charcoal and iron coated charcoal, there seems to be some release of adsorbed arsenic at 2 h. Iron-coated sand material showed a different pattern and was found to be the best among the three materials examined. Having found a better result in this case it was intended to examine the repeated use of the iron-coated sand material. The same material was consecutively reused on days 2, 3, 4, and 5 for the removal of arsenic from spiked tap water. There was a decrease in removal efficiency, which was found to be 75% on day 3 while it was only 55% on day 5 (Table **6**).

The results thus indicate that iron-coated sand can be used for decontamination of arsenic with efficiency up to 96% on day 1, while it can also be used for arsenic removal with 75% efficiency up to day 3.

Table 1. Removal percentage of As (III) from arsenic spiked (50 ng/mL concentration) tap water sample by Ferric hydroxide suspension, after 2 h of treatment.

| Sample | The Concentration of As (III) (ng/mL) | | |
|---|---|---|---|
| | Spiked Samples | Spiked Samples Treated with Ferric Hydroxide Suspension | Removal Percentage (%) |
| 1 | 54.90 | 2.80 | 94.53 |
| 2 | 55.75 | 3.65 | 93.45 |
| 3 | 54.90 | 3.96 | 92.78 |
| Average | 55.18 | 3.13 | 93.58 |

Table 2. Removal percentage of As (V) from arsenic spiked (50 ng/mL concentration) tap water sample by Ferric hydroxide suspension, after 2 h of the treatment.

| Sample | Concentration of As (V) (ng/mL) | | |
|---|---|---|---|
| | Spiked Samples | Spiked Samples Treated with Ferric Hydroxide Suspension | Removal Percentage (%) |
| 1 | 51.44 | 0.48 | 99.06 |
| 2 | 56.90 | 1.26 | 97.78 |
| 3 | 52.00 | 0.82 | 98.42 |

(Table 2) cont.....

| Sample | Concentration of As (V) (ng/mL) | | |
|---|---|---|---|
| | Spiked Samples | Spiked Samples Treated with Ferric Hydroxide Suspension | Removal Percentage (%) |
| Average | 53.44 | 0.85 | 98.42 |

Table 3. Performance data of arsenic removal capacity of sand at different pH.

| S.No. | pH | % Removal by the Adsorbent | % Removal from Blank |
|---|---|---|---|
| 1 | 4 | 77 | 4 |
| 2 | 6 | 85 | 7 |
| 3 | 7 | 82 | 6 |
| 4 | 9 | 65 | 3 |

Table 4. Effect of concentration on the removal of arsenic.

| S.No. | Concentration (ppb) | With Iron Coated Sand | Control |
|---|---|---|---|
| 1 | 30 | 85 | 08 |
| 2 | 50 | 76 | 04 |
| 3 | 100 | 70 | 02 |

Table 5. Comparative evaluation of removal percentage (%) of As (III) from arsenic spiked (50 ng/mL concentration) tap samples

| Treatment Time (h) | Treatment | | |
|---|---|---|---|
| | Charcoal | Iron Coated Charcoal | Iron Coated Coarse Sand |
| 1 | 85.00 | 94.15 | 94.27 |
| 2 | 78.00 | 74.52 | 90.45 |
| 3 | 86.00 | 78.02 | 86.18 |

Table 6. Removal Percentage of arsenic (III) from arsenic spiked (50 ng/mL concentration) tap water samples by iron-coated sand for 5 days

| Treatment Time (h) | Iron Coated Sand | | | | |
|---|---|---|---|---|---|
| | Day 1 | Day 2 | Day 3 | Day 4 | Day 5 |
| 1 | 90.45 | 81.66 | 66.00 | 55.77 | 39.57 |
| 2 | 90.27 | 74.97 | 70.20 | 56.17 | 48.90 |
| 3 | 90.18 | 87.77 | 74.13 | 55.48 | 46.20 |
| 4 | 90.12 | 87.77 | 75.20 | 64.59 | 55.36 |

## CONCLUDING REMARKS

The results showed that iron hydroxide is successful in removing arsenic from spiked tap water and further demonstrated that the efficiency of ferric hydroxide is more to remove As (V) while that for As (III) its efficiency is relatively less. The outcome of the research is expected to provide and design suitable technologies for arsenic removal from drinking water, which is of low cost so that even the low-income groups can afford to use the technology and have safe drinking water generated at home. Coarse sand is readily available in river deltas in Northern India. It has a stable texture in an aqueous medium for a long time. It has a large particle size ranging from 0.5 to 1.0 mm. Also, in its natural constitution, it has a significant amount of iron in it. The results suggested a sustainable alternative to provide arsenic-free water by using iron-coated sand. Since the adverse effect of arsenic is through drinking water, this issue can be resolved to some extent by just putting iron-coated sand in drinking water and keeping it for one day. Around 95% arsenic will be removed, and water will be safe for drinking purposes. This can be done using simple water buckets or any simple device at the domestic level for arsenic decontamination in rural areas in a cost-effective manner.

## CONSENT FOR PUBLICATION

Not Applicable.

## CONFLICT OF INTEREST

The author confirms that this chapter contents have no conflict of interest.

## ACKNOWLEDGEMENT

Declared none.

## REFERENCES

[1] Miller SE, Rodriguez RA, Nelson KL. Removal and growth of microorganisms across treatment and simulated distribution at a pilot-scale direct potable reuse facility†. Environ Sci Water Res Technol 2020; 6: 1370-87.
[http://dx.doi.org/10.1039/C9EW01087D]

[2] Chebeir M, Chen G. G.; Haizhou, L. Emerging investigators series: frontier review: occurrence and speciation of chromium in drinking water distribution systems. Environ Sci Water Res Technol 2016; 2: 906-14.
[http://dx.doi.org/10.1039/C6EW00214E]

[3] Zoroddu MA, Aaseth J, Crisponi G, Medici S, Peana M, Nurchi VM. The essential metals for humans: a brief overview. J Inorg Biochem 2019; 195: 120-9.
[http://dx.doi.org/10.1016/j.jinorgbio.2019.03.013] [PMID: 30939379]

[4] Sejio AR, Alfaya MC, Andrade ML, Vega FA. Copper, Chromium, Nickel, Lead and Zinc Levels and Pollution Degree in Firing Range Soils. LDD 2016; 27: 1721-30. REMOVED HYPERLINK FIELD

[5]  Velusamy A, Satheesh Kumar P, Ram A, Chinnadurai S. Bioaccumulation of heavy metals in commercially important marine fishes from Mumbai Harbor, India. Mar Pollut Bull 2014; 81(1): 218-24.
[http://dx.doi.org/10.1016/j.marpolbul.2014.01.049] [PMID: 24631401]

[6]  Kim HS, Kim YJ, Seo YR. An Overview of Carcinogenic Heavy Metal: Molecular Toxicity Mechanism and Prevention. J Cancer Prev 2015; 20(4): 232-40.
[http://dx.doi.org/10.15430/JCP.2015.20.4.232] [PMID: 26734585]

[7]  Santos I, Diniz MS, Carvalho ML, Santos JP. Assessment of essential elements and heavy metals content on Mytilus galloprovincialis from river Tagus estuary. Biol Trace Elem Res 2014; 159(1-3): 233-40.
[http://dx.doi.org/10.1007/s12011-014-9974-y] [PMID: 24763710]

[8]  Salnikow K, Zhitkovich A. Genetic and epigenetic mechanisms in metal carcinogenesis and cocarcinogenesis: nickel, arsenic, and chromium. Chem Res Toxicol 2008; 21(1): 28-44.
[http://dx.doi.org/10.1021/tx700198a] [PMID: 17970581]

[9]  Yadav SK. Heavy metals toxicity in plants: An overview on the role of glutathione and phytochelatins in heavy metal stress tolerance of plants. S Afr J Bot 2010; 76: 167-79.
[http://dx.doi.org/10.1016/j.sajb.2009.10.007]

[10]  Takahashi K, Harada S, Higashimoto Y, et al. Involvement of metals in enzymatic and nonenzymatic decomposition of C-terminal α-hydroxyglycine to amide: an implication for the catalytic role of enzyme-bound zinc in the peptidylamidoglycolate lyase reaction. Biochemistry 2009; 48(7): 1654-62.
[http://dx.doi.org/10.1021/bi8018866] [PMID: 19170548]

[11]  Martinez VD, Vucic EA, Becker-Santos DD, Gil L, Lam WL. Arsenic exposure and the induction of human cancers. J Toxicol 2011; 2011431287
[http://dx.doi.org/10.1155/2011/431287] [PMID: 22174709]

[12]  Jaishankar M, Tseten T, Anbalagan N, Mathew BB, Beeregowda KN. Toxicity, mechanism and health effects of some heavy metals. Interdiscip Toxicol 2014; 7(2): 60-72.
[http://dx.doi.org/10.2478/intox-2014-0009] [PMID: 26109881]

[13]  Tchounwou PB, Yedjou CG, Patlolla AK, Sutton DJ. Heavy metal toxicity and the environment. Exp Suppl 2012; 101: 133-64.
[PMID: 22945569]

[14]  Guha Mazumder DN. Chronic arsenic toxicity & human health. Indian J Med Res 2008; 128(4): 436-47.
[PMID: 19106439]

[15]  Guo HR, Wang NS, Hu H, Monson RR. Cell type specificity of lung cancer associated with arsenic ingestion. Cancer Epidemiol Biomarkers Prev 2004; 13(4): 638-43.
[PMID: 15066930]

[16]  Sangvanich T, Morry J, Fox C, et al. Novel oral detoxification of mercury, cadmium, and lead with thiol-modified nanoporous silica. ACS Appl Mater Interfaces 2014; 6(8): 5483-93.
[http://dx.doi.org/10.1021/am5007707] [PMID: 24660651]

[17]  Lęczkowskaa A, Vilar R. Interaction of metal complexes with nucleic acids. Annu. Rep. Prog. Chem. Sect A: Inorg Chem 2013; 109: 299-316.

[18]  Joshi S, Hasan SK, Chandra R, Husain MM, Srivastava RC. Scavanging action of metallothioneine and green tea polyphenols on Cisplatin and Nickel induced nitric oxide generation and lipid peroxidation in rats. Biomed Environ Sci 2004; 17: 402-9.
[PMID: 15745244]

[19]  Hrudey SE, Bull RJ, Cotruvo JA, Paoli G, Wilson M. Drinking water as a proportion of total human exposure to volatile N-nitrosamines. Risk Anal 2013; 33(12): 2179-208.
[http://dx.doi.org/10.1111/risa.12070] [PMID: 23786353]

[20]  Zhang Z, Pratheeshkumar P, Budhraja A, Son Y-O, Kim D, Shi X. Role of reactive oxygen species in arsenic-induced transformation of human lung bronchial epithelial (BEAS-2B) cells. Biochem Biophys Res Commun 2015; 456(2): 643-8.
[http://dx.doi.org/10.1016/j.bbrc.2014.12.010] [PMID: 25499816]

[21]  Lee J-C, Son Y-O, Pratheeshkumar P, Shi X. Oxidative stress and metal carcinogenesis. Free Radic Biol Med 2012; 53(4): 742-57.
[http://dx.doi.org/10.1016/j.freeradbiomed.2012.06.002] [PMID: 22705365]

[22]  Mehrotra A, Mishra A, Tripathi RM, Shukla N. Mapping of arsenic contamination severity in Bahraich district of Ghagra basin, Uttar Pradesh, India. Geomatics. Nat Hazards and Risks 2014; 7: 101-12.
[http://dx.doi.org/10.1080/19475705.2013.871354]

[23]  Chakraborti D, Singh SK, Rahman MM, *et al.* Groundwater Arsenic Contamination in the Ganga River Basin: A Future Health Danger. Int J Environ Res Public Health 2018; 15(2): 180.
[http://dx.doi.org/10.3390/ijerph15020180] [PMID: 29360747]

[24]  Ahamed S, Kumar Sengupta M, Mukherjee A, *et al.* Arsenic groundwater contamination and its health effects in the state of Uttar Pradesh (UP) in upper and middle Ganga plain, India: a severe danger. Sci Total Environ 2006; 370(2-3): 310-22.
[http://dx.doi.org/10.1016/j.scitotenv.2006.06.015] [PMID: 16899281]

[25]  Rani A, Kumar A, Lal A, Pant M. Cellular mechanisms of cadmium-induced toxicity: a review. Int J Environ Health Res 2014; 24(4): 378-99.
[http://dx.doi.org/10.1080/09603123.2013.835032] [PMID: 24117228]

[26]  Pavesi T, Moreira JC. Mechanisms and individuality in chromium toxicity in humans. J Appl Toxicol 2020; 40(9): 1183-97.
[http://dx.doi.org/10.1002/jat.3965] [PMID: 32166774]

[27]  Wakeel A, Xu M, Gan Y. Chromium-Induced Reactive Oxygen Species Accumulation by Altering the Enzymatic Antioxidant System and Associated Cytotoxic, Genotoxic, Ultrastructural, and Photosynthetic Changes in Plants. Int J Mol Sci 2020; 21(3): 728.
[http://dx.doi.org/10.3390/ijms21030728] [PMID: 31979101]

[28]  Flora G, Gupta D, Tiwari A. Toxicity of lead: A review with recent updates. Interdiscip Toxicol 2012; 5(2): 47-58.
[http://dx.doi.org/10.2478/v10102-012-0009-2] [PMID: 23118587]

[29]  Bernhoft RA. Mercury toxicity and treatment: a review of the literature. J Environ Public Health 2012; 2012460508
[http://dx.doi.org/10.1155/2012/460508] [PMID: 22235210]

[30]  Jomova K, Jenisova Z, Feszterova M, *et al.* Arsenic: toxicity, oxidative stress and human disease. J Appl Toxicol 2011; 31(2): 95-107.
[http://dx.doi.org/10.1002/jat.1649] [PMID: 21321970]

[31]  Ratnaike RN. Acute and chronic arsenic toxicity. Postgrad Med J 2003; 79(933): 391-6.
[http://dx.doi.org/10.1136/pmj.79.933.391] [PMID: 12897217]

[32]  Mandal P. An insight of environmental contamination of arsenic on animal health. Emerg Contam 2017; 3: 17-22.
[http://dx.doi.org/10.1016/j.emcon.2017.01.004]

[33]  Shi Y-L, Chen W-Q, Wu S-L, Zhu Y-G. Anthropogenic Cycles of Arsenic in Mainland China: 1990-2010. Environ Sci Technol 2017; 51(3): 1670-8.
[http://dx.doi.org/10.1021/acs.est.6b01669] [PMID: 28043121]

[34]  Mochizuki H. Arsenic Neurotoxicity in Humans. Int J Mol Sci 2019; 20(14): 3418.
[http://dx.doi.org/10.3390/ijms20143418] [PMID: 31336801]

[35] Minatel BC, Sage AP, Anderson C, *et al.* Environmental arsenic exposure: From genetic susceptibility to pathogenesis. Environ Int 2018; 112: 183-97.
[http://dx.doi.org/10.1016/j.envint.2017.12.017] [PMID: 29275244]

[36] Chen X, Hong Y, Zheng P, *et al.* The economic research of arsenic trioxide for the treatment of newly diagnosed acute promyelocytic leukemia in China. Cancer 2020; 126(2): 311-21.
[http://dx.doi.org/10.1002/cncr.32519] [PMID: 31714584]

[37] Firkin F. Carcinogenic risk of retained arsenic after successful treatment of acute promyelocytic leukemia with arsenic trioxide: a cause for concern? Leuk Lymphoma 2014; 55(5): 977-8.
[http://dx.doi.org/10.3109/10428194.2013.856429] [PMID: 24144311]

[38] Alimoghaddam K. A review of arsenic trioxide and acute promyelocytic leukemia. Int J Hematol Oncol Stem Cell Res 2014; 8(3): 44-54.
[PMID: 25642308]

[39] McArthur JM, Banerjee DM, Hudson-Edwards KA, *et al.* Natural organic matter in sedimentary basins and its relation to arsenic in anoxic ground water: the example of West Bengal and its worldwide implications. Appl Geochem 2004; 19: 1255-93.
[http://dx.doi.org/10.1016/j.apgeochem.2004.02.001]

[40] Moncur MC, Dogan P, Birks SJ, Ptacek CJ, Weish B, Thibaut Y. Source and distribution of naturally occurring arsenic in groundwater from Alberta's Southern Oil Sands Regions. Appl Geochem 2015; 62: 171-85.
[http://dx.doi.org/10.1016/j.apgeochem.2015.02.015]

[41] Ng JC, Wang J, Shraim A. A global health problem caused by arsenic from natural sources. Chemosphere 2003; 52(9): 1353-9.
[http://dx.doi.org/10.1016/S0045-6535(03)00470-3] [PMID: 12867164]

[42] Chowdhury S, Mazumder MAJ, Al-Attas O, Husain T. Heavy metals in drinking water: Occurrences, implications, and future needs in developing countries. Sci Total Environ 2016; 569-570: 476-88.
[http://dx.doi.org/10.1016/j.scitotenv.2016.06.166] [PMID: 27355520]

[43] Singh R, Singh S, Parihar P, Singh VP, Prasad SM. Arsenic contamination, consequences and remediation techniques: a review. Ecotoxicol Environ Saf 2015; 112: 247-70.
[http://dx.doi.org/10.1016/j.ecoenv.2014.10.009] [PMID: 25463877]

[44] Shankar S, Shanker U, Shikha . Arsenic contamination of groundwater: a review of sources, prevalence, health risks, and strategies for mitigation. ScientificWorldJournal 2014; 2014304524
[http://dx.doi.org/10.1155/2014/304524] [PMID: 25374935]

[45] Cho KH, Sthiannopkao S, Pachepsky YA, Kim KW, Kim JH. Prediction of contamination potential of groundwater arsenic in Cambodia, Laos, and Thailand using artificial neural network. Water Res 2011; 45(17): 5535-44.
[http://dx.doi.org/10.1016/j.watres.2011.08.010] [PMID: 21917287]

[46] Saha D, Sahu S. A decade of investigations on groundwater arsenic contamination in Middle Ganga Plain, India. Environ Geochem Health 2016; 38(2): 315-37.
[http://dx.doi.org/10.1007/s10653-015-9730-z] [PMID: 26116052]

[47] Xia Y, Liu J. An overview on chronic arsenism *via* drinking water in PR China. Toxicology 2004; 198(1-3): 25-9.
[http://dx.doi.org/10.1016/j.tox.2004.01.016] [PMID: 15138026]

[48] Kim K-W, Chanpiwat P, Hanh HT, Phan K, Sthiannopkao S. Arsenic geochemistry of groundwater in Southeast Asia. Front Med 2011; 5(4): 420-33.
[http://dx.doi.org/10.1007/s11684-011-0158-2] [PMID: 22198754]

[49] Shewale M, Bhandari D, Garg RK. A Review of Arsenic in Drinking Water: Indian Scenario (2007-2017). IJSRSET 2017; 3: 300-4.

[50]  Rasheed H, Slack R, Kay P. Human health risk assessment for arsenic: A critical review. Crit Rev Environ Sci Technol 2016; 19-20: 1529-83.
[http://dx.doi.org/10.1080/10643389.2016.1245551]

[51]  Ranjan RK, Ramanathan AL, Parthasarathy P, Kumar A. Hydrochemical characteristics of groundwater in the plains of Phalgu River in Gaya, Bihar, India. Arab J Geosci 2013; 6: 3257-67.
[http://dx.doi.org/10.1007/s12517-012-0599-1]

[52]  Yano Y, Ito K, Kodama A, *et al.* Arsenic Polluted Groundwater and Its Countermeasures in the Middle Basin of the Ganges, Uttar Pradesh State, India. J Environ Prot (Irvine Calif) 2012; 3: 856-62.
[http://dx.doi.org/10.4236/jep.2012.328100]

[53]  Vishnoi N, Dixit S, Sharma YK, Singh DP. Arsenic occurrence in Ground Water and Soil of Uttar Pradesh, India and its Phytotoxic Impact on Crop Plants. Res J Pharm Biol Chem Sci 2018; 4: 338-46.

[54]  Bhardwaj V, Singh DS, Singh AK. Hydrogeochemistry of groundwater and anthropogenic control over dolomitization reactions in alluvial sediments of the Deoria district: Ganga plain, India. Environ Earth Sci 2010; 59: 1099-09.
[http://dx.doi.org/10.1007/s12665-009-0100-y]

[55]  Sanjay Kumar S, Pandey G, Sharma A. Assessment of Arsenic in Groundwater of Gorakhpur District in Uttar Pradesh (India). IJERT 2014; 03: 12.

[56]  Singh AL, Singh VK. Arsenic contamination in Ground water of Ballia, Uttar Pradesh State, India. Appl Water Sci 2018; 8: 95.
[http://dx.doi.org/10.1007/s13201-018-0737-3]

[57]  timesofindia.indiatimes.com/articleshow/38339052.cms?utm

[58]  Gorchev HG, Ozolins G. WHO guidelines for drinking-water quality. WHO Chron 1984; 38(3): 104-8.
[PMID: 6485306]

[59]  Naujokas MF, Anderson B, Ahsan H, *et al.* The broad scope of health effects from chronic arsenic exposure: update on a worldwide public health problem. Environ Health Perspect 2013; 121(3): 295-302.
[http://dx.doi.org/10.1289/ehp.1205875] [PMID: 23458756]

[60]  Abernathy CO, Liu YP, Longfellow D, *et al.* Arsenic: health effects, mechanisms of actions, and research issues. Environ Health Perspect 1999; 107(7): 593-7.
[http://dx.doi.org/10.1289/ehp.99107593] [PMID: 10379007]

[61]  Hughes MF, Beck BD, Chen Y, Lewis AS, Thomas DJ. Arsenic exposure and toxicology: a historical perspective. Toxicol Sci 2011; 123(2): 305-32.
[http://dx.doi.org/10.1093/toxsci/kfr184] [PMID: 21750349]

[62]  Chen Y, Ahsan H. Cancer burden from arsenic in drinking water in Bangladesh. Am J Public Health 2004; 94(5): 741-4.
[http://dx.doi.org/10.2105/AJPH.94.5.741] [PMID: 15117692]

[63]  Applebaum KM, Karagas MR, Hunter DJ, *et al.* Polymorphisms in nucleotide excision repair genes, arsenic exposure, and non-melanoma skin cancer in New Hampshire. Environ Health Perspect 2007; 115(8): 1231-6.
[http://dx.doi.org/10.1289/ehp.10096] [PMID: 17687452]

[64]  Shen S, Li XF, Cullen WR, *et al.* Arsenic binding to proteins. Chem Rev 2013; 113(10): 7769-92.
[http://dx.doi.org/10.1021/cr300015c] [PMID: 23808632]

[65]  Flanagan SV, Johnston RB, Zheng Y. Arsenic in tube well water in Bangladesh: health and economic impacts and implications for arsenic mitigation. Bull World Health Organ 2012; 90(11): 839-46.
[http://dx.doi.org/10.2471/BLT.11.101253] [PMID: 23226896]

[66]  Cho KH, Sthiannopkao S, Pachepsky YA, Kim K-W, Kim JH. Prediction of contamination potential

of groundwater arsenic in Cambodia, Laos, and Thailand using artificial neural network. Water Res 2011; 45(17): 5535-44.
[http://dx.doi.org/10.1016/j.watres.2011.08.010] [PMID: 21917287]

[67] Kim K-W, Chanpiwat P, Hanh HT, Phan K, Sthiannopkao S. Arsenic geochemistry of groundwater in Southeast Asia. Front Med 2011; 5(4): 420-33.
[http://dx.doi.org/10.1007/s11684-011-0158-2] [PMID: 22198754]

[68] Hare V, Chowdhary P, Baghel VS. Influence of bacterial strains on Oryza sativa grown under arsenic tainted soil: Accumulation and detoxification response. Plant Physiol Biochem 2017; 119: 93-102.
[http://dx.doi.org/10.1016/j.plaphy.2017.08.021] [PMID: 28850869]

[69] Kumar M, Kumar P, Ramanathan AL, Prosun B. Arsenic enrichment in groundwater in the middle Gangetic Plain of Ghazipur District in Uttar Pradesh, India. J Geochem Explor 2010; 105: 83-94.
[http://dx.doi.org/10.1016/j.gexplo.2010.04.008]

[70] Kumar P, Thakur PK, Bansod BKS, Debnath SK. Groundwater: a regional resource and a regional governance. Environ Dev Sustain 2018; 20: 1133-51.
[http://dx.doi.org/10.1007/s10668-017-9931-y]

[71] Nicomel NR, Leus K, Folens A, Voort P, Laing GD. Technologies for Arsenic Removal from Water: Current Status and Future Perspectives. Int J Environ Res Public Health 2015; 13- ijerph13010062.

[72] Mondal P, Bhowmick S, Chatterjee D, Figoli A, Van der Bruggen B. Remediation of inorganic arsenic in groundwater for safe water supply: a critical assessment of technological solutions. Chemosphere 2013; 92(2): 157-70.
[http://dx.doi.org/10.1016/j.chemosphere.2013.01.097] [PMID: 23466274]

# CHAPTER 5

# Studies on Polymeric Ceramic Composite Membranes for Water Treatment

**Fakhra Jabeen**[1,*], **Qazi Inamur Rahman**[2] and **Miad Ali Siddiq**[1]

[1] *Jazan University, Jazan, Saudi Arabia*
[2] *Integral University, Lucknow, India*

**Abstract:** Environmental chemistry is the study of chemical processes occurring in the environment for understanding the diverse issues related to human health and resource conservation. These significant effects may be felt on a global scale, through the presence of water pollutants or toxic substances arising from chemical waste. The increasing world population, rapid industrialization, and human activities have resulted in higher water demand throughout the world. The fast spread of contamination problems worldwide and their effects on the natural resources of water led to the evolution of environmental chemistry. This evolution relies on the different membranes technology to facilitate the scientific investigations on the contamination extent and optimize remediation efforts. Polymeric ceramic composite membranes comprise a captivating field of membrane separation technology. Rapid development and innovation have been done in the modification of these membranes. These membranes have superiority in terms of high temperature and chemical resistance, higher chemical, and mechanical stability, and have higher longevity. All these outstanding features have made these membranes ideal for water treatment and desalination applications. This chapter is a review of the development, and the use of polymer composite membranes in treating wastewater. A brief description of synthesizing these membranes through different routes is given and is reviewed critically.

**Keywords:** Ceramic membrane, Desalination, Polymer, Sol-gel process, Water treatment.

## INTRODUCTION

Water is the main source of life; from being the basis of human survival to the economic development of a country. One of the major issues our society faces today is the shortage of fresh water. Increasing global population, periodical

---

* **Corresponding author Fakhra Jabeen:** Jazan University, Jazan, Saudi Arabia; E-mail: fakhrajabeen@gmail.com

Tahmeena Khan, Abdul Rahman Khan, Saman Raza, Iqbal Azad and Alfred J. Lawrence (Eds.)
All rights reserved-© 2021 Bentham Science Publishers

droughts, and rapid industrialization have resulted in higher water demand around the world. To complicate the issue, fresh water resources are reducing year after year throughout the world.

With the increase in world population, the gap between the supply and demand for water is growing and reached at such an alarming rate in some parts of the world that it is threatening the very existence of humans [1]. With the increase in population, the demand for drinking water has increased seven times [2]. Over the next 30 years, it is estimated that the population will grow by 40% and demand for domestic, agricultural, and industrial water sources will be increased, especially in developing countries where the need for water is greater than the population [3]. Lack of fresh water is a growing problem worldwide as only 1% of the earth's fresh water is available for people to consume [4]. It was reported by the United States geological survey that 96.5% of the earth's water is in oceans and seas and 1.7% in icebergs. The remaining percentage is in the form of salty water, brackish water, which is found as surface water in an outfall, and groundwater in salty sinkhole [5]. Demand for clean water is at the top of the global agenda of critical issues. India has 16% of the world's population and 4% of fresh water reservoirs [6]. Due to rapid urbanization, industrialization and development, there is an increased chance of recycling unsafe water in developing countries such as India. Although India occupies only 3.29 million square kilometres, making up 2.4% of the world's arid land, it carries more than 15% of the world's population.

The population of India as of March 31, 2011, was 1,210,193,422 (Census, 2011). The livestock population grew by 4.6% from 512 million in 2012 to 536 million in 2019 in India, which is about 20% of the world's total livestock. However, the total annual water consumption in the country is 1086 $km^3$, which is only 4% of the world's water assets [7]. The total annual water supply for groundwater and surface water is 396 $km^3$ and 690 $km^3$, respectively [8]. As a result of rapid population growth and increasing water challenge, stress on India's water supply is increasing and the availability of water per capita is declining day by day. In India, surface water levels in 1991 and 2001 were 2300 $m^3$ and 1980 $m^3$ respectively, and this is estimated to have dropped to 1401 $m^3$ and 1191 $m^3$ in 2025 and 2050 respectively [9]. The national water demand by 2050 is estimated to be1450 $km^3$ higher than the current availability of 1086 $km^3$. To overcome the water scarcity and the need for clean drinking water, there is a need for the development of new water resources and the protection of existing water resources through appropriate water treatment strategies [10].

The hydrosphere contains more than 75% of the earth's surface, including all types of water resources such as oceans, seas, rivers, lakes, streams, lakes, glaciers, icebergs, and groundwater. About 97% of the world's fresh water is in the form of oceans, which is inaccessible for human consumption due to its high salt content and total dissolved solids (Table **1-2**). About 2% of water resources are available in ice-cold areas and glaciers, while only 1% is available as clean water for human use and other uses. Fresh water is also available in the form of rain, snow, dew, and so on. The main use of water is in irrigation (30%), thermal power plants (50%), and other uses including domestic (8%) and industrial use (12%). Insecticides, pesticides, fertilizers, humans, animals, as well as industrial wastes, pollute the surface water.

Table 1. Classification of water according to its concentration of solids.

| Description | >Dissolved Solids (mg/L) |
|---|---|
| Drinking water | Less than 1000 |
| Mildly brackish | 1000 – 5000 |
| Moderately brackish | 5000 – 15000 |
| Heavily brackish | 15000 – 35000 |
| Average seawater | 35000 |

Table 2. Palatability of water according to concentration of its total dissolved solids.

| Palatability | Dissolved Solids (mg/L) |
|---|---|
| Excellent | Less than 300 |
| Good | 300 – 600 |
| Fair | 600 – 900 |
| Poor | 900 – 1200 |
| Unacceptable | More than 1200 |

# WATER POLLUTION

Of all-natural resources, water is essential to the existence of living organisms; civilization has polluted it and is facing its consequences. A decrease in physical, chemical, and biological properties of water can be defined as Water pollution caused due to natural weathering of rocks, minerals, soil sediments, nutrients, and organic matters of soil transported by erosion. This deterioration in the quality of water has increased in the last few decades mainly due to human activities in industrial and agricultural processes. In a few decades, there has been a growing global concern about the widespread distribution of pollutants from human activities, industrial and agricultural activities, and the potentially harmful effects of these pollutants on humans or ecosystems.

## PARAMETERS OF POLLUTION

Different parameters determine the nature and extent of pollution in water:

**(i) Physical parameters-** Color or water shade, foul smell, turbidity, density, and temperature.

**(ii) Chemical parameters** - pH, total dissolved solids, cation-anion concentration, total hardness, suspended solids, dissolved oxygen, residual chorine, chemical oxygen demand, redox potential, radioactive substances, organic matter, metallic ions, oxides, and industrial by-products.

**(iii) Biological parameters** - bacteria, various microorganisms, algae, small animals such as protozoans and crustaceans.

## Water Pollutants

Different types of water pollutants come under the following categories:

**(i) Inorganic pollutants** - These pollutants include inorganic salt, metals, heavy metals, ionic compounds, trace elements, complexes of metals with organic moiety, and mineral acids.

**(ii) Organic pollutants** - Pathogens, plant nutrients, sewage, organic compounds, biodegradable, and non-degradable substances come under these pollutants.

**(iii) Sediments** - Due to soil erosion, insoluble soil components enter water bodies thereby contaminating them. Several factors such as agricultural processes, construction activities, and mining activities have great influences on solid erosion rates.

**(iv) Radioactive substances** - Out of all the pollutants, they pose the greatest dangers to human health. Radioisotopes are classified as natural and artificial. Natural radioisotopes are produced by cosmic rays and are absorbed into the soil through precipitation, whereas artificial radioisotopes enter the environment mainly through nuclear fusion and research organizations. Some radioisotopes such as K-40, Ra-222, Ra-226, Ra-228, Pb-210, and C-14 enter the human body in incompatible ways.

**(v) Thermal pollutants** – These pollutants contain coal fires and nuclear fuel used by power stations as only a small fraction of the heat generated using nuclear fuel is successfully converted to operation and the remainder is wasted. At present even in coal-fired plants, the efficiency is not more than 40%. The condensers used in these power plants use water from a nearby river, lake, or municipal well and discharge the wastewater back to the source, the temperature of which rises by 10 °C in the process. This reduces the oxygen demand of the water and affects aquatic life.

Water pollution is caused by various sources, particularly point sources which are identifiable and non-point sources that are non-identifiable.

## MAIN SOURCES OF POLLUTANTS

Water pollution is caused by different sources mainly point sources that are identifiable at a single location and non-point sources whose location cannot be identified (Table **3**).

**Table 3. Major point and non-point sources.**

| Point Sources | Non-point Sources |
| --- | --- |
| Industrial waste disposal | Pollution due to industrial chemicals |
| Industrial effluent | Pollution due to agricultural activities |
| Water treatment plant | Oil pollution |
| Municipal sewage leakage | Eutrophication |
| Combined sewer overflows | Consequences of eutrophication |
| Raw sewage disposal | Control of eutrophication |
| Leaching residue tips | Heavy metals like Arsenic, Aluminium, Boron, Cadmium, Copper, Iron, Lead, Manganese, Molybdenum, Mercury Nickel and Zinc |
| Sanitary landfills | Insecticides and pesticides |
| Aerial fallout | Radioactive waste |

Industrial waste disposal pollution due to industrial chemicals and agricultural activities like water treatment plant generate oil pollution, municipal sewage leakage eutrophication combined sewer overflows, control of eutrophication leaching, heavy metals such as arsenic, aluminium, boron, cadmium, copper, iron, lead, manganese, nickel, and zinc, *etc.*

Water is used by people for various domestic purposes such as drinking, cooking, bathing, washing clothes, amusement, car-washing, and toilet cleaning, *etc.* Water is also used for various agricultural purposes, industrial objectives, transportation objectives, electricity generation, fishing, and so on. The amount of freshwater available is not enough to meet the growing needs of the people. Traditional water resources, such as rivers, lakes, lakes, *etc.*, at the surface level, are not completely reliable because most of them contain rainwater, and these days, rainfall is usually below normal. This causes the failure of water resources such as rivers, lakes, and ponds from being restored.

Due to the given reasons, the surface resources also fail to reach a certain level. Experts have estimated that more than a billion people do not have access to safe drinking water. Each year more than 5 million people die from water-related problems; 4 million of them are children. The growing demand for water and the ever-increasing population means that water supply is becoming a serious issue [5, 9].

Membranes appeared as a functional means for water treatment in the 1960s. With the emergence of high-performance membranes, the use of membranes for water treatment using highly advanced membranes made of new materials and monitored in various configurations has become quite popular.

The increase in the shortage of clean water resources led towards alternative resources such as seawater. In the 1970s, research began with a water-repellent membrane. Demonstrating its effectiveness in producing freshwater from brackish water, the membrane became an effective alternative to evaporation technology in the commercial water treatment industry. Over the years, the standards for treated water have become definite, and several new applications have emerged. However, the membrane has risen to the challenge and continues to function accurately and successfully [11]. These days, membrane-based separation processes are generally used in our daily lives in the food, biotechnology, petrochemical, and pharmaceutical industries. Their simplicity and cost-effectiveness, compared to other common segregation procedures such as absorption, adsorption, and distillation, have made them more popular and widespread in their usage. Lower energy consumption, potential to mix with other processes, effortlessness of scaling up, uninterrupted operations, high intensity, and electronic performance are some of the main advantages of membrane processes, whereas the limitations of membrane processes include membrane fouling, restricted chemical stability, and short lifespan.

A good membrane should produce high flux at low pressure, should be of practical use, produce lesser by-products, enhance water quality, and has an easy treatment [12]. The membrane method provides approximately 53% of the world's total production of pure water through an effective water treatment due to its simplicity in operation, no chemical additives, lower cost, phase stability, high yields, and greater volume removal. Due to its expanded properties, membrane formation plays an important role in the treatment of saltwater, wastewater, desalination of seawater, dairy industry, and waste treatment, *etc* [13 - 16].

## MEMBRANES AND THEIR CLASSIFICATION

A membrane is a physical barrier that selectively allows unwanted substances to pass through and useful substances to retain on its surface [17].

Membranes are classified on basis of nature and geometry (Fig. **1**).

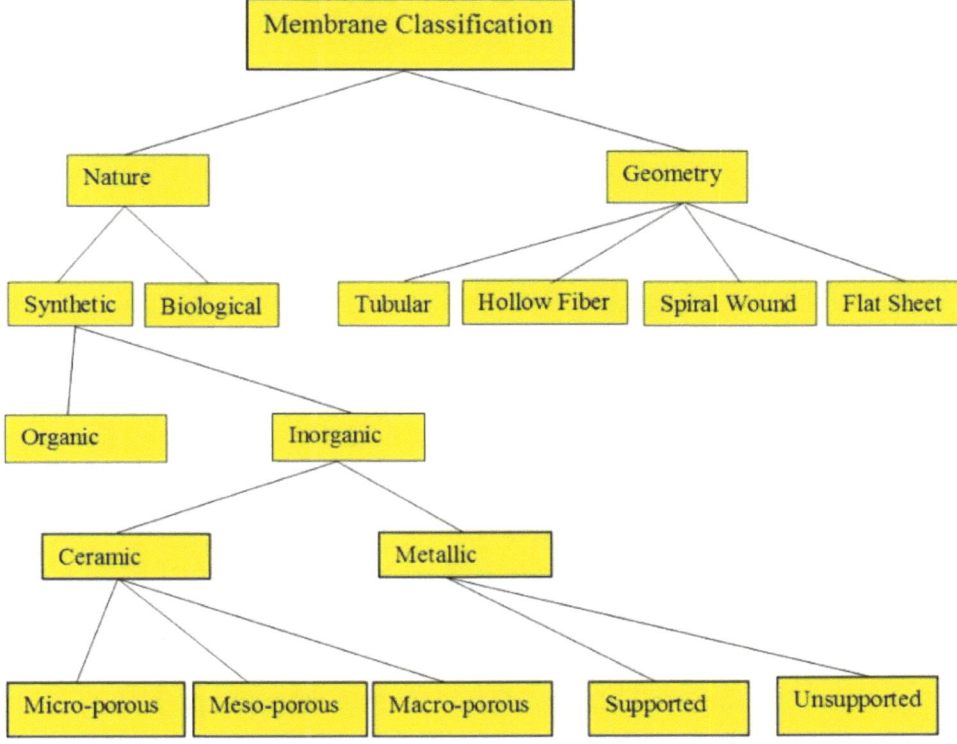

**Fig. (1).** Classification of Membranes.

### Synthetic Membrane

A synthetic membrane is designed artificially for separation purposes in industries. These membranes are used successfully in small and large industrial operations since the mid-twentieth century.

They are used for various separation processes such as dialysis, electrodialysis, hemodialysis, hemofiltration, ultrafiltration, and pervaporation, where the process is usually selected according to the driving force used such as concentration, electrical power, and pressure variation respectively, throughout the membrane. On the other hand, the membranes used for separation processes are highly durable, with excellent tolerability, temperature resistance, and high durability.

Much work has been done in the last 25 years to build the best membrane, using almost every available polymeric material.

**Biological Membrane**

A biological membrane acts as a selective barrier within living organisms. It can be bounding or separating and separates the interior of the cells from their outer space. It is composed of lipid bilayers, which are mainly two layers of amphipathic phospholipids, with their hydrophobic tail regions turned inwards, their polar head regions form intracellular and extracellular faces of the bilayer. Semi permeability is the selective permeability of these membranes. Partial and differential permeability means that different molecules may diffuse and pass by passive, active, and other types of transport, which are controlled by proteins.

**Organic Membranes**

These membranes are made up of synthetic or natural polymers and useful due to their characteristics, such as porosity, surface area, excellent selection, high penetration, good processability for potential fluids in the gas separation of $H_2$, $CO_2$, $O_2$, and other industrial-suitable gases. Gas separation is one of the most critical and important step for industrial processes. Metal-organic framework (MOF) membranes are generally used for this purpose. Common materials used to form organic membranes are polyethylene (PE), polytetrafluoroethylene (PTFE), polypropylene (PP), cellulose acetate (CA), polyacrylonitrile (PAN), polyimide (PI), polysulfone (PS), and polyethersulfone (PES).

The choice of the membrane depends on a variety of factors, including constituents of the feed solution, performance parameters, application type, and separation objectives.

**Inorganic Membrane**

It consists of materials such as ceramic, carbon, silica, zeolite, various oxides and metals such as palladium, silver, and their alloys, oxides, or metals, and can be found in multilayer supporting structures or as self-supporting structures. For $O_2$ or $H_2$ molecules, these membranes can be selectively permeable, energy-efficient, and bear extreme conditions.

Common inorganic membranes are dynamic membranes, liquid membranes, ceramic membranes, silica membranes, zeolite membranes, carbon membranes, and hybrid inorganic membranes. Slip casting, sol-gel method, chemical vapour deposition, and pyrolysis methods are widely used in the manufacture of inorganic membranes. Commonly used membranes for water treatment and desalination are alumina ($Al_2O_3$), titania ($TiO_2$), zirconia ($ZrO_2$), silica ($SiO_2$), and carbon membrane.

## Metallic Membranes

Metallic membranes are made by mixing metal powders such as tungsten, palladium, or stainless steel and placing them on a permeable surface. The main use of a metal membrane is to separate hydrogen, with palladium and its alloys being the most preferred. Surface poisoning is the main disadvantage of the metallic membrane.

Metallic membranes are formed on the cross-linked surface either as a sintered layer of fine powder or as metal oxide, having dense sheets or films used for hydrogen gas separation, because they offer the highest selectivity due to their dense structure, which prevents the overflow of large atoms and molecules such as $CO$, $CO_2$, $O_2$, and $N_2$ and can also work at high operating temperatures. Metals are deposited as thin layers on various supports (*e.g.*, mirrors, pottery, or other metals) to enhance the transmembrane flux, mechanical and thermal stability, and reliability.

The distinctive advantages of these membranes are robust surface, high flux rate, ease to connect to the module, metal embedded structure, low fouling rate, and lower cost. These membranes are useful in food and beverage industries, water filtration, process engineering, as cleaning agents, for fine filtration of catalyst particles, medical and environmental technologies.

## Ceramic Membranes

The ceramic membranes contain metals (aluminium or titanium), non-metals (oxide, nitride, or carbide) and are often used in highly acidic or basic areas media due to their inert nature, having pore size varying from 0.01 to 10 μm which includes microfiltration and ultra-filtration, high sensitivity to the temperature gradient, leading to cracking of the membrane which is the major drawback. Zeolite membranes are highly selective for gas separation due to the same pore size and catalytic activity, which are useful in membrane reactor applications.

## Micro-Porous Memeberanes

Microporous membranes are made up of polymeric films having a pore size from 0.03 μm to 10 μm in diameter. Polymers such as polyolefins, polyamides, polyester, and polyether-based copolymer with appropriate chemical, mechanical, and filler properties essential for high performance are used.

With the advancement in technology strong, thin, defect-free, cost-effective, high-performance membranes can be designed, having larger volume and open pores.

Membrane features are important from the point of view of pore chemistry because it plays a major role in the process of cell division. Pore structures can be determined using pore size, configuration, dimension, and performance.

**Meso-Porous Membranes**

Meso-porous membranes consist of a solid matrix with pores having diameters in the intermediate range between 2 and 50 nm. Having ordered, equal distribution of pores of various geometries, they are suitable for use in many processes. They consist of a silica-based matrix having a larger surface area and pore volume. Sintering, phase inversion, stretching, track etching, template leaching, slip casting, and sol-gel process are the methods used for membrane fabrication.

Meso porous silica membranes have potential applications including separation, adsorption, and cellular transport, due to their large surface area, high pore volume, and adaptable pore size. Specifically, the mesoporous silica membranes consisting of two or three pore structures are the most promising molecular sieve material for gas separation.

**Macro Porous Membranes**

Macroporous membranes have a solid matrix with a pore size diameter of more than 50 nm. They are prepared through phase inversion, decontamination, stretching, track etching, template leakage, slide distribution, and the sol-gel process. These types of membranes are useful in microfiltration and ultra-filtration. These membranes could be organic (polymers), inorganic (glass, metal, and clay), uniform, or non-uniform. Depending on their preparation they can be of flat sheet, fibre-free, capillary, and tubular type.

They have a fine hydrogen-permeable palladium layer coated on metallic support, either with dense support made of material with high hydrogen permeability or porous metal support with low hydrogen transport resistance.

The dense membrane based on palladium is used in the reactor and hydrogen purification.

Unsupported membranes are used to separate hydrogen from gas mixtures. Palladium and its alloys are high-performance materials, due to their high solubility and hydrogen permeability. Since palladium is expensive, one way is to apply a thin layer of palladium to the tantalum or vanadium support film that is also hydrogen-permeable. A composite palladium membrane is applied in the catalytic reactor formed by a thin layer of palladium or palladium alloy, embedded on the porous material, such as ceramic or stainless steel.

These membranes allow selective removal of hydrogen from the catalytic source.

Membrane modules may have different configurations and geometries, such as tubular, hollow fibre, spiral wound, and flat sheets. Tubular membranes (Fig. **2**) have tube-like structures, are 5 to 15 mm wide, with porous walls. They contain two tubes; the inner tube, called the membrane of the tube, and the outer tube, called the shell.

They are found inside a tube made of a distinctive type of material (polymer, ceramic) which is the supporting layer of the membrane. Because the location of the tubular membrane is inside the tube, the flow of the tubular membrane is usually inside out. The feed stream flows across the membrane tube and is separated from the outer shell while the concentrate is collected at the opposite end of the membrane tube.

**Fig. (2).** Tubular Membrane.

Tubular modules operate with the tangential cross flow and are often used to process heavy feed streams such as those with high dissolved solids, high suspended solids, oil or grease, and wastewater management. These layers have less contamination compared to hollow fibre, wind wound and flat sheet systems. They can work more strongly with emulsified oil compared to other membranes and can be physically cleaned with sponge balls.

The disadvantages of tubular layers are the compactness of the lower packaging and the larger size compared to the empty fibre and spiral-wound elements. Due to the wide internal diameter of the tubular modules, the flow requirements are much higher than other system configurations. The cost of the operation of the tubular membranes is often very high.

Tubular membrane filtration serves as a part of the industrial water filtration process to provide good pollution reduction. This is a pressure-driven process, which crosses the flow using a thin layer to separate solid particles from the liquid. After passing through the tubular membrane filter system, the water is sent for postponement of osmosis processing, or re-treated. The use of tubular filters can accurately remove the contents from the feed water sources to produce high-quality water. This water is incorporated into a reverse osmosis plant that produces high-quality clean water, allowing the industrial processing plants to follow the discharge rules and helps keep the environment safe.

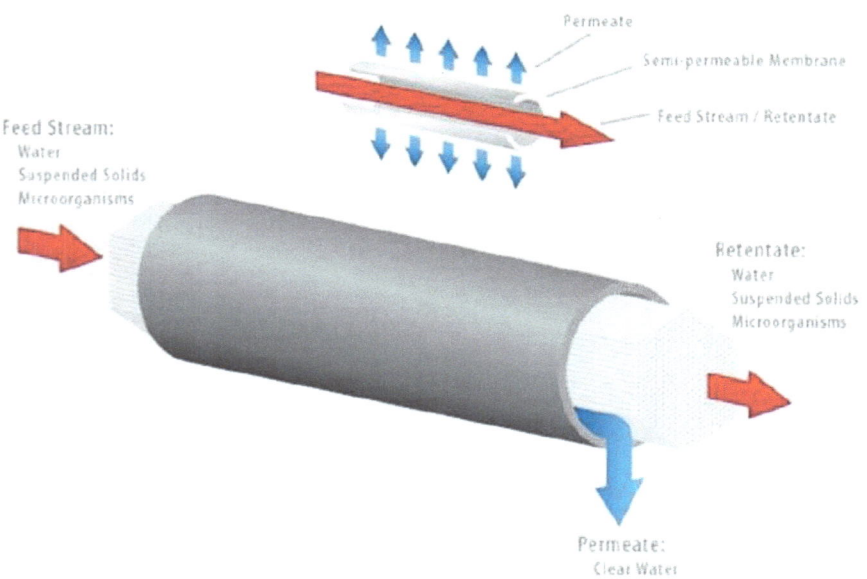

**Fig. (3).** Hollow Fiber Membrane.

These systems provide an effective and inexpensive solution for filtering wastewater. The membrane used in the tubular filter membrane can withstand heavy chemical processing, cleaning, and abrasion for many years.

Hollow fibre membranes (Fig. **3**) are made of semi-permeable barriers using thousands of long, fine filaments ranging from 1-3.5 mm in diameter, embedded in polyvinylchloride shell. Each thread is very small in diameter and flexible, hollow fibres filtering works on equivalent principles such as tubular and capillary adjustment but uses a smaller tube diameter that allows more flexibility.

These membranes have high packaging density due to smaller diameter and flexibility and having a configuration that cannot be achieved in other filters. They can also be backflushed from the permeate side and cleaned of air, and they can process feed streams with high total suspended solids.

Hollow fibre membrane is used in all forms of filtration, from microfiltration to osmosis conversion, water treatment, salt distribution, tissue engineering, cell culture, medicine, and fluid and gas separation. These membranes offer an inexpensive way to filter large volumes of substances and are made in different diameters; these membrane technologies provide great flexibility in the management of various streams, from groundwater filtration to wastewater treatment. Due to their hollow configuration, they can be used to separate drinking water, beer, and wine filters in the beverage industry, juice processing, biotech applications, reversing osmosis pretreatment, industrial water, and contaminated water. Irreparable fouling and fibre tear are major problems related to hollow fibre filtration. Due to their plasticity, they split under high pressure as compared to other filters such as tubular or spiral-wound elements. These membranes have higher operating costs in comparison to other configurations.

Spiral-wound membrane (Fig. **4**) comprises membranes, feed spacers, permeate spacers, and a permeate tube. Initially, the membrane is fabricated and doubled in half with the membrane facing inwards. The feed spacer is then placed between the folded layers, forming a membrane sandwich. The objective of the feed spacer is to provide a space for water to flow between the membranes, and to allow a constant flow between the membrane leaves. After that, the permeate spacer is attached to the permeate tube, using adhesive membrane sandwich is attached to the permeate spacer. The next permeate layer is laid down and sealed with adhesive, and the whole process is repeated until all the necessary permeate spacers have been attached to the membrane. The finished layers of membranes are then wrapped around the tube to form a spiral structure.

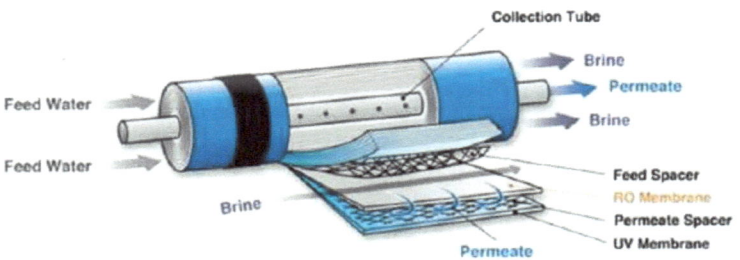

**Fig. (4).** Spiral-Wound Membrane.

Spiral-wound membranes act as a crossflow system, which is perfect for scaling up from flat sheets, sufficient for small production, feasibility studies, and development process. These materials can be used under a wide range of feed flow rates, permeate flux, and operating pressure or temperature to maintain the membrane filtering system. Spiral-wound membranes occur in different shapes with different spacers, membrane types, lengths, and widths that allow them to be used in a variety of applications. These membranes are used in various applications including whey protein concentrations, cathode/anode paint retrieval, lactose concentrations, desalting dye and concentration, sulfate removal, and oil separation in polluted water.

Membrane Spiral-wound contamination is much greater than fouling in tubular membrane filtration processes. Due to the higher packaging density, total suspended solids should drop to a minimum in the feed stream to prevent clogging of the membranes. These membranes are not capable of handling mechanical cleaning like tubular membranes which have less packing density than hollow fibre.

Flat sheet frames (Fig. **5**) are mounted on plate-and-frame or cassette devices. The distance between each membrane is usually 0.5 to 2.0 mm. They have a planar arrangement and are mostly rectangular. The sheet, shell, or panel can be rigid, semi-rigid, or flexible.

Rigid panels have a supporting plastic plate of 4 to 6 mm where the membrane boundary is welded on both sides. The membrane is divided into a rigid plate supported by a thin spacer. Water flows from the outside to the inside of the panel through the spacer material of space and is collected from the permeate outlet tubes. The panels are loaded onto a cassette that holds many panels, permeate outlet tubes, and are attached to one or two manifolds.

**Fig. (5).** Flat Sheet Membrane.

The main features of flat sheets membrane are durability, high porosity, high flux, low fouling, chemical stability, and resistance. These membranes are used for a variety of purposes including the dairy industry, food and beverage industries, recycled water, landfill leachate treatment, industrial recycling and wastewater, and municipal sewage. The membrane is separated from the ceramic membrane by the polymeric membrane.

Ceramic membranes are made from metals or ceramics exhibiting high structural, mechanical, and thermal stability. Among the ceramic membranes, microporous silica membrane is widely used in molecular sieving. Size-based filtering processes require pore size in the nanometer range which is still a problem for the ceramic microporous silica membranes and has not yet been resolved. However, zeolite membranes have high thermal stability, mechanical stability, and hydrophilic nature. These membranes are used as excellent filters on the cellular scale compared to other inorganic elements. Although the ceramic zeolite membrane is widely used in various filtration processes, factors such as ion exchange, acidity, extraction or adsorption capacity and catalytic property reduce their effectiveness [18].

The main features of the ceramic membrane are:

- Consistent and compatible pore size.
- Very narrow and significant pore-size distribution on the surface.
- Prolonged and reliable lifetime.
- Excessive permeate flux, elevated concentration factor, and reasonable waste discharge.
- High stability to organic medium.
- High-temperature resistance, high-pressure resistance, high chemical resistance.
- Inflexible and stiff with no deformation.
- Membranes bonded to the substance by strong and stiff ceramic bonds.

- Deterioration and corrosion-resistant.
- Unresponsive to bacterial action.
- Can be frequently sterilized by steam and other chemicals.
- Ability to backflow and be revived.
- Rejuvenation after fouling.
- Can process extremely viscous fluids.
- Larger void area per unit area of filtration surface and applicable for filtration at high temperature.
- Steady over a broad pH range.

## APPLICATIONS OF CERAMIC MEMBRANES

The main features of ceramic membranes are compatible pore size, narrow pore size distribution on the surface, prolonged lifetime, high flux, high concentration factor, and appropriate waste disposal. They have high stability, high temperature, pressure and chemical resistance, resistance to corrosion, and are non-responsive to bacterial action. They can be sterilized by steam and other chemicals, possess the ability to be renewed, and can be restored after fouling. They can process excessive viscous liquids, and have a larger void per unit area for filtration at high temperature, and possess stability over a wide pH range.

Ceramic membranes are generally used in industries for water purification, biotechnology, pharmaceuticals, dairy, food and beverages, chemicals and petrol, microelectronics, metallurgical processes, and energy production.

Ceramic membranes are suitable for cleaning chemicals at high temperatures while using alkalies, chlorine, ozone, hydrogen peroxide and strong inorganic acids or using steam. Initially, the ceramic membranes were used in the treatment of water.

### Chemical Industry

- Recycling and cleaning of organic solvents
- Desalination of many products
- Recovery of pigments and dyes
- Separation of catalysts
- Product cleaning and separation
- The concentration of metal hydroxide solutions and polymer suspensions.

### Metal Industry/Surface Engineering

- Treatment of water/oil emulsions
- Recovery of many metals

- Recycling and disposal of degreasing
- Cleaning of wastewater
- Treatment of wastewater from glass and glass fibre production

**Textiles/Pulp and Paper Industry**

- Fractionation, concentration, isolation, and sterilization of antibiotics, enzymes, proteins, amino acids, and vitamins.
- Concentration, separation, and dewatering of biomass and algae.
- Separation of yeast
- Desalination.
- Disposal of fat emulsions

**Food and Beverages**

- Purification of pure drinking water.
- Tea and other filtration processes for clarification under high temperature.
- Clarification of juice, wines, draft beer, and yellow wine, *etc*.
- Sterilization and clarification of whey and milk.
- Separation and fractionation of whey and milk ingredients.
- Desalination of whey
- Dewatering of many products.
- Sterilization and clarification of soy sauce, vinegar, and other derivatives.

**Recycling and the Environment**

- Water/oil separation
- Recovery of pesticides and pharmaceuticals.
- Retention of microorganisms.
- Retention of radioactive substances and heavy metals.

- Recycling of water from swimming pools.

- Purification of the drain of sewage plants.

On the other hand, polymeric membranes have higher flexibility, film-forming property, chemical stability, mechanical strength, selective transport of chemical species, inexpensive materials for their manufacture. They have the required pore size for a variety of filtration processes and find their use in pressure-driven processes such as ultrafiltration, nano-filtration, reverse osmosis, gas separation, drug delivery, and wastewater treatment (Fig. **6**) [19 - 22]. Materials used for the production of polymeric layers include Polyamide (PA), Cellulose acetate (CA), Polyvinyl alcohol (PVA), Polysulfone (PSF), Polyethersulfone (PES), Polyvinylidene fluoride (PVDF), Polyvinyl chloride (PVC), Polypropylene (PP), Polyacrylonitrile (PAN), Polyimide (PI), Polyethylene (PE), and chitosan [23].

According to the pore size and filtering process, the membranes are divided into four different categories as, Microfiltration (MF), Ultrafiltration (UF), Nanofiltration (NF), and Reverse Osmosis (RO).

In microfiltration, components have a pore size of 1-0.1 micrometre used for the separation of larger particles (macromolecules, colloids, bacteria, and other particles). The ultrafiltration membranes have a pore size of 0.1-0.01 micrometre. They can separate viruses, macromolecules, solutes, and high molecular weight substances are retained whereas, water and low molecular weight substances permeate.

Nanofiltration membrane is a condensed membrane structure with small pores as compared to ultrafiltration and microfiltration membranes and removes (divalent ions) and allows monovalent ions to pass, with a pore size domain of 0.1-0.001 mm. Reverse Osmosis membranes are more condensed as compared to nanofiltration and require a high pressure-driven filtration process with a narrow pore size domain (< 0.001 mm) thus almost separate all monovalent ions or impurities and recover clean water (Fig. **1**) [24, 25].

A nanofiltration membrane is a condensed membrane structure with smaller pores compared to ultrafiltration and microfiltration that removes (divalent ions) and allows the monovalent ions to pass through, with a pore size of 0.1- 0.001 mm.

Compared to nanofiltration, reverse osmosis requires condensed membranes and is a pressure-driven filter with a small pore size (<0.001 mm) thus recovering fresh water and separating all monovalent ions and contaminants. (Fig. **1**) [24, 25].

## TYPE OF MEMBRANES AND CHARACTERISTICS

Fig. (6). Types of membranes and characteristics.

## POLYETHERSULFONE MEMBRANE CHARACTERISTICS AND ITS TYPES

(PES) membranes, due to their excellent oxidative, hydrolytic, thermal, and mechanical properties are widely used in various separation processes. PES membranes are produced by phase-inversion procedure and subsequent membrane formation depending on the concentration of solvents, non-solvents, additives, coagulation baths, and solution temperature [26]. The PES membrane has important functions in the bio-medical field because of its compatibility with blood, and is used in various blood-related diseases (such as plasmapheresis, hemodialysis, hemofiltration, *etc.*), and artificial organs for blood purification. These membranes are hydrophobic and familiar with membrane fouling [27]. Polysulfone (PSF) is one of the most widely used polymers for membrane formation because of its mechanical strength, chemical inertness, thermal stability, and pH resistance. PSF membranes are used in a wide range of applications, such as hemodialysis, hemofiltration, water treatment, and in the purification of proteins, *etc.* Damage to the membrane caused by PSF's hydrophobic signals, sufficiently reduces the function of the membrane and its function, thereby resulting in a significant barrier to the application of the membrane [28]. Therefore, many efforts have been made to improve the membrane hydrophilicity and filtration properties by incorporating hydrophilic

Polyethylene glycol (PEG) or polyvinyl pyrrolidone (PVP) into the membrane solution. Polyethylene glycol or polyvinyl pyrrolidone can promote pore formation in polymeric components and increase their saturation properties [29]. Jeonghyun et al. [30] remoulded the polysulfone membrane with tetrahydrofuran, compliant with water pressure (8 bars) for sodium alginate separation. They concluded that the polysulfone recast membrane showed a high-water flow rate of 259 L / m$^2$ and 98.8% rejection of sodium alginate in the filtration process.

Polyvinylidene fluoride (PVDF) is a polymer used in the preparation of membranes for ultrafiltration and microfiltration because of its improved thermal and mechanical stability and supreme chemical resistance [31]. Despite their hydrophobic nature, it is receptive to fouling during wastewater treatment. Contamination can be reduced by surface tempering of the PVDF membrane by attaching a hydrophilic layer of nanoparticles to its surface by linking [32 - 34]. Nevstrueva et al. [35] synthesized $TiO_2$ amalgamated cellulose membrane for the process of ultrafiltration and investigated the effect of nanoparticle size on membrane function. They found that the $TiO_2$ nanoparticle with a diameter of 10 nm offered better antifouling properties compared to using a larger particle size of 26-30 nm. Liu et al. [36] modified the separation performance of PVDF membrane by PVDF / $\gamma$-$Al_2O_3$ with a 2% weight. $\gamma$- $Al_2O_3$ exhibited better performance. Oh et al. [37] synthesized PVDF membranes using $TiO_2$ nanoparticles dispersed into fabricating solution utilizing PET films. The resulting nanocomposite membrane exhibited improved resistance to fouling.

Cellulose acetate (CA) membranes are used commercially in many gas separations processes, due to their high solubility in terms of carbon dioxide and hydrogen sulfide within the cellulose acetate polymer matrix [38]. Ahmad et al. [39] synthesized a hybrid matrix membrane made from cellulose acetate, with multi-walled carbon nanotubes and examined the higher performance ability of the hybrid matrix membrane for carbon dioxide and nitrogen gas separation. Lee et al. [40] analysed the features of cellulose acetate membranes altered by chitin nanocrystals through the surface coating. They found that a modified hydrophobic cellulose acetate matrix with a contact angle of 132° was reduced to 0° after transformation into a hydrophilic membrane.

Cellulose Triacetate (CTA) is used as a polymeric membrane material in ultrafiltration membrane for dialysis, reverse osmosis, and forward osmosis due to its excellent desalting properties with reasonable stiffness, higher biocompatibility, and cost-efficient. CTA membranes have hydrophilic properties and prevent fouling [41].

Polymeric nanocomposite membranes are improved by the distribution of nanomaterial in their polymeric membrane matrices. Polymer nanocomposite membranes are active in solvent nanofiltration, pervaporation, methanol fuel cells, sensors, gas segregation, water treatment, proton exchange, and membrane fuel cells. On the other hand, these membranes are widely used for the separation of liquid-solid, gas-gas, and liquid-liquid.

Nanocomposite membranes use organic, inorganic, hybrid, bio-organic nanomaterials. The process of phase modification is used to produce polymer nanocomposite membranes either in form of hollow fibre or flat sheets by dispersing nano-materials into polymer solutions [42]. Different ways of fabrication such as dip coating, sintering, track-etching, stretching, template leaching method, and phase inversion method are used for the fabrication of the desired framework and features. The phase transformation process is a significant and common process of polymeric membrane formation [43 - 45].

Natural polymers such as cellulose and cellulose derivatives are used for the manufacture of membranes such as collodion and cellulose acetate. These polymeric materials are used in the manufacture of desalination membranes embedded into hyperfiltration modules, to produce pure water free from saline and seawater.

## DESALINATION FOR WATER TREATMENT

Water Treatment by desalination is the process of removal of excessive salt and other minerals from the sea to get clean water suitable for human and animal use or irrigation [46].

This process is used all over the world to provide people with a reliable supply of clean water. In old days, many communities utilized this process to convert seawater into drinking water. Naturally, the sun causes water to evaporate from the earth's surface such as lakes, seas, rivers, and streams. Steam, at last, meets cool air, where it gets re-condensed and form dew or rain. Artificially this process is much faster, using man-made techniques for heating and cooling and desalination, whereas solar energy is used in a natural process to form rain which is the largest source of fresh water on earth (Fig. 7). Artificial distillation systems are duplicates of this natural process on a small scale. Desalination is also a major concern in agriculture [47]. Desalination methods use reverse-osmosis in which saline water is forced through a membrane allowing water molecules to pass through but retention of salts and other minerals. Another potential product of this process is salt. The main objective in this process is focused on creating more affordable ways to provide fresh water for human consumption. Along with recycled wastewater, this is one of the few rainfall-independent sources of water

[48]. Desalination processes use chemical engineering technology in which saline water is fed into operating machinery, heat energy, water pressure or electricity is applied, and two streams are produced; a stream of pure water and a stream of concentrated brine which must be decomposed.

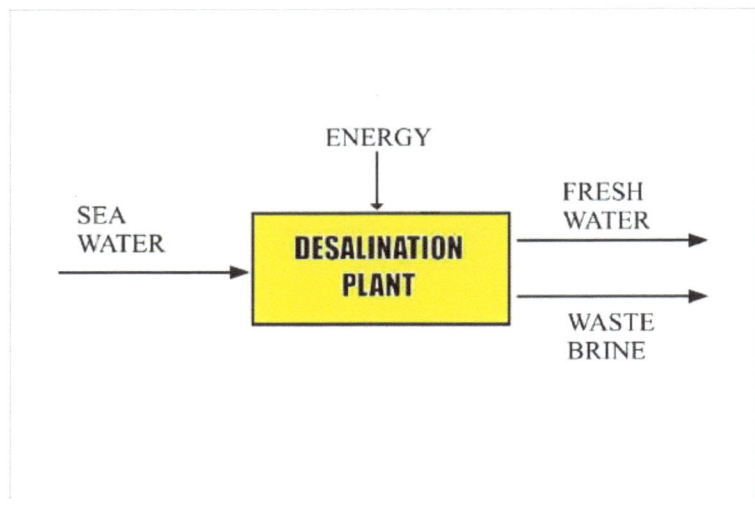

**Fig. (7).** Basic principle of desalination.

In the meantime, there are more than 12,500 industrial desalination plants (Table 4) around the world that produce fresh water from the sea and brackish water [49]. As a standard, Seawater contains approximately 35,000 mg/L of salt, but these estimates can vary from 24,000 to 42,000 mg/L depending on the area [50]. Brackish water is less salty than seawater, which is between 2,000-5,000 mg/L. However, drinking water requirements require salinity levels below 250 mg/L [51]. In seawater containing 35,000 mg/L of salt, a 99.3% low salt rejection layer is required to produce drinking water in seawater.

**Table 4. The world top most desalination plants.**

| Desalination Plants | Locations | Capacity (m³/day) |
|---|---|---|
| Ras Al Khair | Saudi Arabia | 1,036,000 |
| Taweelah | United Arab Emirates | 909,200 |
| Shuaiba 3 | Saudi Arabia | 880,000 |
| Sorek | Israel | 624,000 |
| Rabigh 3 IWP | Saudi Arabia | 600,000 |
| Fujairah 2 | United Arab Emirates | 591,000 |

Various processes used for desalination are Multi-Stage Flash Distillation (MSF), Multi-Effect Distillation (MED), Thermal Vapor Compression (TVC), solar distillation, ice, hydrate formation, membrane distillation, Mechanical Vapor Compression (MVC), Reverse Osmosis (RO), Electrodialysis (ED) and ionic exchange (Fig. **8**). Among these, the most common are multi-stage flash distillation, multi-effect distillation, and reverse osmosis [52].

Multi-effect distillation is the earliest known technique, used in the middle of the 19th century to make freshwater. The basic principle involves the transfer between steam and seawater in different phases. This usually results in a higher proportion of water produced by steam. A large scale means that the process is good to produce a high amount of water with a low amount of energy input [52]. Multi-stage flash distillation is an improved version of the multi-effect distillation is a standard procedure used today; having a series of flash chambers where seawater is heated and pure condensate is collected. While this process is simple and reliable, it requires more energy than multi-effect distillation, which makes it more expensive [52]. A reversal of osmosis emerged in the 1960s with saltwater treatment and in the 1980s reverse osmosis became very competitive with current seawater purification systems for seawater. The main advantage of this process is the low power consumption due to the lack of evaporation action [52].

## TYPES OF DESALINATION PROCESSES

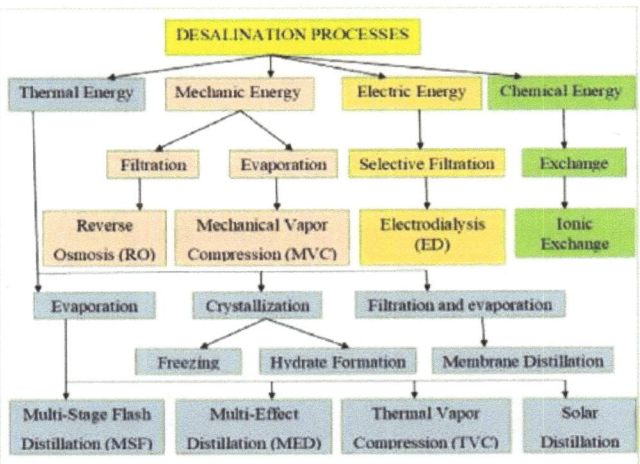

**Fig. (8).** Types of desalination processes.

# ADVANTAGES AND DISADVANTAGES OF DIFFERENT DESALINATION PROCESSES

## 1. Multi-Stage Flash Distillation (MSF)

Advantages

• MSF plants have large capacities and the salinity of the feed water does not have much influence on the process.

• There is a minimal requirement for the pretreatment of the feed water and produces very high-quality water as the product.

• MSF plants do not have very strict operational and maintenance procedures.

• It can be combined with other processes.

Disadvantages

• They are costly to build and need a high level of technical knowledge.

• High energy is required and the recovery ratio is low.

## 2. Multi-Effect Distillation (MED)

Advantages

• Small energy is consumed as compared to other thermal processes. The process operates at low temperature and concentration to avoid corrosion and scaling.

• Highly reliable, simple to operate, and does not need pretreatment.

• Small to medium-sized plant size, low maintenance cost, continuous operation with minimum supervision.

• Can be adapted for any energy source, including hot water, waste heat from power generation, industrial processes, or solar heating.

Disadvantages

• Operates at high temperature that increases corrosion and scale formation

• Water formed as a product usually has a very high temperature and requires cooling before it can be used as potable water. The recovery ratio is low.

## 3. Thermal Vapor Compression (TVC)

Advantages

- Small to large plant size, low investment costs
- Minimum corrosion and scaling risk
- Lower thermal energy consumption
- Low operating costs, efficient use of plant volume.

Disadvantages

- High maintenance of compressors and heat exchangers.
- Energy consumption is high.
- Capital cost is high

## 4. Solar Distillation

Advantages

- Produces pure water, no conventional energy required, no skilled operator is required.
- Local manufacturing and repairing
- Can purify highly saline water, no energy costs due to the use of solar panels
- Very low maintenance.

Disadvantages

- Needs a lot of space and investment cost due to the price of land.
- The current costs of panels are high.

## 5. Freezing

Advantages

- Small energy consumption
- Produce very pure potable water

Disadvantages

- The slow growth of ice crystals
- Handling ice and water mixtures is complicated

**6. Hydrate Formation**

Advantages

- No fuel needed, no waste produced
- Not expensive to operate
- It produces a great deal of energy

Disadvantages

- Depends on the waves, needs a suitable site.
- Some designs are noisy

**7. Membrane Distillation**

Advantages

- Requires low pressure, temperatures, and vapour space than the boiling point of feed solution.
- High osmotic pressure, high separation factor of nonvolatile solutes.
- Production of high-quality water
- Production of potable water from brackish water in arid regions.

Disadvantages

- Feedwater should be free from organic pollutants
- Sediment clogging
- High initial cost, low performance with time

**8. Mechanical Vapor Compression (MVC)**

Advantages

- Economically better as requires a low maintenance cost

- Eco-friendly, waste steam recycled to generate electricity

- Compact size, reduced corrosion risk, and minimized scaling risk

- Process stability, no need for external thermal energy

Disadvantages

- High maintenance and spare parts requirements, highly skilled operators required.

- Capital cost is high

## 9. Reverse Osmosis (RO)

Advantages

- The processing system is simple, but a clean supply of feedwater needed to minimize the need for frequent cleaning of the membrane

- Low installation costs, low maintenance

- Plants have an efficient space/production capacity ratio, ranging from 25 000 to 60 000 L/day/m$^2$

- The process uses an unlimited and reliable water source

- The process removes organic and inorganic contaminants

- Negligible environmental impact, minimal use of chemicals

Disadvantages

- Feedwater usually needs to be pretreated to remove particulates

- High-quality materials and equipment needed

- Risk of bacterial contamination of the membranes

## 10. Electrodialysis (ED)

Advantages

- High water recovery design and generates low concentrate volume for disposal

• Targets arsenic, fluoride, radium, and nitrate removal. Easy to control salt removal and product quality by adjusting the amount of electricity applied to membrane stack.

• Long membrane life, low installation costs.

• Does not require any of the solutions to change its state of matter, simply separates the solute from the solvent.

Disadvantages

• A large amount of electricity required to produce the desired effect.

• Chemical usage for pre-treatment,

• Sometimes leaks occur in the membrane stacks.

## 11. Ionic exchange

Advantages

• Purifies safe drinking water for homes and communities

• Quality product water is obtained. Efficient removal of pollutants from wastewater.

• Recycle and recover water and valuable products for reuse.

• Low installation costs, low maintenance and green technology.

Disadvantages

• Require pretreatment

• Inefficient for treating high salinity streams.

Advantages of desalination

• It provides people with potable water

• It provides water to the agricultural industry

• It uses tried-and-tested technology

Disadvantages of desalination

• It consumes a large amount of energy

- It can be costly

- It can be harmful to the environment

## CONCLUDING REMARKS

Water treatment is required for the purification of water. Desalination procedures create new sources of freshwater from seawater or brackish waters. This chapter has summarized the basic features of membranes required for water treatment. Global population growth, rapid industrialization, and human activity have deteriorated the quality of the existing water supply and raised the global demand for pure water. Desalination of seawater has become an increasingly important source of usable water. An environmentally friendly approach, energy use, and environmental impact must address the effect of desalination methods on the purification of seawater to freshwater. In the meantime, desalination has become a very cost-effective solution to address the shortage of freshwater, especially in tropical and coastal areas. In recent years, there has been significant progress in membrane desalination techniques, especially in the form of the membrane module, energy retrieval, pre-and post-treatments methods that have made it inexpensive compared to the thermal procedures. The benefits of using renewable energy for desalination processes have become a more sensible and superior alternative to stressing energy situations and a sustainable solution to water scarcity.

The lack of clean drinking water hinders social and economic development in many areas, where solar resources are plentiful. Therefore, the use of solar energy for desalination in African countries and the Middle East has emerged as a promising solution to meet the water demand. It can play a role in solving water shortage problems and reducing carbon dioxide emissions through environmentally friendly processes. For further research and development, stakeholders, governments, industries, universities, and research institutes all around the world need to develop and improve the seawater desalination process to make it more affordable globally, especially in countries lacking common forms of energy and suffering from water scarcity. Similarly, various water desalination systems require extensive research and analysis to assess their potential for development, use, and efficiency. Besides, a combination of renewable energy and desalination procedures needs to be done together. The benefits of seawater desalination are important to human needs, but the economic, cultural, and geographical costs of the practice remain high. In many parts of the world, alternatives offer the same benefits of fresh seawater at lower economic and environmental costs. These alternatives include the treatment of low-quality local water resources, promoting district water transfers, improving conservation and efficiency, promoting wastewater recycling, and implementing effective land-use planning.

## CONSENT FOR PUBLICATION

Not Applicable.

## CONFLICT OF INTEREST

The author confirms that this chapter contents have no conflict of interest.

## ACKNOWLEDGEMENT

Declared none.

## REFERENCES

[1] Li Y, Tian K. Application of vacuum membrane distillation in water treatment. J Sustain Dev 2009; 2: 183-6.
[http://dx.doi.org/10.5539/jsd.v2n3p183]

[2] Watkins K. Watkins, K. Human development report: United Nations development programme, 2006.

[3] Water U. Coping with water scarcity: Challenge of the twenty-first century. Prepared for World Water Day 2007.

[4] Pangarkar BL, Thorat PV, Parjane SB, Abhang RM. Performance evaluation of vacuum membrane distillation for desalination by using a flat sheet membrane. Desalination Water Treat 2010; 21: 328-34.
[http://dx.doi.org/10.5004/dwt.2010.1400]

[5] Greenlee LF, Lawler DF, Freeman BD, Marrot B, Moulin P. Reverse osmosis desalination: water sources, technology, and today's challenges. Water Res 2009; 43(9): 2317-48.
[http://dx.doi.org/10.1016/j.watres.2009.03.010] [PMID: 19371922]

[6] Abraham T, Luthra A. Socio-economic and technical assessment of photovoltaic powered membrane desalination processes for India. Desalination 2011; 268: 238-48.
[http://dx.doi.org/10.1016/j.desal.2010.10.035]

[7] Kumar R, Singh RD, Sharma KD. Water resources of India. Curr Sci 2005; 89: 794-811.

[8] Ministry of Water Resources, Integrated water resources development a plan for action, Report for the National Commission for Integrated Water Resource Development, 1999.. 1999.

[9] Emerson G. Every Drop is Precious: Greywater as an Alternative Water Source, 98 of Research Bulletin no.4, Queensland Parliamentary Library, 1998, 98.

[10] Pendergast MM, Hoek EM. A review of water treatment membrane nanotechnologies. Energy Environ Sci 2011; 4: 1946-71.
[http://dx.doi.org/10.1039/c0ee00541j]

[11] Amjad Z. Reverse Osmosis: Membrane Technology, Water Chemistry and Industrial Applications. New York: Van Nostrand Reinhold 1993.

[12] Kochkodan V, Hilal N. A comprehensive review on surface modified polymer membranes for biofouling mitigation. Desalination 2015; 356: 187-207.
[http://dx.doi.org/10.1016/j.desal.2014.09.015]

[13] Li D, Yan Y, Wang H. Recent advances in polymer and polymer composite membranes for reverse

and forward osmosis processes. J Prog Polym Sci 2016; 61: 104-55.
[http://dx.doi.org/10.1016/j.progpolymsci.2016.03.003]

[14] Luo J, Cao W, Ding L, Zhu Z, Wan Y. Treatment of dairy effluent by shear-enhanced membrane filtration: The role of foulants. J Seppur 2012; 96: 194-203.
[http://dx.doi.org/10.1016/j.seppur.2012.06.009]

[15] Yan Y, Li D, Wang H. Advances in polymer and polymer composite membranes for forward and reverse osmosis processes. Prog Polym Sci 2016; 16: 104-25.

[16] Mezher T, Fath H, Abbas Z, Khaled A. Techno-economic assessment and environmental impacts of desalination technologies. J Desal 2011; 266: 263-73.
[http://dx.doi.org/10.1016/j.desal.2010.08.035]

[17] Shi W, He B, Ding J, Li J, Yan F, Liang X. Preparation and characterization of the organic-inorganic hybrid membrane for biodiesel production. Bioresour Technol 2010; 101(5): 1501-5.
[http://dx.doi.org/10.1016/j.biortech.2009.07.014] [PMID: 19656676]

[18] Joshi R, Alwarappan S, Yoshimura M, Sahajwalla V, Nishina Y. Graphene oxide: The new membrane material. J Apmt 2015; 1: 1-12.

[19] Ulbricht M. Advanced functional polymer membranes. J Polym 2006; 47: 2217-62.
[http://dx.doi.org/10.1016/j.polymer.2006.01.084]

[20] Yampolskii Y. Polymeric gas separation membranes. Macromolecules 2012; 45: 3298-311.
[http://dx.doi.org/10.1021/ma300213b]

[21] Rao M, Sircar S. Performance and pore characterization of nanoporous carbon membranes for gas separation. J Membr Sci 1996; 110: 109-18.
[http://dx.doi.org/10.1016/0376-7388(95)00241-3]

[22] Farrell S, Sirkar KK. Mathematical model of a hybrid dispersed network-membrane-based controlled release system. J Control Release 2001; 70(1-2): 51-61.
[http://dx.doi.org/10.1016/S0168-3659(00)00336-9] [PMID: 11166407]

[23] Jhaveri JH, Murthy ZA. Comprehensive review on anti-fouling nanocomposite membranes for pressure driven membrane separation processes. J Desal 2016; 379: 137-54.
[http://dx.doi.org/10.1016/j.desal.2015.11.009]

[24] Kochkodan V, Johnson DJ, Hilal N. Polymeric membranes: surface modification for minimizing (bio)colloidal fouling. Adv Colloid Interface Sci 2014; 206: 116-40.
[http://dx.doi.org/10.1016/j.cis.2013.05.005] [PMID: 23777923]

[25] Drioli E. Fontananova, EIntegrated membrane processes. Membrane Operations: Innovative Separations and Transformations 2009; 265-83.

[26] Barth C, Goncalves M, Pires A, Roeder J, Wolf B. Asymmetric polysulfone and polyethersulfone membranes: Effects of thermodynamic conditions during formation on their performance. J Membr Sci 2000; 169: 287-99.
[http://dx.doi.org/10.1016/S0376-7388(99)00344-0]

[27] Zhao C, Xue J, Ran F, Sun S. Modification of polyethersulfone membranes a review of methods. Prog Mater Sci 2013; 58: 76-150.
[http://dx.doi.org/10.1016/j.pmatsci.2012.07.002]

[28] Ravishankar H, Roddick F, Navaratna D, Jegatheesan V. Preparation, characterisation and critical flux determination of graphene oxide blended polysulfone (PSf) membranes in an MBR system. J Environ Manage 2018; 213: 168-79.
[http://dx.doi.org/10.1016/j.jenvman.2018.02.063] [PMID: 29494933]

[29] Gao H, Sun X, Gao C. Antifouling polysulfone ultrafiltration membranes with sulfobetaine polyimides as novel additive for the enhancement of both water flux and protein rejection. J Membr Sci 2017; 542: 81-90.

[http://dx.doi.org/10.1016/j.memsci.2017.07.053]

[30] Hwang J, Choi J, Kim JM, Kang SW. Water treatment by polysulfone membrane modified with tetrahydrofuran and water pressure. Macromol Res 2016; 24: 1020-3.
[http://dx.doi.org/10.1007/s13233-016-4145-y]

[31] Xiao Y, Liu X, Wang D, Lin Y, Han Y. Feasibility of using an innovative PVDF MF membrane prior to RO for reuse of a secondary municipal effluent. J Desal 2013; 311: 16-23.
[http://dx.doi.org/10.1016/j.desal.2012.10.022]

[32] Hou J, Dong G, Ye Y, Chen V. Enzymatic degradation of bisphenol-A with immobilized laccase on $TiO_2$ sol-gel coated PVDF membrane. J Membr Sci 2014; 469: 19-30.
[http://dx.doi.org/10.1016/j.memsci.2014.06.027]

[33] Kang G, Cao Y. Application and modification of poly (vinylidene fluoride) (PVDF) membranes-a review. J Membr Sci 2014; 463: 145-65.
[http://dx.doi.org/10.1016/j.memsci.2014.03.055]

[34] Zaviska F, Drogui P, Grasmick A, Azais A, Heran M. Nanofiltration membrane bioreactor for removing pharmaceutical compounds. J Membr Sci 2013; 429: 121-9.
[http://dx.doi.org/10.1016/j.memsci.2012.11.022]

[35] Nevstrueva D, Pihlajamaki A, Manttari M. Effect of a $TiO_2$ additive on the morphology and permeability of cellulose ultrafiltration membranes prepared *via* immersion precipitation with ionic liquid as a solvent. Cellulose 2015; 22: 3865-76.
[http://dx.doi.org/10.1007/s10570-015-0746-4]

[36] Liu F, Abed MM, Li K. Preparation and characterization of poly (vinylidene fluoride) (PVDF) based ultrafiltration membranes using nano $\gamma$-$Al_2O_3$. J Membr Sci 2011; 366: 97-103.
[http://dx.doi.org/10.1016/j.memsci.2010.09.044]

[37] Oh SJ, Kim N, Lee YT. Preparation and characterization of PVDF/$TiO_2$ organic-inorganic composite membranes for fouling resistance improvement. J Membr Sci 2009; 345: 13-20.
[http://dx.doi.org/10.1016/j.memsci.2009.08.003]

[38] Bernardo P, Drioli E, Golemme G. Membrane gas separation: A review/ state of the art. Ind Eng Chem Res 2009; 48: 4638-63.
[http://dx.doi.org/10.1021/ie8019032]

[39] Ahmad A, Jawad Z, Low S, Zein S. A cellulose acetate/multi-walled carbon nanotube mixed matrix membrane for $CO_2$/$N_2$ separation. J Membr Sci 2014; 451: 55-66.
[http://dx.doi.org/10.1016/j.memsci.2013.09.043]

[40] Goetz LA, Jalvo B, Rosal R, Mathew AP. Superhydrophilic anti-fouling electrospun cellulose acetate membranes coated with chitin nanocrystals for water filtration. J Membr Sci 2016; 510: 238-48.
[http://dx.doi.org/10.1016/j.memsci.2016.02.069]

[41] Yu Y, Wu Q Y, Liang H Q, Gu L, Xu Z K. Preparation and characterization of cellulose triacetate membranes via thermally induced phase separation. J. App. Polym. Sci., 2017, 134.

[42] Yin J, Deng B. Polymer-matrix nanocomposite membranes for water treatment. J Membr Sci 2015; 479: 256-75.
[http://dx.doi.org/10.1016/j.memsci.2014.11.019]

[43] Jung JT, Kim JF, Wang HH, di Nicolo E, Drioli E. Understanding the non-solvent induced phase separation (NIPS) effect during the fabrication of microporous PVDF membranes *via* thermally induced phase separation (TIPS). J Membr Sci 2016; 514: 250-63.
[http://dx.doi.org/10.1016/j.memsci.2016.04.069]

[44] Rezakazemi M, Sadrzadeh M, Mohammadi T, Matsuura T. Methods for the preparation of organic-inorganic nanocomposite polymer electrolyte membranes for fuel cells.Organic-inorganic composite polymer electrolyte membranes. Springer 2017; pp. 311-25.
[http://dx.doi.org/10.1007/978-3-319-52739-0_11]

[45]   Rezakazemi M, Sadrzadeh M, Matsuura T. Thermally stable polymers for advanced high-performance gas separation membranes. Pror Energy Combust Sci 2018; 66: 1-41.
[http://dx.doi.org/10.1016/j.pecs.2017.11.002]

[46]   Desalination,The American Heritage Science Dictionary, Houghton Mifflin Company, *via* dictionarycom 2007.

[47]   english.people.com.cn

[48]   Fischetti M. Fresh from the sea. Sci Am 2007; 297(3): 118-9.
[http://dx.doi.org/10.1038/scientificamerican0907-118] [PMID: 17784633]

[49]   Ettouney HM, El-Sessouky HT, Faibish RS, Gowin PJ. Evaluating the Economics of Desalination. Chem Eng Prog 2002; 98: 32-9.

[50]   Riley RL, Baker RW, Cussler EL, Eds., *et al.* In Membrane Separation Systems: Recent Developments and Future Directions. Park Ridge: Noyes Data Corporation 1991; pp. 276-328.

[51]   Mallevialle J, Odendaal PE, Wiesner MR, Eds. Water Treatment: Membrane Processes. New York: McGraw-Hill 1996.

[52]   Vander Bruggen B, Vandecasteele C. Distillation *vs.* membrane filtration: overview of process evolutions in seawater desalination. Desalination 2002; 143: 207-18.
[http://dx.doi.org/10.1016/S0011-9164(02)00259-X]

# CHAPTER 6

# Chemosensors For Anions Of Biological and Environmental Relevance

**Shweta Agarwal**[1,*]

[1] *Isabella Thoburn College, Lucknow, India*

**Abstract:** Anions are prevalent in nature and have important roles in many biological, medical, industrial, and environmental processes. These processes lead to the release of anions in the environment, which act as pollutants at higher concentrations. The proper management of these anions requires adequate detection techniques. Anion sensing, a branch of supramolecular chemistry, deals with chemosensors that are capable of selective recognition and detection of anions through optical or electrochemical response. Further, these compounds are also used for the construction of sensory devices and the extraction and separation of anions. Chemosensors are very useful for the detection of potentially toxic (*e.g.*, fluoride, cyanide) and environmentally hazardous (*e.g.*, phosphate, nitrate) anions as well as in medical diagnostics. Consequently, anion sensing has become one of the most active areas of supramolecular chemistry. The design and synthesis of anion-selective receptors and sensors are challenging, as compared to cation counterparts, due to their different sizes, shapes, high hydration energies, and pH-dependent properties. Three approaches have been used for the detection of anions by chemosensors *viz.* binding site-signalling subunit approach, displacement approach, and chemodosimeter approach. This chapter focuses on small molecular optical chemosensors and the mechanisms adopted for the detection of anions.

**Keywords:** Anion-π interaction, Calix [4] pyrrole, Chemodosimeter, Colourimetric, Displacement approach, Electrostatic interaction, Fluorescence sensors, Halogen-bonding, Hydrogen-bonding, Hydrophobic interaction, Lewis acid, Metal complexes, Molecular assembly, Naked-eye detection, Non-covalent interactions, Optical sensor, Recognition, Self-assembly.

## INTRODUCTION

Anions play many important roles in biological, chemical, and industrial processes, and find widespread use in catalysis, medical diagnostics, and environmental chemistry. Deviation from the required concentration of anions has

---

* **Corresponding author Shweta Agarwal:** Department of Chemistry, Isabella Thoburn College, Lucknow, U.P., India; E-mail: shwetagupta78@gmail.com

Tahmeena Khan, Abdul Rahman Khan, Saman Raza, Iqbal Azad and Alfred J. Lawrence (Eds.)
All rights reserved-© 2021 Bentham Science Publishers

adverse effects on biological systems as well as on the environment [1]. Anions are released from the living systems and industrial effluents in the environment and cause pollution; if not managed properly, this can have devastating effects. Certain anions are important for plant growth and development and are added in fertilizers *e.g.*, nitrate, sulfate, phosphate, and chloride ions. Overuse of fertilizers and soil erosion leads to the release of these anions into the environment. Phosphate ions, along with other ions, are the principal cause of eutrophication [2]. Sulphate ions contribute to the permanent hardness of water, and along with the nitrate ions, are a prominent constituent of acid rain. Likewise, cyanide ion which is used in gold mining, electroplating, production of organic chemicals and polymers, including nitriles, nylon, and acrylic plastics, increases the risk of its unwanted release in the environment [3]. Therefore, efficient techniques which can detect the anions selectively under real-world conditions, in the presence of other anions, are desirable.

**Biological Significance of Anions**

A few important anions which are responsible for maintaining the normal functioning of life are DNA (carrier of genetic code), phospholipids (involved in the formation of the cell membrane), and ATP (the energy currency of biological systems). Carboxylate ions such as citrate, succinate, maleate, *etc.* are present in different steps of Kreb's Cycle.

Many fruits and vegetables contain oxalate, which is an essential nutrient. However, overconsumption of oxalate is associated with the development of kidney stones [2a]. Citrate salts are used as flavouring agents and preservatives in the food industry and as local anticoagulants in clinical practices [4]. Citrate ions inhibit the crystallization of calcium ions; therefore, secretion of less than 320 mg of citrate per day in urine is also associated with renal stone [2b].

Iodide anion is an essential micronutrient and is involved in the synthesis of thyroid hormones $T_3$ and $T_4$. These hormones are responsible for cell differentiation, cell growth, and metabolism. Deficiency of dietary iodide results in the enlarged thyroid gland (goitre) and permanent brain damage in foetuses and children. These disorders can be prevented by adequate iodine intake. However, excessive intake of iodine is also detrimental, triggering goitre and Hashimoto disease [5].

Fluoride is a widespread, non-biodegradable, and biologically important anion and is a comparatively persistent pollutant [2d, 6]. Fluoride is added to drinking water to prevent dental caries and is used for the treatment of osteoporosis. The recommended concentration of fluoride in water should not exceed 2 ppm [2c, 2d]. Fluoride assists in maintaining the teeth' health by inhibiting the acid-assisted demineralisation, promoting remineralisation, interference in the functioning of plaque microorganisms, and alteration in tooth morphology. The primary and most important action of fluoride is topical. The main constituent of tooth enamel is hydroxyapatite (HAP). In the presence of fluoride ions (present in saliva), HAP is converted into fluorohydroxyapetite (FHAP) by substitution of OH$^-$ by F$^-$. Owing to the decreased solubility of FHAP compared to HAP, the process of demineralization is inhibited, while during the formation of the tooth, FHAP incorporates into the tooth enamel and assists in remineralisation of tooth enamel and alters the tooth morphology. In addition to mineralization and demineralization, fluoride ions interact with oral plaque bacteria and inhibit the production of lactic acid formed by the fermentation of carbohydrates. So, in this way, under recommended doses, fluoride assists in maintaining the health of skeletal tissues. However, excessive levels of fluoride in water lead to the diseased condition fluorosis, caused by the accumulation of fluoride in teeth and bones, along with other clinical consequences [2c, 2d].

Other anions of biological significance are chloride and bicarbonate which, in combination with sodium ions, are responsible for the regulation of Anion Gap (AG) in serum and maintaining cellular pH [7]. In addition to regulating electrical neutrality, chloride helps to regulate the distribution of water in the body while bicarbonate plays an important role in the transportation of $CO_2$. Any variation in AG levels results in acute illness, which includes mental disorder, acute renal failure, and disorders in lungs, kidneys, and other organs. Furthermore, an elevated level of chloride anions in sweat chloride test is used to diagnose cystic fibrosis (CF), which is a lethal genetic disease characterized by the production of thick and sticky mucus. CF is caused by dysregulation of chloride channels of epithelial cells [2e, 2f]. Cyanide ion is a neurotoxic agent and is lethal in a very small dose. It binds strongly to the active site of the cytochrome oxidase, leading to inhibition of the mitochondrial electron-transport chain, and decreases oxidative metabolism [2g, 3].

Thus, the diagnosis of some diseases can be established by the detection and analysis of certain anions. Therefore, developing novel methods for anion detection will not only improve the management of anion induced environmental hazards but may also offer efficient diagnostic tools. Much research is being focused on finding inexpensive, reliable, and simple ways of detecting anions in solution and biological samples.

The present chapter gives a brief introduction of the important techniques available for detection of anions, along with a detailed study of small, molecular optical chemosensors, the use of which is one of the most explored and practically applicable techniques for detection of anions of biological and environmental significance. Fundamental challenges in designing the chemosensors and the mechanisms by which they operate, accompanied by different classes of discrete optical chemosensors, have been discussed. A brief introduction of molecular ensembles to understand the displacement approach is also covered. Coordination and chemical reactions taking place between the sensors and anions have been shown by diagrams, in some examples, for better understanding.

## Important Techniques for Detection of Anions

Anions can be detected by easy, traditional, analytical methods in laboratories, which involve colour tests. These methods require a high concentration of the analyte and expensive reagents of high-grade purity which are often toxic. Other modern techniques for detection and analysis of anions involve the use of Ion Chromatography (IC), Capillary Electrophoresis (CE), and Chemosensors.

## Ion Chromatography (IC)

Ion Chromatography is a well-established technique for the detection, quantitative analysis, and separation of a wide range of water-soluble analytes including inorganic and organic anions, cations, charged organic compounds (such as DNA and proteins), and metal complexes [8, 9]. In IC, the separation of ions is based on the interaction of ions with the stationary phase and the mobile phase (eluent). Different types of IC techniques are reported in the literature, such as ion-exchange chromatography (IEC), ion-exclusion chromatography, ion-pair chromatography, and chelation-ion chromatography. Amongst these, ion-exchange chromatography (IEC) remains the predominant method of choice. In IEC, the ionizable functional groups are present on the stationary phase, which bind the oppositely charged ions passing through the column. The bound target ions can be eluted and collected as eluent. An important part of IC is the detector, and its selection is based on the nature of the analyte being detected. The most frequently used detector is a conductivity detector. Other detectors used for IC are UV-Visible detector, amperometric detector, and spectrometric detector [10].

## Capillary Electrophoresis (CE)

Capillary electrophoresis is another effective technique for the separation and analysis of ions, based on their electrophoretic mobility, by applying a voltage. Electrophoretic mobility depends on the charge and size of ions and the viscosity of the medium [11, 12]. CE has high-resolution efficiency compared to IC. There are different methods to perform CE *viz.* capillary zone electrophoresis (CZE), capillary gel electrophoresis (CGE), micellar electrokinetic chromatography (MEKC), capillary electro-chromatography (CEC), capillary isoelectric focusing (CIEF), and capillary isotachophoresis (CITP). Amongst these, CZE is the prevailing method used for anion detection. Similar to IC, different detector modes are attached to CE which are an integral part of CE. These are indirect UV detector, conductivity detector, indirect fluorescence detector, mass spectrometry detector, and potentiometric detector. Amongst these, indirect UV detector and conductivity detector are widely used [13].

## Chemosensors

Chemosensing is used as one of the extensively explored techniques for selective detection and analysis of specific anions, even in the trace amount. Since the last three decades, the development of chemosensors for anion detection has become one of the most active areas of supramolecular chemistry. Chemosensors are chemical compounds that interact selectively with the target analyte (here anions) in the presence of other competitive analytes and produce a readily observable response, which can be assessed qualitatively and/or quantitatively [14]. Based on the responses produced by the sensors on interaction with anions, they are broadly classified into two classes: optical chemosensors and electrochemical sensors.

Optical chemosensors include colourimetric chemosensors and fluorescent chemosensors and produce changes in the optical properties, on exposure to anions. Good examples of sensor-analyte interactions are visible colour changes observed in hormone monitoring strips and pH paper. An improved understanding of sensor-anion interaction can be achieved by monitoring luminescent or fluorescent changes by using a UV lamp, and observing the change in intensity of emitted light (on/off). Generally, the sensors in which the intensity of absorption or emission is increased in the presence of target anion (turn-on) are preferred. Quantitative estimation of anions, which is more important from a scientific point of view, can be carried out by using instruments, such as UV-visible spectrophotometer, spectrofluorometer, NMR, and X-ray diffractometer.

The other types of sensors used for anion detection are electrochemical sensors, which give an electrochemical response on perturbation in redox properties upon complexation with the anion. Two strategies are followed for the construction of electrochemical sensors [15]. In the first strategy, a redox-active group is attached to the anion binding site and the change in the reduction or oxidation potential on complexation with an anion is observed. This approach is analogous to the approach used for the construction of classical optical sensors. The second strategy is the construction of ion-selective electrodes (ISEs) which have an ion-selective membrane, which is formed by immobilization of receptor molecules and can bind the anions reversibly [16].

Changes in physical states, such as rheological properties of matter, on interaction with an analyte, may also be considered as a readily observable sensing response *eg.* a sol may solidify to a gel or *vice-versa*. Change in rheological properties, need no optical or redox change, and can be determined by physical state; this offers new routes for the construction of responsive material.

In summary, many techniques are available for the detection of anions in solution. Typical instruments such as IC and CE are sophisticated (so do not offer onsite sample monitoring), highly expensive and require complicated procedure and data analysis. Moreover, these methods could not be used for the study of biological samples, and their detection limit is up to micromolar level. Additionally, among various types of chemosensors discussed above, the electrochemical sensors are very sensitive, but their cost, lifetime, and cautious handling limit their wide applicability. On the other hand, optical chemosensors offer accurate, efficient, selective, and cost-effective detection of a variety of anions which is either difficult or not possible to detect by utilizing any other technique, especially when they are in extreme and complicated environmental or biological systems. The colourimetric sensors signal the binding event as a colour change, and low-cost or no equipment is required for the detection of anions. Moreover, the fluorescence sensors are highly sensitive and a low detection limit of the anion can be reached. The sensitivity of chemosensors can be understood from their detection limit, which is as low as 2ppb [17].

## Optical Chemosensors for Anions

### Challenges in Development of Chemosensors

Initially, the development of receptors for selective binding of anions was challenging and relatively slow compared to the receptors for cations, and required a robust designing strategy. This is due to some physical and chemical properties of anions [18]. The most common are:

a. Ionic size: Ionic radii of anions are greater than that of cations having the same number of electrons. Therefore, the charge to size ratio is relatively small for anions. Consequently, the effective nuclear charge of anions is less which results in less effective electrostatic interaction of anions with the receptor.

b. Geometry: Anions are of different geometrical shapes, as shown in Fig. **1**. The diversified structures of the anions require a receptor that can have interactions with the anions of varying geometries.

c. Anions exist in a narrow pH range and their ionization states are also variable at different pH values. Therefore, the receptors must be able to recognize the anion at the given pH.

d. The solvation energy of anions is very high. Therefore, the receptor must be able to compete effectively with the surrounding solvent molecules for binding the anion.

| Geometry | Spherical | Linear | Trigonal Planar | Tetragonal Planar | Tetrahedral | Octahedral | Complex shape |
|---|---|---|---|---|---|---|---|
| Examples | $F^-$, $Cl^-$, $Br^-$, $I^-$ | $SCN^-$, $N_3^-$ | $CO_3^{2-}$, $NO_3^-$ | $PtCl_4^-$ | $BF_4^-$, $SO_4^{2-}$, $PO_4^{3-}$, $ClO_4^-$ | $PF_6^-$, $[Fe(CN)_6]^{3-}$ | DNA. Proteins |

**Fig. (1).** Geometry of different anions

e. The selectivity of a receptor is influenced by the hydrophobicity of the anion. It is known that hydrophobic anions bind more strongly in hydrophobic binding sites.

Despite these challenges, plenty of anion receptors present in nature bind selectively with the anion of choice [19 - 23]. These systems inspired chemists to design and synthesize synthetic receptors for anions. For designing anion receptors, consideration of geometry and basicity of the anion and the nature of the solvent medium is important. Selectivity can be achieved by the complementarity between the receptor and the anion. Most of the receptors bind the anions through non-covalent interactions [24] involving hydrogen-bonding, electrostatic interaction, halogen bonding, metal coordination, anion-π interactions, and a combination of many of these interactions together. All receptors cannot act as a sensor. For a receptor to act as a sensor, three components are essential. The first component is a binding site capable of coordination to target anion. This is usually associated with a binding event between the receptor and an anion. The second important element that must be present, is a signalling subunit or signal transducer. It translates the chemical information of the binding events, converts it into a signal, such as a change in the optical spectrum or electronic state, and the last component is a method of measuring the physical change detected at the signalling unit and converting it into information.

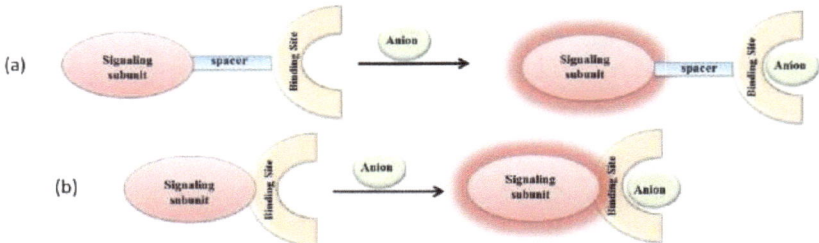

**Fig. (2).** Schematic representation of Binding-site signalling subunit approach (a) binding site and signalling subunit are linked through a spacer (b) signalling subunit is integrated part of the binding site.

## Sensing Mechanisms of Chemosensors

Anion chemosensors display the sensing mechanism by any of the following three approaches:

## Binding Site-Signalling Subunit Approach

This is the most common approach deployed by many sensors [25 - 27]. In this approach, the signalling subunit is either attached to the binding site through a spacer or is an integrated part of the binding site (Fig. 2). The binding site interacts with the anion through electrostatic interaction, hydrogen bonds, and metal coordination. The electrostatic interactions with anions are found when using the positively charged receptors.

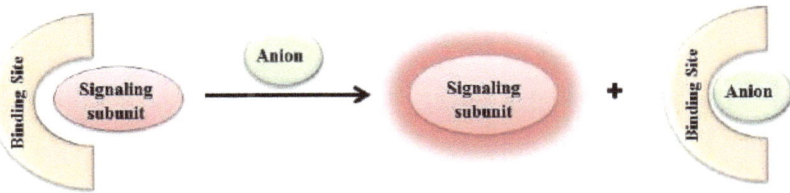

**Fig. (3).** Schematic representation of Displacement assay approach.

A hydrogen bond can be established when a hydrogen atom is covalently attached to a highly electronegative atom and interacts with another electronegative atom (of the same or different molecule) having lone pair of electrons. Sensors deploying the hydrogen-bonding groups have been widely used in binding sites for anion recognition. Metal complexes have also been used as anion binding sites. Metal complexes can bind anions, forming stronger bonds than those generally observed using electrostatic or hydrogen-bonding interactions. In some cases, a certain binding site may combine several of those binding groups in a proper spatial distribution, for the selective coordination of target anions. On the coordination of the anion with the binding site, the chemical information is translated by the signalling subunit, leading to a change in the optical properties and thus allows the detection of the anion.

## Displacement Approach

This method was first reported by Eric Anslyn [28]. In this approach, the binding site and signalling subunit are not covalently linked but exist as a molecular ensemble [29]. In the presence of the target anion, the signalling subunit is displaced from the binding site into the solution (Fig. **3**). If the spectroscopic properties of the free signalling subunit are different from that in a molecular ensemble, the anion binding process leads to a signalling event. Here, it is important to note that the target anion must have a higher affinity (high value of binding constant) for the binding unit than the signalling subunit.

## Chemodosimeter Approach

In this case, the word chemodosimeters or chemoreactants is used instead of chemosensors because target anion induces an irreversible chemical reaction resulting in the formation of new species with different optical properties [30 - 32]. As shown in Fig. **4**, the target anion reacts with chemodosimeter to produce a new product, either through the covalent bond formation or by acting as a catalyst.

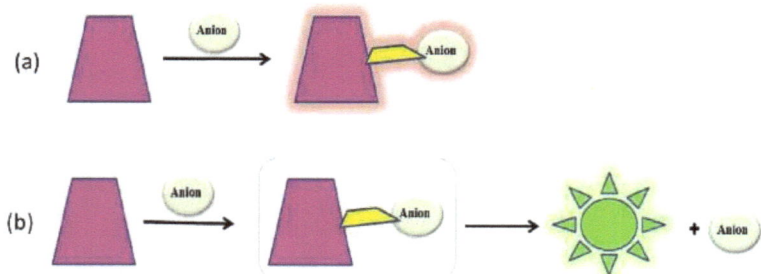

**Fig. (4).** Schematic representation of Chemodosimeter approach (a) formation of a bond between the anion and chemodosimeter (b) anion catalyzes the conversion of chemodosimeter into a new product.

## Optical (Colourimetric and Fluorescence) Chemosensors for Anions

Sensors that change their optical properties in the presence of target anions are of particular importance. The colour (absorption of light in the visible region) and luminescence (emission of light in UV-visible region) of a compound depend on the electronic energy level of the compound which in turn, depends on the chemical structure. In conjugated compounds, the energy gap between the HOMO and LUMO is critical in determining the colour. In compounds with extended conjugated systems, the absorption band is shifted towards the longer wavelength; therefore, such compounds appear dark in colour. The presence of auxochromes further deepens the colour, due to the involvement of nonbonding electrons in resonance. On absorption of light, the molecule gets excited and reaches the excited state from the ground state. The spontaneous emission of radiation from an excited state is known as luminescence. Luminescence involves fluorescence when the emission from the excited singlet state happens and phosphorescence when emission from the excited triplet state occurs. In fluorescence, on the absorption of light, immediate emission of radiation at a longer wavelength takes place. Upon interaction of anion with the sensor, the quenching or enhancement of fluorescence emission of the sensor takes place through photo-induced electron transfer (PET), electronic energy transfer (EET) and monomer-excimer formation.

## Optical Anion Sensing by Discrete molecules

In discrete molecular sensors, the binding site and signalling subunits are connected either by a covalent bond or by π-bond. Upon interaction of the receptor with the anion, the optical properties of the receptor change. This class includes integrated receptors as well as chemodosimeters. This section is subdivided based on the types of non-covalent interactions involved in the sensing of anions.

## Hydrogen Bond Chemosensors

The binding of the anion with the receptor through a hydrogen bond is one of the commonly used strategies for the detection of anions. The formation of a hydrogen bond between the anion and the receptor changes the optical properties of the signalling subunit. A hydrogen bond is directional; therefore, it is important for the recognition of polyatomic anions, where shape complementarity is necessary. Compounds having various types of hydrogen-bond forming groups like amines, amides, urea, thiourea, pyrrole, indoles, and benzimidazole, *etc.*, are commonly used for the synthesis of anion sensors employing hydrogen-bond, whereas sensors using OH and CH groups for detection of anions through H-bond are less.

A series of 'off-the-shelf' organic compounds, having two or more donor groups and a chromophore (Fig. **5**), such as 1,2-diaminoanthraquinone, 1,8-diaminoanthraquinone, 4-nitroaniline, 4-nitro-1,2-phenylenediamine, L-leucine-4-nitroanilide, 1-(4-nitrophenyl)-2-thiourea, 4-nitrophenol, alizarin, 2,2'-bi(-hydroxy-1,4-naphthoquinone), acid blue 45, naphthol AS, 9(10H)-acridone, and Direct Yellow 50, were used as colourimetric sensors for the detection of anions in organic solvents [33]. These compounds formed coloured charge-transfer complexes with anions through hydrogen-bond, in solution. For instance, on exposure of a solution of 1,2-diaminoanthraquinone in dichloromethane to fluoride, chloride, bromide, iodide, phosphate, and sulfate anions (100 equiv.

each), the initial yellow color ($\lambda_{max}$ = 478 nm) became dark purple ($\lambda_{max}$ = 555 nm), red ($\lambda_{max}$ = 519 nm), reddish-orange ($\lambda_{max}$ = 513 nm), orange ($\lambda_{max}$ = 499 nm), purple ($\lambda_{max}$ = 548 nm), and orange ($\lambda_{max}$ = 493 nm), respectively. Similar observations were noted with 1, 8- diaminoanthraquinone but to a lesser extent. In addition to anthraquinones, 2,2-bis(3-hydroxy-1,4-naphthoquinone) was found to be a potential chloride sensor, under extraction type of conditions involving a two-phase aqueous-dichloromethane system, in the presence of [15] crown-5 in the organic phase as a phase-transfer catalyst. The receptor can act as a colourimetric indicator for chloride anion originally present in the aqueous phase. 2, 2-bis (3-hydroxy-1, 4-naphthoquinone) is not only able to detect chloride ion in laboratory samples but it has also been used successfully for samples obtained from the Gulf of Mexico.

Compound **1** (Fig. **6**) contains naphthalene linked to two urea groups, relating to *p*-nitrophenyl groups [34]. When receptor **1** was exposed to a strongly basic fluoride anion, it formed a hydrogen-bonded complex and the UV-visible spectrum displayed a bathochromic shift from 350 nm to 498, with a concomitant colour change from pale yellow to red. Receptor **1** displayed pale yellow colour due to charge-transfer between the electron-rich donor units and the electron-deficient *p*-nitrophenyl groups and it gets enhanced on complexation of the receptor with the F⁻ ion, resulting in a visible colour change.

**Fig. (5).** Off-the-shelf reagents used for anion sensing through Hydrogen bonding.

Compounds **2a** and **2b** (Fig. **6**) are like compound **1** and showed a colour change from pale yellow to purple and pale yellow to pink respectively, on binding with both F⁻ and AcO⁻ ions in DMSO [35]. In **2a**, the electron-withdrawing nitro group is at the *para* position which makes the -OH group more acidic. This results in deprotonation of **2a** by strongly basic F⁻ and AcO⁻ ions in an aqueous medium. The detection limit of **2a** was found to be 0.40 and 0.35 ppm for F⁻ and AcO⁻ ions, respectively.

**Fig. (6).** Colorimetric anion sensors employing Hydrogen bond.

## Halogen Bond Chemosensors

Halogen bond is analogous to hydrogen bond. Halogen atom, when covalently bonded to an electron-withdrawing group, is electrophilic and can accept an electron-pair from a Lewis base. Heavier halogens such as iodine and bromine can be easily polarized and so these are suitable for the formation of halogen bonds. In comparison to a hydrogen bond, the halogen bond is more hydrophobic and directionally confined [36 - 38] and therefore is capable of geometrically precise recognition of chaotropic anions in water. Beer *et al.* for the first time reported macrocyclic halo-imidazolium based fluorescent anion sensors **3a** and **3b** (Fig. **7**), deploying halogen bond [39].

**Fig. (7).** Turn-on fluorescence sensors employing halogen bond.

Anion recognition was confirmed by an increase in fluorescence of a naphthalene unit on binding with an anion. Anion selectivity is achieved by the complementarity between the cavity size of the sensor and the size of the anion. Sensor **3a,** having bulky iodo groups, binds the smaller bromide anion, whereas the less bulky sensor **3b,** having smaller bromo groups, recognized iodide anion selectively in 9:1 methanol: water. A pyrene appended acyclic triazolium based anion sensor **4** has been reported, which senses dihydrogenphosphate anion in acetone [40]. Sensing action involves bringing of pyrene nuclei together on anion binding, which results in the enhancement of excimer fluorescence of pyrene termini.

## Boron Based Lewis Acid Chemosensors

Compounds containing boron atom as Lewis acid centre [41, 42] have been used for sensing the strongly basic anions by the formation of Lewis acid-base complex. The interaction of the anion with boron atom changes the geometry of boron from trigonal planar to tetrahedral. The loss of planarity at the boron atom changes the electronic energy levels of the molecule. Thus, the variation in the electronic properties of the system may produce variation in luminescence and/or fluorescence.

Based on this concept, James *et al.* have reported a fluorescence sensor 2-naphthylboronic acid (**5**) for recognition of fluoride anion (Fig. **8**) [43]. Tetrahedralization of **5** with fluoride anion, in 1:1 $CH_3OH$: $H_2O$ solvent at pH 5.5, was signalled by quenching of fluorescence involving PET mechanism.

Compound **5** can sense fluoride anion in the 50-70 mM range.

Boronic acids can recognize F⁻ ion in an organic solvent, but the use of these compounds for the detection of other anions is very limited. Few exceptional examples are described here. Compound **6** (Fig. **8**) has selectivity for cyanide anion over fluoride anion in aqueous solvents, and the anion binding event was confirmed by quenching of fluorescence [44, 45]. Likewise, selective detection of pyrophosphate anion is made possible by **7** (Fig. **8**) due to the assistance of a nearby zinc centre. The binding of pyrophosphate anion in aqueous conditions was accompanied by turn-on fluorescence [46]. Very recently, a triaryl borane, functionalized bis-benzimidazole with a phenyl bridge (**8**), has been reported as a turn-off sensor having a high affinity for cyanide and fluoride anions over other anions [47].

## Metal Complexes as Chemosensors

The idea of using metal complexes for anion recognition and sensing is the most useful strategy and is conceived from metalloenzymes. These enzymes have metal cations, such as $Zn^{2+}$, $Mg^{2+}$, $Cu^{2+}$ *etc.*, as the anion binding sites and exhibit selective binding with anions. The metal ions play pivotal roles [48] in deciding the nature of anionic substrates, structure enforcement and all steric switching.

**Fig. (8).** Boron based fluorescence sensors.

Many transition metal complexes and lanthanide complexes [24, 27] have been reported in the literature, which are capable of selective detection of target anions in an aqueous medium. Metal complexes elicit a binding response, either by metal-mediated preorganization of ligands attached to it for effective binding with the target anion or by changes in the optical properties of the metal centre when the metal centre itself acts as a site for anion binding. Metal complexes having transition metals have partially filled d-orbitals. Various types of transitions are observed on the absorption of light which include d-d transitions, the ligand to metal charge transfer (LMCT) or metal to ligand charge transfer (MLCT). If coordination of anions to the metal centre of metal complexes generates intense CT bands in the visible region, then it assists in colourimetric detection of anions.

Among the transition metal complexes, zinc complexes of dipicoylamino (DPA) [49] ligands, separated by different spacers, are the most exploited complexes used as sensors for anions. Yoon *et al.* have reported Zn(II)-DPA receptor **9** (Fig. **9**) which recognized the pyrophosphate anion over others like ATP, ADP, AMP, $H_3PO_4^-$, $F^-$, $Cl^-$, $Br^-$, $I^-$, $AcO^-$, and $HSO_4^-$, in aqueous medium at physiological pH [50]. The binding output depends upon the excimer formation due to the 2+2 type of complex between the receptor and pyrophosphate anion.

**Fig. (9).** Metal complexes as sensors.

Another example of a transition metal complex is Cu (II) complex of quinolone with (9) ane-$N_3$ pendant arm **10** (Fig. **9**) which acts as a colourimetric sensor for iodide and cyanide anions [51]. The iodide and cyanide anions can bind the copper centre of the complex and the binding response was recognized by a colour change from light blue to dark green for 1 equivalent of $I^-$ ion, and to dark blue and then pink for 1 equivalent and 2 equivalents of CN⁻ ions, respectively, in acetonitrile. It is interesting to note that sensor **10** exclusively recognized the cyanide anion in water and produced blue colour with one or higher equivalents of cyanide anion.

## Anion-π Chemosensors

The non-covalent interaction between an anion and the aromatic or unsaturated compounds, containing electron-withdrawing groups, is called anion-π interaction. Very few sensors that employ anion-π interactions [52, 53] are reported in the literature, to date. Organic chromophores and fluorophores are unsaturated conjugated systems, and their interactions with anions directly alter their spectral properties. The change in spectral properties arises either through the formation of a charge-transfer complex or electrostatic influence on energy levels.

One of the earliest examples of sensors following anion-π interaction is the binding of halides with the neutral π-acceptor 1, 2, 4, 5-tetracyanobenzene (**11**) (Fig. **10**), in acetonitrile [54]. When **11** comes in contact with an anion, a charge transfer complex is formed and the sensor shows a significant colour change from colourless to dark red or purple, depending on the anion. In another case, the π electron-deficient *s*-tetrazine, incorporated into a macrocycle (**12**) (Fig. **10**), was used as a selective and colourimetric sensor for fluoride anion, in DMSO and aqueous solution [55]. The change in colour of the solution from red to green was attributed to the formation of F⁻-tetrazine (anion-π) CT-complex. Another fluoride sensor based on electron-deficient naphthalenediimide arene (**13**) (Fig. **10**) has been reported which formed a charge transfer complex exclusively with fluoride, resulting in colourless to orange and then pink hue in DMSO, along with fluorescence enhancement [56].

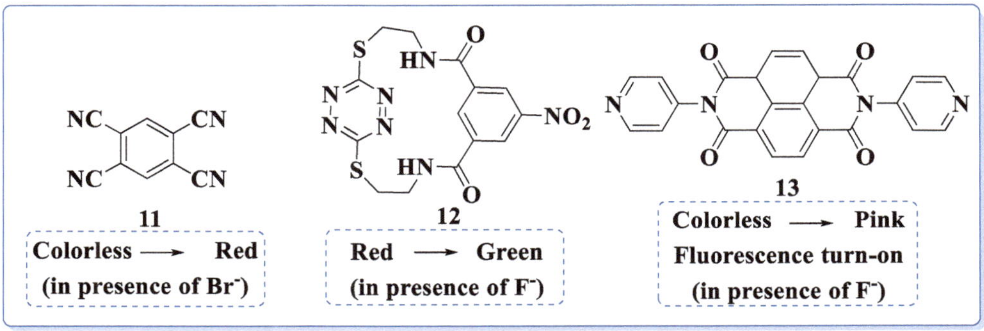

**Figure 10**: Optical sensors based on Anion-π Interaction.

**Fig. (10).** Optical sensors based on Anion-π Interaction.

## Chemosensors Based on Electrostatic Interactions

The use of charged chemosensors for anion detection is also extensively studied. Positively charged sensors bind the anions through strong electrostatic interactions [15, 18, 24]. Czarnik *et al.* have reported anthrynylpolyamines **14** (Fig. **11**) as a fluorescent chemosensor [57] for $HPO_4^{2-}$ ion, at pH 6. In an acidic medium, the receptor **14** was converted into its protonated form, which binds with the $HPO_4^{2-}$ ion and exhibited chelation-induced enhanced fluorescence (CHEF). Gellini *et al.* have reported a fluorescent macrocyclic polyamine **15** (Fig. **11**) for sensing halides [58]. At pH 6 the receptor **15** exists in bi- and tri- protonated forms, which exhibited different behaviour for different halides. In the presence of fluoride and chloride anions, enhancement of fluorescence was observed. On the other hand, in the presence of bromide and iodide anions, protonated sensor **15** showed quenching of fluorescence.

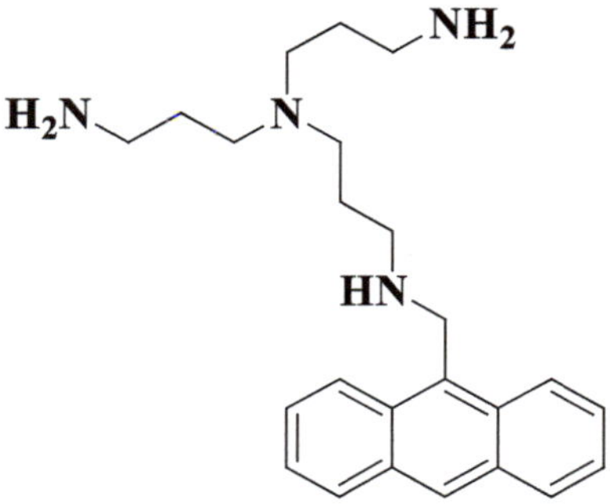


**14**

**Fluorescent turn-on**
**(in presence of $HPO_4^{2-}$ at pH 6)**

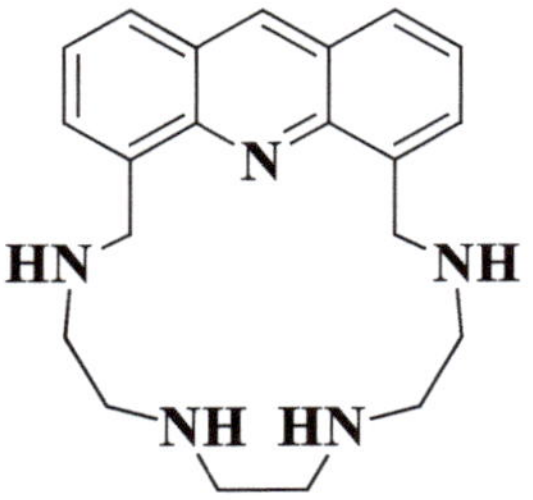


**15**

**Fluorescence turn-on**
**(in presence of $F^-$ and $Cl^-$ at pH 6)**
**Fluorescence turn-off**
**(in presence of $Br^-$ and $I^-$ at pH 6)**

**Figure 11**: Optical sensors based on Electrostatic Interaction.

**Fig. (11).** Optical sensors based on Electrostatic Interaction.

## Chemodosimeters

Detection of anions can be achieved when the anion reacts irreversibly with a compound or catalyzes a reaction and generates a new species, having different spectral properties as compared to the starting material, thus allowing the detection of anions. Despite the attractiveness of this approach, its use is limited to the detection of fluoride and cyanide anions [59 - 62]. This is because of the two properties of anions: basicity and nucleophilicity. Strongly basic anions promote deprotonation of the molecules whereas strongly nucleophilic anions form a new covalent bond by nucleophilic attack on the molecule. Another limitation of these compounds is that they cannot be reused as the anions form covalent bonds. In a few cases, the nucleophilic attack is followed by bond cleavage, as in desilylation reactions.

The hydrogen atoms of hydrogen-bond donors are slightly acidic, which undergo deprotonation by more basic anions. Therefore, these compounds may act as sensors and can be used for the detection of more basic anions. Sensors incorporating a naphthalimide framework are excellent examples for this category of sensors. Deprotonation of **16** (Fig. **12**) by fluoride anion, in DMSO, is characterized by a colour change, from yellow to deep purple, and substantial change in fluorescence [63]. Similar changes in colour and fluorescence for **16** were noted with strong bases but not with other basic anions (dihydrogenphosphate anion and other halides).

Recently, a chemodosimeter based on a selenourea scaffold **17** (Fig. **12**), for the recognition of cyanide and sulfide in an aqueous solution, has been reported [64]. Spectroscopic studies revealed the formation of the diselenides (for $CN^-$, and $S^{2-}$, respectively) by nucleophilic addition of these anions in $H_2O/CH_3CN$ (75:25, v/v). Binding of $CN^-$ and $S^{2-}$ with **17** is also detected by a visible colour change, from dark yellow to light yellow, and colourless, for $CN^-$ and $S^{2-}$ anions, respectively.

**Fig. (12).** Chemodosimeters.

Deprotection of alcohols/phenols, containing silyl groups, by tetra butyl ammonium fluoride (TBAF), is an example of a desilylation reaction where the fluoride anion catalyzes the removal of a silyl group. This approach has been used to develop fluoride-sensitive reporter molecules. The first example of this type of sensors, involving desilylation reaction, is compound **18** which gave a coloured response in presence of fluoride anion due to the formation of a dye, resorufin, in 1:1 $CH_3CN: H_2O$ [65].

## Optical Sensing by Molecular Assemblies

Preassembled supramolecular assemblies, formed by non-covalent interaction of binding unit and signalling subunit (which may be chromophores, fluorophores, or quenchers), can be used to simulate an optical response instead of using discrete molecules [17, 66, 67]. These molecular ensembles act as hosts for anions and produce on/off optical response. The newly formed receptor-anion assembly can be analysed to identify the anion.

A 1:1 complex of *meso*-octamethylcalix [4] pyrrole (**18a**) and *p*-nitrophenolate (**19**) ion acts as a turn-on colourimetric sensor for naked-eye detection of fluoride anion (Fig. **13**). The intense yellow colour of **19**, in dichloromethane, dissipated on gradual addition of a solution of *meso*-octamethylcalix [4] pyrrole (**18a**) and was restored upon addition of fluoride anion (in the form of tetra butyl ammonium salt). The regeneration of the yellow colour is attributed to the displacement of *p*-nitrophenolate ion by fluoride anion and the formation of a hydrogen bond complex. UV-Visible spectrum displayed a significant decrease in the absorbance of *p*-nitrophenolate ion (**19**) at 432 nm, upon the addition of a solution of **18a**; the colour was regenerated upon addition of fluoride anion [66].

**Fig. (13).** Chemosensors based on Displacement approach.

Methyl pyridinium appended calix [4] pyrrole (**18b**) formed a non-fluorescent complex with caromenolate ion (**20**) in acetonitrile (Fig. **13**). The fluorescence of the resulting supramolecular complex was restored upon the addition of pyrophosphate anion ($HP_2O_7^{3-}$) due to the release of **20**. The selectivity of the complex for pyrophosphate anion ($HP_2O_7^{3-}$) (Ka = $2.55 \times 10^7$ M$^{-1}$) over the phosphate and fluoride anions is due to the well-matched size and charge distribution of $HP_2O_7^{3-}$ ion with the cavity of the receptor [67]. Replacement of methyl pyridinium groups in **18b** by benzimidazolium groups (**18c**) gave selectivity for bicarbonate ($HCO_3^-$) anion in $CH_3CN$ [68]. Complex **18c:20** was used to measure the concentration of $CO_2$ in carbonated drinks and the detection limit for $HCO_3^-$ was found to be 4nM. The sensor system can be regenerated by the addition of $Na^+$ ions; thus, this system can be reused for detection purposes.

## CONCLUDING REMARKS

Due to the prevalent role of anions in our lives, techniques that offer selectivity, sensitivity, and applicability, in multiple environments simultaneously, are required. The present chapter highlights the different mechanisms and systems that have been used for optical recognition and sensing of anions by small molecules and molecular ensembles. Many portable devices from these chemosensors have been innovated which offer on-site monitoring of samples. Besides the challenges, the field of chemosensors is continuously growing and the future of chemosensors for anion sensing is very promising.

## LIST OF ABBREVIATIONS

| | |
|---|---|
| **AG** | Anion Gap |
| **CF** | Cystic Fibrosis |
| **IC** | Ion Chromatography |
| **IEC** | Ion Exchange Chromatography |
| **CE** | Capillary Electrophoresis |
| **CZE** | Capillary Zone Electrophoresis |
| **CGE** | Capillary Gel Electrophoresis |
| **CIEF** | Capillary Isoelectric Focusing |
| **CITP** | Capillary Isotacophoresis |
| **UV** | Ultra-Violet |
| **ISE** | Ion Selective Electrode |
| **CT** | Charge Transfer |
| **PET** | Photo-induced Electron Transfer |
| **CHEF** | Chelation Enhanced Fluorescence |

## CONSENT FOR PUBLICATION

Not Applicable.

## CONFLICT OF INTEREST

The author confirms that this chapter contents have no conflict of interest.

## ACKNOWLEDGEMENT

Declared none.

## REFERENCES

[1] aFinkielstein VA, Goldfarb DS. Strategies for preventing calcium oxalate stones. CMAJ 2006; 174(10): 1407-9.
[http://dx.doi.org/10.1503/cmaj.051517] [PMID: 16682705] bAbdulhadi MH, Hall PM, Streem SB. Hypocitraturia and its role in renal stone disease. Cleve Clin J Med 1988; 55(3): 242-5.
[http://dx.doi.org/10.3949/ccjm.55.3.242] [PMID: 3416412] cYadav KK, Kumar S, Pham QB, et al. Fluoride contamination, health problems and remediation methods in Asian groundwater: A comprehensive review. Ecotoxicol Environ Saf 2019; 182109362
[http://dx.doi.org/10.1016/j.ecoenv.2019.06.045] [PMID: 31254856] dSrivastava S, Flora SJS. Srivastava, Sakshi; Flora, S. J. S. Fluoride in drinking water and skeletal fluorosis: A review of the global impact. Curr Environ Health Rep 2020; 7(2): 140-6. [Case, M.].
[http://dx.doi.org/10.1007/s40572-020-00270-9] [PMID: 32207100] eCase M. Physiology. Chloride ions and cystic fibrosis. Nature 1986; 322(6078): 407.
[http://dx.doi.org/10.1038/322407a0] [PMID: 2426596] fBrown SD, White R, Tobin P. Keep them breathing: Cystic fibrosis pathophysiology, diagnosis, and treatment. JAAPA 2017; 30(5): 23-7.
[http://dx.doi.org/10.1097/01.JAA.0000515540.36581.92] [PMID: 28441669] gHassan DMA, Farghali MRF. Cyanide pollution in different water sources in assiut, Egypt: levels, distributions and health risk assessment. Res J Environ Sci 2018; 12: 213-9.
[http://dx.doi.org/10.3923/rjes.2018.213.219]

[2] H3eisler, J.; Glibert, P.; Burkholder, J.; Anderson, D.; Cochlan, W.; Dennison, W.; Gobler, C.; Dortch, Q.; Heil, C.; Humphries, E.; Lewitus, A.; Magnien, R.; Marshall, H.; Sellner, K.; Stockwell, D.; Stoecker, D.; Suddleson, M. Eutrophication and harmful algal blooms: A scientific consensus. Harmful Algae 2008; 8: 3-13.

[3] Abedi-Orang B, Seifpanahi-Shabani K, Kakaie R. Mathematical modeling of fate and transport of cyanide pollutant in the gold mine tailings: With emphasis on physico-chemical process. Environ Earth Sci 2020; 79: 189.
[http://dx.doi.org/10.1007/s12665-020-08927-2]

[4] Honore PM, De Bels D, Preseau T, Redant S, Spapen HD. Citrate: How to get started and what, when, and how to monitor? J Transl Int Med 2018; 6(3): 115-27.
[http://dx.doi.org/10.2478/jtim-2018-0026] [PMID: 30425947]

[5] Lucia M, Navarro AM. About iodide: A friendly and necessary ion. Annals. Thyroid Res 2019; 5: 223-8.

[6] Mukherjee I, Singh UK. Groundwater fluoride contamination, probable release, and containment mechanisms: a review on Indian context. Environ Geochem Health 2018; 40(6): 2259-301.
[http://dx.doi.org/10.1007/s10653-018-0096-x] [PMID: 29572620]

[7] Kraut JA, Madias NE. Serum anion gap: its uses and limitations in clinical medicine. Clin J Am Soc Nephrol 2007; 2(1): 162-74.
[http://dx.doi.org/10.2215/CJN.03020906] [PMID: 17699401]

[8]   Cummins PM, Rochfort KD, O'Connor BF. Ion-Exchange Chromatography: Basic Principles and Application.Protein Chromatography: Methods in Molecular Biology, Walls, D; Loughran, S. NY 2017; Vol. 1485: pp. 209-23.
[http://dx.doi.org/10.1007/978-1-4939-6412-3_11]

[9]   Buchberger WW. Detection techniques in ion analysis: what are our choices? J Chromatogr A 2000; 884(1-2): 3-22.
[http://dx.doi.org/10.1016/S0021-9673(00)00283-1] [PMID: 10917418]

[10]  Michalski R. Ion chromatography applications in waste water analysis. Separation 2018; 5: 16.
[http://dx.doi.org/10.3390/separations5010016]

[11]  Soga T, Imaizumi M. Capillary electrophoresis method for the analysis of inorganic anions, organic acids, amino acids, nucleotides, carbohydrates and other anionic compounds. Electrophoresis 2001; 22(16): 3418-25.
[http://dx.doi.org/10.1002/1522-2683(200109)22:16<3418::AID-ELPS3418>3.0.CO;2-8] [PMID: 11669520]

[12]  Kaniansky D, Masár M, Marák J, Bodor R. Capillary electrophoresis of inorganic anions. J Chromatogr A 1999; 834(1-2): 133-78.
[http://dx.doi.org/10.1016/S0021-9673(98)00789-4] [PMID: 10189691]

[13]  Pacakova V, Stulık K. Capillary electrophoresis of inorganic anions and its comparison with ion chromatography. J Chromatogr A 1997; 789: 169-80.
[http://dx.doi.org/10.1016/S0021-9673(97)00830-3]

[14]  Patil SR, Jain PR, Sahoo SK, *et al.* A thesaurus comprising brief introduction of chemosensor for recognition of metal cations and anions. World J Pharm Res 2017; 6: 296-340.

[15]  aDavis F, Collyer SD, Higson SPJ. The construction and operation of anion sensors: Currentstatus and future perspectives. Top Curr Chem 2005; 255: 97-124.
[http://dx.doi.org/10.1007/b101164] bBusschaert N, Caltagirone C, Van Rossom W, Gale PA. Applications of supramolecular anion recognition. Chem Rev 2015; 115(15): 8038-155.
[http://dx.doi.org/10.1021/acs.chemrev.5b00099] [PMID: 25996028] cHein R, Beer PD, Davis JJ. Electrochemical anion sensing: Supramolecular approaches. Chem Rev 2020; 120(3): 1888-935.
[http://dx.doi.org/10.1021/acs.chemrev.9b00624] [PMID: 31916758]

[16]  Radecka H, Hanna R. Redox-active monolayers deposited on gold electrode surface-universal platforms for electrochemical sensing. Sens Mater 2020; 32: 1065-78.
[http://dx.doi.org/10.18494/SAM.2020.2662]

[17]  Kaur S, Wang HH, Tae Lee J, Lee C. Displacement-based, chromogenic calix[4]pyrrole-indicator complex for selective sensing of pyrophosphate anion. Tetrahedron Lett 2013; 54: 3744-7.
[http://dx.doi.org/10.1016/j.tetlet.2013.04.129]

[18]  Beer PD, Gale PA. Anion recognition and sensing: The state of the art and future perspectives. Angew Chem Int Ed Engl 2001; 40(3): 486-516.
[http://dx.doi.org/10.1002/1521-3773(20010202)40:3<486::AID-ANIE486>3.0.CO;2-P] [PMID: 11180358]

[19]  Pflugrath JW, Quiocho FA. Sulphate sequestered in the sulphate-binding protein of Salmonella typhimurium is bound solely by hydrogen bonds. Nature 1985; 314(6008): 257-60.
[http://dx.doi.org/10.1038/314257a0] [PMID: 3885043]

[20]  Neurath H. Proteolytic enzymes, past and future. Proc Natl Acad Sci USA 1999; 96(20): 10962-3.
[http://dx.doi.org/10.1073/pnas.96.20.10962] [PMID: 10500108]

[21]  Mangani S, Ferraroni M. Natural Anion Receptors: Anion Recognition by Proteins.Supramolecular Chemistry of Anions. New York 1997; pp. 63-78.

[22]  Dutzler R, Campbell EB, Cadene M, Chait BT, MacKinnon R. X-ray structure of a ClC chloride channel at 3.0 A reveals the molecular basis of anion selectivity. Nature 2002; 415(6869): 287-94.

[http://dx.doi.org/10.1038/415287a] [PMID: 11796999]

[23] Ledvina PS, Yao N, Choudhary A, Quiocho FA. Negative electrostatic surface potential of protein sites specific for anionic ligands. Proc Natl Acad Sci USA 1996; 93(13): 6786-91.
[http://dx.doi.org/10.1073/pnas.93.13.6786] [PMID: 8692896]

[24] Molina P, Zapata F, Caballero A. Anion recognition strategies based on combined non-covalent interactions. Chem Rev 2017; 117(15): 9907-72.
[http://dx.doi.org/10.1021/acs.chemrev.6b00814] [PMID: 28665114]

[25] Santos-Figueroa LE, Moragues ME, Climent E, Agostini A, Martínez-Máñez R, Sancenón F. Chromogenic and fluorogenic chemosensors and reagents for anions. A comprehensive review of the years 2010-2011. Chem Soc Rev 2013; 42(8): 3489-613.
[http://dx.doi.org/10.1039/c3cs35429f] [PMID: 23400370]

[26] Amendola V, Esteban-Gómez D, Fabbrizzi L, Licchelli M. What anions do to N-H-containing receptors. Acc Chem Res 2006; 39(5): 343-53.
[http://dx.doi.org/10.1021/ar050119l] [PMID: 16700533]

[27] Aletti AB, Gillen DM, Gunnlaugsson T. Luminescent/colorimetric probes and (chemo-) sensors for detecting anions based on transition and lanthanide ion receptor/binding complexes. Coord Chem Rev 2018; 354: 98-120.
[http://dx.doi.org/10.1016/j.ccr.2017.06.020]

[28] Wiskur SL, Ait-Haddou H, Lavigne JJ, Anslyn EV. Teaching old indicators new tricks. Acc Chem Res 2001; 34(12): 963-72.
[http://dx.doi.org/10.1021/ar9600796] [PMID: 11747414]

[29] Gale PA, Caltagirone C. Anion sensing by small molecules and molecular ensembles. Chem Soc Rev 2015; 44(13): 4212-27.
[http://dx.doi.org/10.1039/C4CS00179F] [PMID: 24975326]

[30] Mohr GJ. A chromoreactand for the selective detection of HSO3- based on the reversible bisulfite addition reaction in polymer membranes. Chem Commun (Camb) 2002; 22(22): 2646-7.
[http://dx.doi.org/10.1039/B207621G] [PMID: 12510279]

[31] Caballero A, Zapata F, Beer PD. Interlocked host molecules for anion recognition and sensing. Coord Chem Rev 2013; 257: 2434-55.
[http://dx.doi.org/10.1016/j.ccr.2013.01.016]

[32] Ferreira NL, de Cordova LM, Schramm ADS, Nicoleti CR, Machado VG. Chromogenic and fluorogenic chemodosimeter derived from Meldrum's acid detects cyanide and sulfide in aqueous medium. J Mol Liq 2019; 282: 142-53.
[http://dx.doi.org/10.1016/j.molliq.2019.02.129]

[33] Miyaji H, Sessler JL. Off-the-shelf colorimetric anion sensors. Angew Chem Int Ed 2001; 5: 154-7.
[http://dx.doi.org/10.1002/1521-3773(20010105)40:1<154::AID-ANIE154>3.0.CO;2-G]

[34] Cho EJ, Ryu BJ, Lee YJ, Nam KC. Visible colorimetric fluoride ion sensors. Org Lett 2005; 7(13): 2607-9.
[http://dx.doi.org/10.1021/ol0507470] [PMID: 15957902]

[35] Singh A, Sahoo SK, Trivedi DR. Colorimetric anion sensors based on positional effect of nitro group for recognition of biologically relevant anions in organic and aqueous medium, insight real-life application and DFT studies. Spectrochim Acta A Mol Biomol Spectrosc 2018; 188: 596-610.
[http://dx.doi.org/10.1016/j.saa.2017.07.051] [PMID: 28779621]

[36] Tepper R, Schubert US. Halogen bonding in solution: Anion recognition, templated self-assembly, and organocatalysis. Angew Chem Int Ed Engl 2018; 57(21): 6004-16.
[http://dx.doi.org/10.1002/anie.201707986] [PMID: 29341377]

[37] Pancholi J, Beer PD. Halogen bonding motifs for anion recognition. Coord Chem Rev 2020; 416213281

[http://dx.doi.org/10.1016/j.ccr.2020.213281]

[38] Mako TL, Racicot JM, Levine M. Supramolecular Luminescent Sensors. Chem Rev 2019; 119(1): 322-477.
[http://dx.doi.org/10.1021/acs.chemrev.8b00260] [PMID: 30507166]

[39] Zapata F, Caballero A, White NG, *et al.* Fluorescent charge-assisted halogen-bonding macrocyclic halo-imidazolium receptors for anion recognition and sensing in aqueous media. J Am Chem Soc 2012; 134(28): 11533-41.
[http://dx.doi.org/10.1021/ja302213r] [PMID: 22703526]

[40] Zapata F, Caballero A, Molina P, Alkorta I, Elguero J. Open bis(triazolium) structural motifs as a benchmark to study combined hydrogen- and halogen-bonding interactions in oxoanion recognition processes. J Org Chem 2014; 79(15): 6959-69.
[http://dx.doi.org/10.1021/jo501061z] [PMID: 25020191]

[41] Wu X, Chen XX, Jiang YB. Recent advances in boronic acid-based optical chemosensors. Analyst (Lond) 2017; 142(9): 1403-14.
[http://dx.doi.org/10.1039/C7AN00439G] [PMID: 28425507]

[42] Galbraith E, James TD. Boron based anion receptors as sensors. Chem Soc Rev 2010; 39(10): 3831-42.
[http://dx.doi.org/10.1039/b926165f] [PMID: 20820463]

[43] Cooper CR, Spencer N, James TD. Selective fluorescence detection of fluoride using boronic acids. Chem Commun (Camb) 1998; 1365-6.
[http://dx.doi.org/10.1039/a801693c]

[44] Hudnall TW, Gabbaï FP. Ammonium boranes for the selective complexation of cyanide or fluoride ions in water. J Am Chem Soc 2007; 129(39): 11978-86.
[http://dx.doi.org/10.1021/ja073793z] [PMID: 17845043]

[45] Bhat HR, Jha PC. Selective complexation of cyanide and fluoride ions with ammonium boranes: A theoretical study on sensing mechanism involving intramolecular charge transfer and configurational changes. J Phys Chem A 2017; 121(19): 3757-67.
[http://dx.doi.org/10.1021/acs.jpca.7b00502] [PMID: 28443335]

[46] Hudnall TW, Chiu CW, Gabbaï FP. Fluoride ion recognition by chelating and cationic boranes. Acc Chem Res 2009; 42(2): 388-97.
[http://dx.doi.org/10.1021/ar8001816] [PMID: 19140747]

[47] Brazeau AL, Yuan K, Ko SB, Wyman I, Wang S. Anion sensing with a blue fluorescent triarylboron-functionalized bisbenzimidazole and its bisbenzimidazolium salt. ACS Omega 2017; 2(12): 8625-32.
[http://dx.doi.org/10.1021/acsomega.7b01631] [PMID: 31457395]

[48] Kral V, Rusin O, Shishkanova T, Volf R, Matejka P, Volka K. Anion binding: From supramolecules to sensors. Chem Listy 1999; 93: 546-53.

[49] Ngo HT, Liu X, Jolliffe KA. Anion recognition and sensing with Zn(II)-dipicolylamine complexes. Chem Soc Rev 2012; 41(14): 4928-65.
[http://dx.doi.org/10.1039/c2cs35087d] [PMID: 22688834]

[50] Lee HN, Xu Z, Kim SK, *et al.* Pyrophosphate-selective fluorescent chemosensor at physiological pH: formation of a unique excimer upon addition of pyrophosphate. J Am Chem Soc 2007; 129(13): 3828-9.
[http://dx.doi.org/10.1021/ja0700294] [PMID: 17348656]

[51] Aguado Tetilla M, Aragoni MC, Arca M, *et al.* Colorimetric response to anions by a "robust" copper(II) complex of a [9]aneN$_3$ pendant arm derivative: CN$^-$ and I$^-$ selective sensing. Chem Commun (Camb) 2011; 47(13): 3805-7.
[http://dx.doi.org/10.1039/c0cc04500d] [PMID: 21290057]

[52] Giese M, Albrecht M, Rissanen K. Anion-π interactions with fluoroarenes. Chem Rev 2015; 115(16):

8867-95.
[http://dx.doi.org/10.1021/acs.chemrev.5b00156] [PMID: 26278927]

[53] Rather IA, Wagay SA, Ali R. Emergence of anion-π interactions: The land of opportunity in supramolecular chemistry and beyond. Coord Chem Rev 2020; 415213327
[http://dx.doi.org/10.1016/j.ccr.2020.213327]

[54] Rosokha YS, Lindeman SV, Rosokha SV, Kochi JK. Halide recognition through diagnostic "anion-π" interactions: molecular complexes of Cl⁻, Br⁻, and I⁻ with olefinic and aromatic π receptors. Angew Chem Int Ed Engl 2004; 43(35): 4650-2.
[http://dx.doi.org/10.1002/anie.200460337] [PMID: 15352195]

[55] Zhao Y, Li Y, Qin Z, Jiang R, Liu H, Li Y. Selective and colorimetric fluoride anion chemosensor based on s-tetrazines. Dalton Trans 2012; 41(43): 13338-42.
[http://dx.doi.org/10.1039/c2dt31641b] [PMID: 23001332]

[56] Guha S, Saha S. Fluoride ion sensing by an anion-π interaction. J Am Chem Soc 2010; 132(50): 17674-7.
[http://dx.doi.org/10.1021/ja107382x] [PMID: 21114330]

[57] Huston ME, Akkaya EU, Czarnik AW. Chelation enhanced fluorescence detection of non-metal ions. J Am Chem Soc 1989; 111: 8735-7.
[http://dx.doi.org/10.1021/ja00205a034]

[58] Chelli R, Pietraperzia G, Bencini A, et al. A fluorescent receptor for halide recognition: clues for the design of anion chemosensors. Phys Chem Chem Phys 2015; 17(16): 10813-22.
[http://dx.doi.org/10.1039/C5CP00131E] [PMID: 25814174]

[59] Pati PB. Organic chemodosimeter for cyanide: A nucleophilic approach. Sens Actuators B Chem 2016; 222: 374-90.
[http://dx.doi.org/10.1016/j.snb.2015.08.044]

[60] Zhang L, Zou LY, Guo JF, Ren AM. Theoretical investigation on the one- and two-photon responsive behavior of fluoride ion probes based on diketopyrrolopyrrole and its π-expanded derivatives. New J Chem 2016; 40: 4899-910.
[http://dx.doi.org/10.1039/C6NJ00432F]

[61] Dhiman S, Ahmad M, Singla N, et al. Chemodosimeters for optical detection of fluoride anion. Coord Chem Rev 2020; 405213138
[http://dx.doi.org/10.1016/j.ccr.2019.213138]

[62] Kaur K, Saini R, Kumar A, et al. Chemodosimeters: An approach for detection and estimation of biologically and medically relevant metal ions, anions and thiols. Coord Chem Rev 2012; 256: 1992-2028.
[http://dx.doi.org/10.1016/j.ccr.2012.04.013]

[63] Gunnlaugsson T, Kruger PE, Jensen P, Pfeffer FM, Hussey GM. Simple naphthalimide based anion sensors: deprotonation induced colour changes and $CO_2$ fixation. Tetrahedron Lett 2003; 44: 8909-13.
[http://dx.doi.org/10.1016/j.tetlet.2003.09.148]

[64] Casula A, Llopis-Lorente A, Garau A, et al. A new class of silica-supported chromo-fluorogenic chemosensors for anion recognition based on a selenourea scaffold. Chem Commun (Camb) 2017; 53(26): 3729-32.
[http://dx.doi.org/10.1039/C7CC01214D] [PMID: 28300250]

[65] Kim SY, Hong JI. Chromogenic and fluorescent chemodosimeter for detection of fluoride in aqueous solution. Org Lett 2007; 9(16): 3109-12.
[http://dx.doi.org/10.1021/ol0711873] [PMID: 17629289]

[66] Gale PA, Gale PA, Twyman LJ, Handlin CI, Sessler JL. A colourimetric calix[4]pyrrole–4-nitrophenolate based anion sensor. Chem Commun (Camb) 1999; 1851-2.
[http://dx.doi.org/10.1039/a905743i]

[67] Sokkalingam P, Kim DS, Hwang H, Sessler JL, Lee CH. A dicationic calix[4]pyrrole derivative and its use for the selective recognition and displacement-based sensing of pyrophosphate. Chem Sci (Camb) 2012; 3: 1819-24.
[http://dx.doi.org/10.1039/c2sc20232h]

[68] Mulugeta E, He Q, Sareen D, et al. Recognition, sensing, and trapping of bicarbonate anions with a dicationic meso-bis(benzimidazolium) calix[4]pyrrole. Chem 2017; 3: 1008-20.
[http://dx.doi.org/10.1016/j.chempr.2017.10.007]

# CHAPTER 7

# Antibiotic Pollution: Challenges and Strategies

**Saman Raza**[1,*] **and Tahmeena Khan**[2]

[1] *Isabella Thoburn College, Lucknow, India*
[2] *Integral University, Lucknow, India*

**Abstract:** Antibiotics have been used as antimicrobial agents to fight a variety of infectious diseases, for the past more than 100 years. Apart from this, they are also extensively used in animal farming, agriculture, and aquaculture, all over the world. However, this frequent and large-scale overuse and incorrect use lead to the excessive dispersal of antibiotics in water and soil, resulting in their accumulation in the environment, which is known as antibiotic pollution. The removal of antibiotics from water and soil is complicated due to their non-biodegradable nature, and special techniques must be used for the same. This pollution has serious implications on both human health and the ecological balance. The major adverse effect is antibiotic resistance, wherein, microbes become less susceptible to treatment with antibiotics, posing problems for both the patient and the physician. This chapter describes the causes and consequences of antibiotic pollution, the challenges it presents, and the strategies to counter them.

**Keywords:** Adjuvant-therapy, Antibiotic, AOP, Beta-lactam, Efflux, Inhibitors, Non-biodegradable, Non-target, Permeabilizers, Pollution, Resistance, Wastewater.

## INTRODUCTION

Ever since Alexander Fleming discovered penicillin in 1929, antibiotics have become the most significant medical discovery of the twentieth century. This is because antibiotics were able to treat a variety of microbial infections, many of which were fatal and had no treatment earlier [1]. By the 1970s, over 160 new antibiotics and their semi-synthetic derivatives were introduced for the treatment of infectious diseases [2].

---

[*] **Corresponding author Saman Raza:** Department of Chemistry, Isabella Thoburn College, Lucknow, India; E-mail: samanmahek@gmail.com

Tahmeena Khan, Abdul Rahman Khan, Saman Raza, Iqbal Azad and Alfred J. Lawrence (Eds.)
All rights reserved-© 2021 Bentham Science Publishers

Antibiotics can be classified based on their structure, mode of action, spectrum of activity, etc. Table **1** gives the classification of antibiotics based on their structure. The structure and mode of action of the important classes of antibiotics have also been described in the table.

**Table 1. Different classes of antibiotics based on the structure.**

| Class of Antibiotics | Examples | Structure | Mode of Action |
|---|---|---|---|
| ß-Lactams | Penicillins such as amoxicillin and flucloxacillin | | Inhibit bacterial cell wall biosynthesis |
| | Cephalosporins such as cefalexin. | | Inhibit bacteria cell wall biosynthesis |
| Aminoglycosides | Streptomycin, neomycin, kanamycin, paromomycin. | | Inhibit the synthesis of proteins by bacteria |
| Chloramphenicol | Chloramphenicol | | Inhibits the synthesis of proteins by bacteria |

(Table 1) cont.....

| Class of Antibiotics | Examples | Structure | Mode of Action |
|---|---|---|---|
| Glycopeptides | Vancomycin, teicoplanin. | | Inhibit bacterial cell wall biosynthesis |
| Quinolones | Ciprofloxacin, levofloxacin, trovafloxacin. | | Interfere with bacterial DNA replication and transcription |
| Oxazolidinones | Linezolid, posizolid, tedizolid, cycloserine. | | Inhibit the synthesis of proteins by bacteria |
| Sulfonamides | Prontosil, sulfanilamide, sulfadiazine, sulfisoxazole. | | Do not kill bacteria but prevent their growth and multiplication. |
| Tetracyclines | Tetracycline, doxycycline, limecycline, oxytetracycline. | | Inhibit the synthesis of proteins by bacteria |

(Table 1) cont.....

| Class of Antibiotics | Examples | Structure | Mode of Action |
|---|---|---|---|
| Macrolides | Erythromycin, clarithromycin, azithromycin. | | Inhibit the synthesis of proteins by bacteria |

## MECHANISM OF ACTION OF ANTIBIOTICS

Antibiotics may be bactericidal (kill bacteria) or bacteriostatic (stop bacterial growth) in nature. Mostly all antibiotics have one out of the known five mechanisms of action, as shown in Fig. (1).

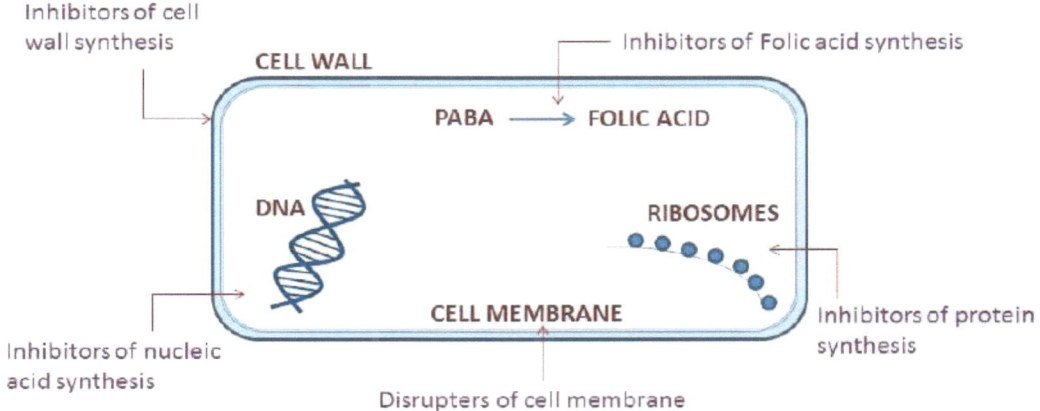

**Fig. (1).** Mechanisms of action of antibiotics.

### i. Inhibition of Bacterial Cell Wall Synthesis

This is the most common mechanism in which the antibiotic inhibits or disrupts the synthesis of peptidoglycan, the polymer that makes up the cell wall. Beta-lactams like penicillin bind with the transpeptidase enzymes due to their structural similarity to the D-alanyl-D-alanine group found in the peptidoglycan structure, thereby preventing the assembly of the peptidoglycan layer [3]. On the other hand, vancomycin disrupts cell wall synthesis by binding with C-terminal acyl--alanyl-D-alanine containing residues in the peptidoglycan precursors, and peptidoglycan cross-linkage is prevented [4]. Antibiotics of this type are bactericidal in nature.

### ii. Inhibition of Bacterial Protein Synthesis

This is the second common mode of action and the target is bacterial RNA. Aminoglycosides act by irreversibly binding to the bacterial 30s ribosomal subunit and disrupting peptide synthesis; this effect is bactericidal. The tetracyclines interact with 30s ribosomal subunits and prevent the binding of incoming aminoacyl-tRNA. However, this appears to occur after the initial binding of the elongation factor Tu–aminoacyl-tRNA complex to the ribosome [5]. On the other hand, chloramphenicol binds to the peptidyl transferase component of 50s ribosomal subunit, blocking peptide elongation [6]. In this way, the elongation of the polypeptide chain is prevented by these two types of antibiotics, and so these antibiotics are predominantly bacteriostatic.

### iii. Disruption of Cell Membranes

Polymyxin disrupts the cytoplasmic membrane of the bacteria through a detergent-like action, by interaction with membrane phospholipids and increasing cellular permeability. This effect is bactericidal [7].

### iv. Inhibition of Nucleic Acid Synthesis

Quinolones bind with DNA gyrases or topoisomerases which are required for DNA transcription and replication. Complex formation with gyrase is followed by the rapid and reversible inhibition of DNA synthesis which stops growth. At higher drug concentrations, cell death occurs. Such antibiotics are bactericidal in nature [8].

### v. Antimetabolite Activity

Sulfonamides are structurally like *p*-aminobenzoic acid (PABA), which is required in the synthesis of folic acid, an essential nutrient needed for correct protein synthesis and in other essential biosynthetic pathways. This competitive inhibition of PABA disrupts the folate cycle. This causes the reduction of folic acid levels leading to errors in DNA synthesis. The overall effect is bacteriostatic [9].

## USES OF ANTIBIOTICS

Antibiotics are used to treat or prevent infections caused by Gram-positive and Gram-negative bacteria, such as urinary tract infections, strep throat, sinusitis, pneumonia, cellulitis, plague, etc. They are also used as prophylactic agents in surgeries and chemotherapy.

Although originally developed as drugs to treat and prevent microbial diseases in humans, with time, antibiotics found use in other fields as well, such as animal farming, agriculture, and fishery.

## ANIMAL FARMING

Antibiotics are routinely used in the treatment of diseases, increasing feed productivity in livestock [10], and as growth promoters. The addition of low levels of certain antibiotics, such as penicillin, aureomycin, streptomycin, terramycin, and bacitracin, to the diet of livestock, increases the rate of growth of young animals. This 'growth effect' may be attributed to the better absorption of nutrients, reduction of toxins produced by gut bacteria, or by the reduction in the frequency of subclinical intestinal infections [11]. The practice of feeding antibiotics to farm animals is widespread and does not cause any harmful effects on public health; for example, even when the animals are fed antibiotics at high levels, they are not found in a detectable amount in the meat obtained.

## AGRICULTURAL PURPOSES

Antibiotics are also used to control certain bacterial diseases in plants of high economic value, such as fruit and vegetable bearing plants as well as ornamental plants. As antibiotics do not stay active on plants for more than a week and no significant residues are found on the harvest, they are widely used and have become an essential part of agriculture [12]. The antibiotics mostly used on plants are streptomycin and oxytetracycline while oxalinic acid and gentamicin are also used in a few countries.

## AQUACULTURE

Antibiotics are used for therapeutic purposes and as prophylactic agents in the fishing industry as well [13]. Aquaculture is the fastest-growing food industry in the world [14], supplying more than 60 million tons of seafood per year; the extensive use of antibiotics to prevent microbial diseases is a major contributor to this notable growth. For instance, the salmon farms in Chile which produced more than 700 thousand tons of salmon in 2016, used more than 300 tons of antibiotics that year. Florfenicol and oxytetracycline are the most frequently used antibiotics in aquaculture [14].

## ANTIBIOTIC POLLUTION

With the growing practice of using antibiotics in different fields, the overuse and misuse of these agents also increased. Consequently, they made their way into the environment, causing detrimental effects; this phenomenon is now referred to as antibiotic pollution [15]. There are several sources of dispersal of antibiotics into the environment, which are discussed below (Fig. 2).

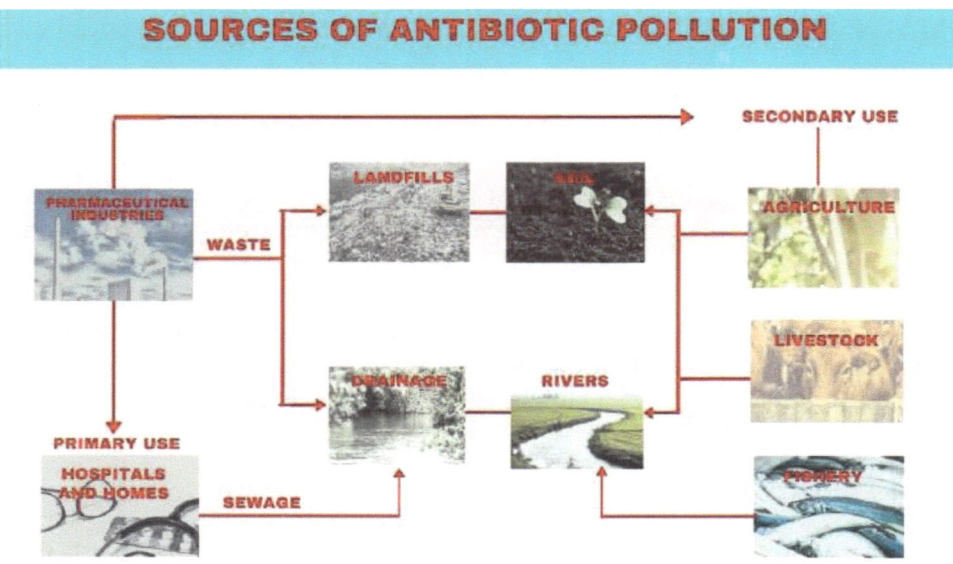

**Fig. (2).** Sources of antibiotic pollution.

i. When a person is treated with an antibiotic, not all the medication is used up inside the body. While hospitals provide relatively controlled environments for the usage of antibiotics, unsupervised in-home antibiotic use (which may include incorrect use of antibiotics or unfinished treatment) results in sub-inhibitory antibiotics concentrations *in situ*. Most of the antibiotic, either in their original form or slightly changed, are released in urine and/or stool and pass into sewage from where they reach surface water. In addition to water pollution, unused antibiotics contaminate the soil when they are disposed of in the trash and end up in landfills.

ii. Antibiotics used in animals end up in manure and enter the soil, water bodies as well as groundwater, along with runoff from the farms and the grazing fields.

iii. Similarly, antibiotics applied as bactericidal agents to crops and fruit trees also get washed off and reach the soil, water bodies, and groundwater. About 300,000 pounds of antibiotics are used in crop production each year [16] which indicates the amount being released into the environment.

iv. Antibiotics used in fisheries are directly added into the water and the surplus keeps accumulating when left untreated.

v. Effluents from pharmaceutical industries are released into the environment *via* wastewater and from landfills. Antibiotics in extremely high concentrations, sometimes close to or above therapeutic levels, have been found in industrial effluents in several countries, such as ciprofloxacin and quinolones in India [17, 18], oxytetracycline and metabolites of penicillin G in China [19, 20], sulfonamides in Croatia [21], lincomycin in South Korea [22], and bacitracin in Norway [23].

Antibiotic usage has been estimated to exceed 100,000 tons per year worldwide and there is increasing concern over the fate of these substances, once released into the environment. Research shows that antibiotic concentration in wastewater, soil, and sediment can range from μg/kg to mg/kg [24]. What makes this situation worrisome is the fact that due to their antimicrobial nature, antibiotics resist biodegradation. They are categorized as pseudo-persistent organic pollutants as they are continually added into the environment, are difficult to remove, and consequently, have a permanent presence. Antibiotics generally cannot be efficiently removed by traditional physicochemical methods applied in wastewater treatment plants, like flocculation, sedimentation and filtering, or biological processes like activated sludge and anaerobic digestion. However, certain newly developed physicochemical techniques (discussed later) have shown success in the remediation of antibiotic contamination in wastewater [25].

## EFFECTS OF ANTIBIOTIC POLLUTION

Unlike other organic pollutants like polychlorinated biphenyls (PCBs), antibiotics not only deteriorate environment quality by disrupting ecosystems but also result in the rapid spread of drug-resistant microbes in the environment [26]. Antibiotic pollution is, therefore, a major cause of concern due to its potential ecological and public health risks [27]. The major adverse effects of antibiotic pollution have been summarized below.

# EFFECT OF ANTIBIOTIC POLLUTION ON HEALTH: ANTIBIOTIC RESISTANCE

The most adverse consequence of antibiotic pollution is antibiotic resistance. As the widespread use of antibiotics increased in medicine and food production, microbial pathogens slowly evolved, developed resistance to antibiotics, and became harder to treat, causing increased morbidity [28]. Antibiotic resistance has become one of the most complex global health challenges and threatens to reverse the extensive progress made against infectious diseases since the discovery of antibiotics. According to the European Centre for Disease Prevention and Control, each year about 25,000 people in Europe die from drug-resistant bacterial infections [29]. Another report, published in Lancet, suggests worldwide mortality of half a million people from the same cause [30].

The transfer of resistant bacteria to humans could occur *via* water or food if food-producing plants are watered with contaminated surface water or sewage sludge, if manure is used as a fertilizer, or if resistant bacteria are present in animal products like dairy, meat, poultry, and fish [31].

Resistance to antibiotics can develop *via* several mechanisms, such as:

i. inactivation of the antibiotic by bacterial enzymes; for example, bacterial beta-lactamase enzyme cleaves the beta-lactam ring of penicillin, making it inactive.

ii. transportation outside of the bacterial cell; for example, bacterial Tetracycline resistant (TetA) proteins force tetracyclines out of the bacterial cell.

iii. modification of the antibiotic's target; for example, point mutations in the bacterial gyrA gene prevent binding by ciprofloxacin [32].

The most prominent medical examples of antibiotic-resistant microbes are vancomycin-resistant *Enterococci*, methicillin-resistant *Staphylococcus aureus*, and multi-resistant *Pseudomonads*.

The implications of antibiotic resistance on health and wellbeing are grave. As a result of resistance, infections can no longer be treated by first-line antibiotics, leading to prolonged illness as well as treatment. Infections caused by resistant bacterial strains usually lead to almost two times higher rates of adverse outcomes than those by susceptible strains. For example, infections caused by carbapenem-resistant *K. Pneumoniae* have two to five times higher mortality rate than carbapenem susceptible strains of the microbe [33].

While death is the most severe adverse effect of antibiotic resistance, there are several others as well, such as complications in the illness, use of stronger medications, prolonged hospital stays, the increased financial burden on the families, *etc*. Also, if not properly controlled, the antibiotic-resistant bacterium may spread to other people. Apart from the problems faced in the treatment of infections, antibiotic prophylaxis is also affected. Surgeries, both simple and complex, as well as chemotherapy, become much more dangerous without effective antibiotics for the prevention and treatment of infections. A study was conducted in the USA to learn about the consequences of increased antibiotic resistance on common surgical procedures and cancer chemotherapy. It was found that 38.7% to 50.9% of pathogens that cause surgical site infections and 26.8% of pathogens that may cause infections after chemotherapy are resistant to standard prophylactic antibiotics. 42% of the infections, after a prostate biopsy, were found to be due to resistance to fluoroquinolones, a prophylactic antibiotic commonly used [34].

Another concern related to antibiotic resistance is unnecessary and overuse of empiric antibiotics, especially the broad-spectrum type. For example, treatment guidelines for pseudomonas infections recommend the use of anti-pseudomonal beta-lactams, even before the microbial test is revealed [35]. In this way, usage of antibiotics, often unnecessary ones, has increased in healthcare, which leads to further pollution and subsequent health problems, turning into a vicious cycle.

## EFFECTS OF ANTIBIOTIC POLLUTION ON THE ENVIRONMENT

Antibiotic contamination not only affects the targeted microbial population but also non-target organisms [36]. Significant levels of antibiotics in surface water are harmful to freshwater organisms like algae, fishes, and zooplankton, as reported in a study on the negative effects of monensin (used to increase animal growth in dairy farms) on species richness, and abundance of zooplankton and phytoplankton biomass, in an artificial pond [37].

It has been shown with the help of bioassays that some of the antibiotics found in surface waters, such as Ciprofloxacin, affect microbes at concentrations below 10 µg/L, although these concentrations might not be deemed harmful to humans. In this context, it may be noted that examples of considerable antibiotic pollution in freshwaters have been reported in the literature, from 15 µg/L up to 450 µg/L [38].

While high concentrations are lethal and may wipe out entire trophic levels in some ecosystems, sub-lethal concentrations of antibiotics are also harmful. This is because antibiotics have direct as well as indirect effects on the affected organisms [39]. Due to their long-term exposure, resistance develops in microbes, even the non-target ones. Other than this, the composition of single-celled communities may also be affected, as demonstrated in laboratory experiments [39].

Similar effects are expected in microbes found in soil as well, which are adversely affected by antibiotic pollution of the landmass. The complex microbial populations found in soil are responsible for cycling nutrients by nitrogen fixation. This is important for maintaining soil quality and fertility of agricultural land. Nitrogen cycling is accomplished by the gram-negative bacteria, *Nitrosomonas* and *Nitrobacter*. Gram-negative and broad-spectrum antibiotics, such as tetracyclines and sulphonamides, could seriously inhibit this nutrient cycling if their concentrations reached high levels. While this result has been observed in laboratory studies, no field studies have found antibiotic concentrations at such levels that would seriously disrupt the nitrification process [40].

Therefore, it can be concluded that antibiotic pollution has implications for the microbial ecosystems as well and by extension, larger organisms, and the ecological balance.

## STRATEGIES TO COUNTER ANTIBIOTIC POLLUTION AND RESISTANCE

Considering the huge risks associated with antibiotic pollution, it is crucial to reduce the overuse and misuse of antibiotics for both medicinal and commercial purposes, as well as to eliminate antibiotic residues in wastes generated from them. The remediation strategies are two-pronged: reduction of antibiotic pollution and countering antibiotic resistance.

### A. Methods for the Reduction of Antibiotic Pollution

#### *1. Removal of Antibiotic Residues from Water*

As discussed earlier, traditional wastewater treatment techniques are not able to degrade and remove antibiotics. However certain newly developed oxidation processes, such as ozonation, Fenton and photo-Fenton reaction, photolysis, and photo-catalysis, have proved successful in the remediation of antibiotic contamination in wastewater [41]. In these Advanced Oxidation Processes (AOPs), hydroxyl radicals are generated by various methods which oxidize and disrupt the active group of the antibiotics, thereby making them inactive. AOPs can be classified into two groups based on the use of light: non-photochemical AOPs and photochemical AOPs, as shown in Table **2** [42].

Table 2. Different types of Advanced Oxidation Processes (AOPs).

| Type of AOP | Name of Process | Requirements | Reaction producing HO· |
|---|---|---|---|
| Non-photochemical | Ozonation | High pH | $3O_3 + H_2O \rightarrow 2\ HO^{\cdot} + 4O_2$ |
| | Ozone/hydrogen peroxide | High pH | $H_2O_2 + 2O_3 \rightarrow 2\ HO^{\cdot} + 3O_2$ |
| | Fenton process | High pH, iron salt as catalyst | $Fe^{+2} + H_2O_2 \rightarrow Fe^{+3} + HO^- + HO^{\cdot}$ |
| Photochemical | Vacuum UV (VUV) photolysis | hv < 190 nm | $H_2O + hv \rightarrow H^{\cdot} + HO^{\cdot}$ |
| | Hydrogen peroxide/UV ($H_2O_2$/UV) process | UV light | $H_2O_2 + hv \rightarrow 2HO^{\cdot}$ |
| | Ozone/UV ($O_3$/UV) process | UV light | $O_3 + hv + H_2O \rightarrow H_2O_2 + O_2$ <br> $H_2O_2 + hv \rightarrow 2HO^{\cdot}$ |
| | Ozone/hydrogen peroxide/UV ($O_3$/$H_2O_2$/UV) process | UV light | $2O_3 + H_2O_2 + UV \rightarrow 2HO^{\cdot} + 3O_2$ |
| | Photo-Fenton process | UV light, Iron salt as catalyst | $Fe(OH)^{+2} + hv \rightarrow Fe^{+3} + HO^{\cdot}$ |
| | Semiconductor-sensitized process | Photocatalysts like Zinc oxide, strontium titanium trioxide, and $TiO_2$ | $TiO_2 + hv \rightarrow e^- + h^+$ <br> $H_2O + h^+ \rightarrow HO^{\cdot} + H^+$ |

Although these advanced methods are useful in removing antibiotics from wastewater, they also have their disadvantages, such as slow rates of degradation and high maintenance cost. To overcome these problems, combinations of existing techniques have also been used with much success. However, as pollution levels are rising rapidly, more research is needed to develop a cost-effective and durable wastewater treatment technique.

## *2. Reduction in the Use of Antibiotics*

The rate of antibiotic resistance is steadily rising, and it is estimated that by 2050, 10 million deaths per year, globally, will be due to antimicrobial resistance [43]. This means that like climate change, an immediate and effective solution for the reduction of the use of antibiotics, at a global level, is required. In 2001, the World Health Organization released a paper named WHO Global Strategy for Containment of Antimicrobial Resistance [44] which laid out 68 recommendations to tackle resistance. The paper recommended key changes to be made by healthcare workers, patients, farmers, and those who sell antibiotics.

Similarly, a Review on Antimicrobial Resistance, which was commissioned by the British government to address the growing global problem of drug-resistant infections, also gave 10 intervention strategies [45], some of which are described below (Fig. 3):

i. An awareness campaign is needed to educate people about the problem of drug resistance.

ii. Development of new drugs to replace the ones that are known to be susceptible to resistance.

iii. Reduction in the unnecessary use of antibiotics in humans and animals.

iv. Development of rapid diagnostic techniques so that unnecessary use of empiric antibiotics may be avoided.

v. Development and use of vaccines and alternative therapy options.

vi. Focus on hygiene and sanitation, and changes in animal housing and husbandry practices, to reduce the spread of infections.

vi. Reduction in the extensive and unnecessary use of antibiotics in agriculture and livestock farming.

The WHO guidelines on the use of antibiotics in livestock recommend that the food industry should stop the routine use of antibiotics in healthy animals, which is done to promote growth and prevent disease. In response to the urgency of the matter, several countries have taken action to reduce the use of antibiotics in food-producing animals, from planning strategies to banning the use of antibiotics for growth promotion. For example, in 2017, the Indian government released a five-year National Action Plan on Antimicrobial Resistance (NAP-AMR), that prioritizes the environment, with increased surveillance for food and animal sectors, regulation, and optimized use of antibiotics. The European Union has banned the use of antibiotics for growth promotion in livestock, since 2006.

As a result of such efforts, educated and aware consumers now demand meat procured without the routine use of antibiotics. Consequently, some major food chains have adopted antibiotic-free policies for their meat supplies. It has been found that interventions that restrict antibiotic use in food-producing animals have reduced antibiotic-resistant bacteria in these animals, by up to 39% [46].

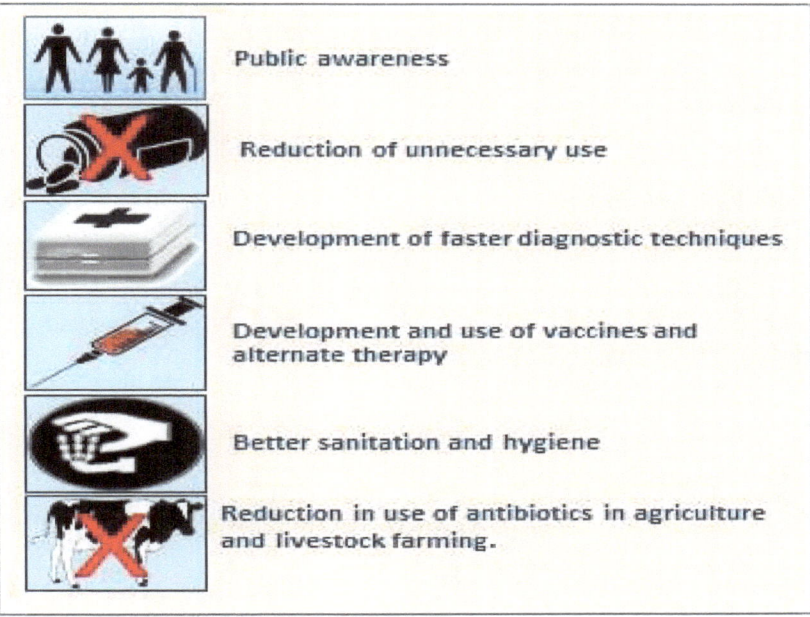

**Fig. (3).** Guidelines to reduce the use of antibiotics.

## B. Methods to Counter Antibiotic Resistance

The treatment options for diseases caused by antibiotic-resistant microbes are limited, posing problems for the doctors. The development of new antibiotics, having novel mechanisms of action or the ability to counter resistance, seems to be the answer to this problem; however, there have been very few new candidates in recent years due to various reasons. An alternative to the development of new antibiotics is to find potentiators of the already existing ones, known as antibiotic adjuvants [47].

### *1. Adjuvant Therapy*

Antibiotic adjuvants are active molecules with little or no antibiotic activity, that when taken in combination with antibiotics, enhance the antimicrobial activity of the latter. Two antibiotics are also considered adjuvants when their effect is synergistic (*i.e.,* the coadministration of the two drugs has a significantly greater effect than that of each antibiotic alone). Antibiotic adjuvants do not directly kill bacteria but enhance antibiotic activity by a) blocking or countering resistance mechanisms, such as enhancing intracellular antibiotic-accumulation, b) inhibiting signalling, and regulatory pathways, or c) boosting the host response to bacterial infection. Based on the mechanism of action, antibiotic adjuvants are of three main types: (i) beta-lactamase inhibitors, (ii) efflux pump inhibitors, and (iii) outer membrane permeabilizers [48].

## *i. Beta-Lactamase Inhibitors*

They are the most used antibiotic adjuvants and are used to protect β-lactam antibiotics from inactivation. Certain bacteria produce β−lactamase enzymes that hydrolyze the β−lactam ring of such antibiotics by an acylation-deacylation based process, making them inactive (Fig. **4**).

**Fig. (4).** Hydrolysis of the beta-lactam ring by beta-lactamase enzymes.

The structure of the β−lactamase inhibitors is like the beta-lactam antibiotics (Fig. **5**). The inhibitors prevent the hydrolysis of the antibiotics by competitive inhibition and in this way the antibiotics remain intact and active. Some examples of β−lactamase inhibitors in clinical use are clavulanic acid, sulbactam and tazobactam [49].

Fig. (5). Structures of some beta-lactamase inhibitors.

## *ii. Efflux Pump Inhibitors*

Microbes have efflux pumps that can eliminate a variety of toxic compounds, such as antibiotics, heavy metals, organic pollutants, *etc.* via active efflux. This efflux mechanism is also responsible for antibiotic resistance because microorganisms force out the drug molecules from the cytoplasm into extracellular spaces, thereby reducing the level of the antibiotic below effective concentration [50]. The inhibition of efflux pumps *via* competitive inhibition is, therefore, a promising strategy to counter antibacterial efflux. In recent years, several efflux pump inhibitors have been discovered and tested. Although no efflux inhibitor has been approved for therapeutic use, they may be beneficial for increasing the susceptibility of different bacterial species to antibiotics [51]. Some examples of efflux inhibitors include certain natural products like the carotenoids capsanthin and capsorubin [52], the flavonoids rotenone and chrysin [52], and the alkaloid lysergol [53], and synthetic molecules like phenylalanine-arginine-β-naphthylamide (PAβN) [54], pyridopyrimidines and arylpiperazines [55] (Fig. 6). Some nanoparticles, for example, zinc oxide [56], have also been found to counter antibiotic resistance, possibly by reducing the efflux of the drug from the microbial cell.

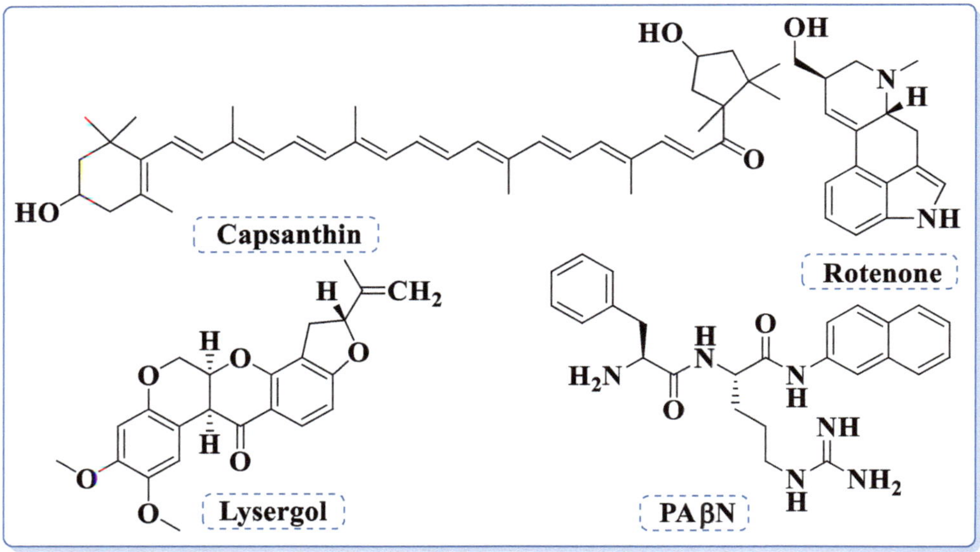

Fig. (6). Structures of some efflux inhibitors.

## *iii. Outer Membrane Permeabilizers*

Gram-negative bacteria are more resistant to antibiotics as they have a unique outer membrane (OM). Hydrophilic antibiotics are prevented from crossing the outer membrane by the lipopolysaccharide (LPS) layer and the underlying phospholipids, whereas hydrophobic ones are kept out by outer membrane proteins [57]. LPS layer is polyanionic, with cation-binding sites which are essential for the integrity of the OM. The naturally occurring polycationic antibiotics of the polymyxin group form a complex with LPS and disorganize the OM, leading to the disruption of the outer membranes (Fig. 7). Recently, it has been shown that de-lipidated polymyxin, as well as other polycations, can also disrupt the outer membranes of Escherichia coli cells [58]. Such compounds are known as permeabilizers and they enhance the permeability of bacterial cells. The use of OM permeabilizers in combination with antibiotics, even those that are not active against Gram-negative bacteria, may provide additional means of controlling their growth [59].

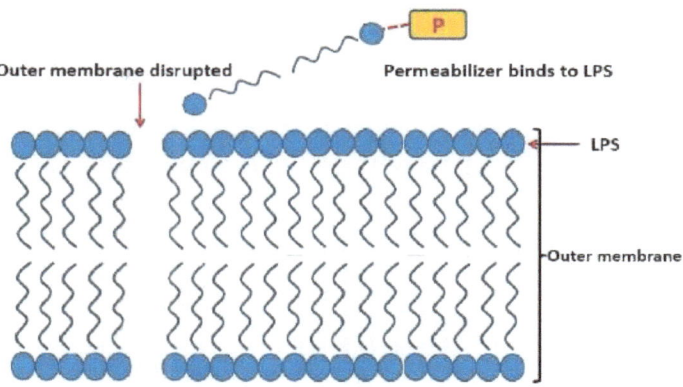

**Fig. (7).** Disruption of the outer membrane by outer membrane permeabilizers.

The antibiotic adjuvant approach provides an alternative and complementary strategy for new antibiotic discovery. The practice of using adjuvant therapy has two main advantages: (i) the use of the available antibiotics can be continued, and (ii) the pressure on the development of novel molecules with a new mechanism of antimicrobial action, can be relaxed. The main disadvantages of the use of antibiotic adjuvants are: (i) there is an increased risk of adverse effects due to potential drug-drug interactions, and (ii) to make the pharmacokinetic and pharmacodynamic properties of the antibiotic and its adjuvant compatible, an effective co-dosing routine must be established.

## 2. Development of New Antibiotics

With the rising trend of resistance, the quest for new antibiotics, having a novel mechanism of action, has become an urgent necessity. Despite this fact, the development of new drugs has not been very successful, especially those targeting Gram-negative bacteria. Some of the new antibiotics and the mode of action which helps them to counter resistance, have been described in Table **3** while their structures are given in Fig. (**8**).

Table 3. Class and mechanism of action of some new antibiotics.

| Name | Class of Antibiotic | Mode of Action/ Mechanism to Circumvent Resistance |
|:---:|:---:|:---:|
| **Delafloxacin** | Fluoroquinolone | i. Interferes with DNA replication<br>ii. Has 10 times more accumulation than Fluoroquinolones [60] |

| Name | Class of Antibiotic | Mode of Action/ Mechanism to Circumvent Resistance |
|---|---|---|
| Meropenem-vaborbactam | Carbapenem and novel boronic acid-based beta-lactamase inhibitor combination | i. Meropenem inhibits the cell wall synthesis<br>ii. Vaborbactam inhibits beta-lactamase [61] |
| Plazomicin | Aminoglycoside | i. Inhibits protein synthesis<br>ii. Is stable towards aminoglycoside-modifying enzymes [62] |
| Eravacycline | Fluorocycline | i. Inhibits protein synthesis<br>ii. Resists tetracycline-specific efflux [63] |
| Cefiderocol | Siderophore cephalosporin | i. Inhibits bacterial cell wall biosynthesis<br>ii. Chelates with iron to bypasses the bacterial porin channels [64] |
| Omadacycline | Aminomethyl-cycline | i. Inhibits protein synthesis<br>ii. Resists tetracycline-specific efflux [65] |

**Fig. (8).** Structure of some new antibiotics.

While only a few new drugs have been approved, several more are in the clinical stage of development. These are mostly derivatives of the major known chemical and functional classes of antibiotics, especially β-lactams, β-lactamase inhibitors and tetracyclines, and have a known mechanism of action. Also, hybrid antibiotics that have more than one component with a synergistic effect are being developed.

Other than these, two new antibiotics, having a new chemical structure, are in clinical development. The first, Afabicin, is an inhibitor of the enzyme FabI which catalyses the bacterial fatty-acid biosynthesis pathway. By disrupting fatty acid biosynthesis, Afabicin exhibits selective antibacterial activity against staphylococcal species. Preliminary studies have indicated that afabicin may not be subject to the rapid emergence of resistance, possibly due to its high-affinity of binding [66]. The second drug is a prodrug, TXA709. It has a benzamide structure and targets the bacterial protein FtsZ which has an essential role in the formation of the septum, which is the new cell wall that forms between two daughter cells because of cell division. As a result, TXA709 prevents bacterial cell division [67].

## CONCLUDING REMARKS

The discovery and subsequent widespread use of antibiotics led to a marked reduction in the spread of infectious diseases, in the twentieth century. However, the overuse and mismanagement of the same have now posed the problem of pollution. The hazards of antibiotic pollution are so grave and so pressing that they require urgent remediation. Steps like spreading awareness, regulations, and legal action to reduce the overuse and improper disposal of antibiotics, as well as proper wastewater treatment, would be useful in reducing the pollution of water and soil with antibiotics. In addition to this, the problem of resistance also needs attention as resistant microbes have become increasingly virulent. While adjuvant therapy and new antibiotics have proved beneficial in the treatment of drug-resistant microbial infections, there is still a crucial need for the development of new drugs, especially against the WHO declared critical-priority microbes. Overall, efforts should focus on- a) spreading awareness, b) innovation of economical techniques to treat wastewater efficiently, and c) development of antibacterial agents that would evade resistance. Sustained and proactive efforts are needed from governments, regulatory bodies, scientists, pharmaceutical companies, healthcare workers, farmers, and common people, to break the cycle of antibiotic pollution and resistance if we wish to lead a healthy life.

## CONSENT FOR PUBLICATION

Not Applicable.

## CONFLICT OF INTEREST

The author confirms that this chapter contents have no conflict of interest.

## ACKNOWLEDGEMENT

Declared none.

## REFERENCES

[1] Levy SB, Bergman MM. The Antibiotic Paradox: How the Misuse of Antibiotics Destroys Their Curative Powers Clinical Infectious Diseases. 2nd ed., Boston, USA: Perseus Publishing 2003.

[2] Davies J. Where have all the antibiotics gone? Can J Infect Dis Med Microbiol 2006; 17(5): 287-90.
[http://dx.doi.org/10.1155/2006/707296] [PMID: 18382641]

[3] Bush K, Bradford PA. β-Lactams and β-Lactamase Inhibitors: An Overview. Cold Spring Harb Perspect Med 2016; 6(8)a025247
[http://dx.doi.org/10.1101/cshperspect.a025247] [PMID: 27329032]

[4] Reynolds PE. Structure, biochemistry and mechanism of action of glycopeptide antibiotics. Eur J Clin Microbiol Infect Dis 1989; 8(11): 943-50.
[http://dx.doi.org/10.1007/BF01967563] [PMID: 2532132]

[5] Kotra LP, Haddad J, Mobashery S. Aminoglycosides: perspectives on mechanisms of action and resistance and strategies to counter resistance. Antimicrob Agents Chemother 2000; 44(12): 3249-56.
[http://dx.doi.org/10.1128/AAC.44.12.3249-3256.2000] [PMID: 11083623]

[6] Bambeke F, Leclercq MM, Glupczynski Y, Tulkens PM. Infect Dis 2017; 2: 1162-80. [Fourth Edition].
[http://dx.doi.org/10.1016/B978-0-7020-6285-8.00137-4]

[7] Velkov T, Roberts KD, Nation RL, Thompson PE, Li J. Pharmacology of polymyxins: new insights into an 'old' class of antibiotics. Future Microbiol 2013; 8(6): 711-24.
[http://dx.doi.org/10.2217/fmb.13.39] [PMID: 23701329]

[8] Chen CR, Malik M, Snyder M, Drlica K. DNA gyrase and topoisomerase IV on the bacterial chromosome: quinolone-induced DNA cleavage. J Mol Biol 1996; 258(4): 627-37.
[http://dx.doi.org/10.1006/jmbi.1996.0274] [PMID: 8636997]

[9] Fernández-Villa D, Aguilar MR, Rojo L. Folic Acid Antagonists: Antimicrobial and Immunomodulating Mechanisms and Applications. Int J Mol Sci 2019; 20(20): 4996.
[http://dx.doi.org/10.3390/ijms20204996] [PMID: 31601031]

[10] Kiser JS. A perspective on the use of antibiotics in animal feeds. J Anim Sci 1976; 42(4): 1058-72.
[http://dx.doi.org/10.2527/jas1976.4241058x] [PMID: 770412]

[11] Feighner SD, Dashkevicz MP. Subtherapeutic levels of antibiotics in poultry feeds and their effects on weight gain, feed efficiency, and bacterial cholyltaurine hydrolase activity. Appl Environ Microbiol 1987; 53(2): 331-6.
[http://dx.doi.org/10.1128/AEM.53.2.331-336.1987] [PMID: 3566269]

[12] Stockwell VO, Duffy B. Use of antibiotics in plant agriculture. Rev Sci Tech 2012; 31(1): 199-210.
[http://dx.doi.org/10.20506/rst.31.1.2104] [PMID: 22849276]

[13] Serrano PH. Responsible use of antibiotics in aquaculture Fisheries Technical Paper 469, Food and Agriculture Organization of the United Nations (FAO), Rome, 2005.

[14] Miranda CD, Godoy FA, Lee MR. Current Status of the Use of Antibiotics and the Antimicrobial Resistance in the Chilean Salmon Farms. Front Microbiol 2018; 9: 1284.
[http://dx.doi.org/10.3389/fmicb.2018.01284] [PMID: 29967597]

[15] Martínez JL. Antibiotics and antibiotic resistance genes in natural environments. Science 2008; 321(5887): 365-7.
[http://dx.doi.org/10.1126/science.1159483] [PMID: 18635792]

[16]  Halling-Sørensen B, Nors Nielsen S, Lanzky PF, Ingerslev F, Holten Lützhøft HC, Jørgensen SE. Occurrence, fate and effects of pharmaceutical substances in the environment--a review. Chemosphere 1998; 36(2): 357-93.
[http://dx.doi.org/10.1016/S0045-6535(97)00354-8] [PMID: 9569937]

[17]  Khan MA. https://timesofindia.indiatimes.com2009.

[18]  Fick J, Söderström H, Lindberg RH, Phan C, Tysklind M, Larsson DGJ. Contamination of surface, ground, and drinking water from pharmaceutical production. Environ Toxicol Chem 2009; 28(12): 2522-7.
[http://dx.doi.org/10.1897/09-073.1] [PMID: 19449981]

[19]  Li D, Yang M, Hu J, Ren L, Zhang Y, Li K. Determination and fate of oxytetracycline and related compounds in oxytetracycline production wastewater and the receiving river. Environ Toxicol Chem 2008; 27(1): 80-6.
[http://dx.doi.org/10.1897/07-080.1] [PMID: 18092864]

[20]  Li D, Yang M, Hu J, Zhang Y, Chang H, Jin F. Determination of penicillin G and its degradation products in a penicillin production wastewater treatment plant and the receiving river. Water Res 2008; 42(1-2): 307-17.
[http://dx.doi.org/10.1016/j.watres.2007.07.016] [PMID: 17675133]

[21]  Babic S, Mutavdzic D, Asperger D, Horvat AJM, Kastelan-Macan M. Determination of veterinary pharmaceuticals in production wastewater by HPTLC-videodensitometry. Chromatographia 2007; 65: 105-10.
[http://dx.doi.org/10.1365/s10337-006-0109-2]

[22]  Sim WJ, Lee JW, Lee ES, Shin SK, Hwang SR, Oh JE. Occurrence and distribution of pharmaceuticals in wastewater from households, livestock farms, hospitals and pharmaceutical manufactures. Chemosphere 2011; 82(2): 179-86.
[http://dx.doi.org/10.1016/j.chemosphere.2010.10.026] [PMID: 21040946]

[23]  Johnning A, Moore ERB, Svensson-Stadler L, Shouche YS, Larsson DGJ, Kristiansson E. Acquired genetic mechanisms of a multiresistant bacterium isolated from a treatment plant receiving wastewater from antibiotic production. Appl Environ Microbiol 2013; 79(23): 7256-63.
[http://dx.doi.org/10.1128/AEM.02141-13] [PMID: 24038701]

[24]  Ma J, Zhai G. Antibiotic Contamination: A Global Environment Issue. J Bioremed Biodeg 2014; 5(): 5- e157.

[25]  Shejale KP, Yadav D, Patil H, Saxena S, Shukla S. Evaluation of techniques for the remediation of antibiotic-contaminated water using activated carbon. Mol Syst Des Eng 2020; 5: 743-56.
[http://dx.doi.org/10.1039/C9ME00167K]

[26]  Arslan-Alaton I, Dogruel S. Pre-treatment of penicillin formulation effluent by advanced oxidation processes. J Hazard Mater 2004; 112(1-2): 105-13.
[http://dx.doi.org/10.1016/j.jhazmat.2004.04.009] [PMID: 15225936]

[27]  Neu HC. The crisis in antibiotic resistance. Science 1992; 257(5073): 1064-73.
[http://dx.doi.org/10.1126/science.257.5073.1064] [PMID: 1509257]

[28]  Davies J, Davies D. Origins and evolution of antibiotic resistance. Microbiol Mol Biol Rev 2010; 74(3): 417-33.
[http://dx.doi.org/10.1128/MMBR.00016-10] [PMID: 20805405]

[29]  European Centre for Disease Prevention and Control Joint Report with EMEA: The Bacterial Challenge: Time to React 2011.

[30]  Davies SC, Fowler T, Watson J, Livermore DM, Walker D. Annual Report of the Chief Medical Officer: infection and the rise of antimicrobial resistance. Lancet 2013; 381(9878): 1606-9.
[http://dx.doi.org/10.1016/S0140-6736(13)60604-2] [PMID: 23489756]

[31] Perreten V, Schwarz F, Cresta L, Boeglin M, Dasen G, Teuber M. Antibiotic resistance spread in food. Nature 1997; 389(6653): 801-2.
[http://dx.doi.org/10.1038/39767] [PMID: 9349809]

[32] Tenover FC. Mechanisms of antimicrobial resistance in bacteria. Am J Med 2006; 119(6) (Suppl. 1): S3-S10.
[http://dx.doi.org/10.1016/j.amjmed.2006.03.011] [PMID: 16735149]

[33] Borer A, Saidel-Odes L, Riesenberg K, et al. Attributable mortality rate for carbapenem-resistant Klebsiella pneumoniae bacteremia. Infect Control Hosp Epidemiol 2009; 30(10): 972-6.
[http://dx.doi.org/10.1086/605922] [PMID: 19712030]

[34] Teillant A, Gandra S, Barter D, Morgan DJ, Laxminarayan R. Potential burden of antibiotic resistance on surgery and cancer chemotherapy antibiotic prophylaxis in the USA: a literature review and modelling study. Lancet Infect Dis 2015; 15(12): 1429-37.
[http://dx.doi.org/10.1016/S1473-3099(15)00270-4] [PMID: 26482597]

[35] Paterson DL. Impact of antibiotic resistance in gram-negative bacilli on empirical and definitive antibiotic therapy. Clin Infect Dis 2008; 47(47) (Suppl. 1): S14-20.
[http://dx.doi.org/10.1086/590062] [PMID: 18713045]

[36] Leung HW, Minh TB, Murphy MB, et al. Distribution, fate and risk assessment of antibiotics in sewage treatment plants in Hong Kong, South China. Environ Int 2012; 42: 1-9.
[http://dx.doi.org/10.1016/j.envint.2011.03.004] [PMID: 21450345]

[37] Hillis DG, Lissemore L, Sibley PK, Solomon KR. Effects of monensin on zooplankton communities in aquatic microcosms. Environ Sci Technol 2007; 41(18): 6620-6.
[http://dx.doi.org/10.1021/es070799f] [PMID: 17948817]

[38] Danner MC, Robertson A, Behrends V, Reiss J. Antibiotic pollution in surface fresh waters: Occurrence and effects. Sci Total Environ 2019; 664: 793-804.
[http://dx.doi.org/10.1016/j.scitotenv.2019.01.406] [PMID: 30763859]

[39] Grenni P, Ancona V, Caracciolo AB. Ecological effects of antibiotics on natural ecosystems: a Review. Microchem J 2018; 136: 25-39.
[http://dx.doi.org/10.1016/j.microc.2017.02.006]

[40] Jensen J. Veterinary medicines and soil quality: The Danish situation as an example.Pharmaceuticals and Personal Care Products in the Environment: Scientific and Regulatory Issues. Washington, D.C.: American Chemical Society 2001; pp. 282-302.
[http://dx.doi.org/10.1021/bk-2001-0791.ch016]

[41] Parsons SA, Williams M. Introduction.Advanced Oxidation Processes for Water and Wastewater Treatment. London, UK: IWA Publishing 2004; pp. 1-6.

[42] Litter MI. Introduction to photochemical advanced oxidation processes for water treatment. Environ Photochem Part II 2005; 2: 325-6.
[http://dx.doi.org/10.1007/b138188]

[43] O'Neill J. Review on Antimicrobial Resistance Antimicrobial Resistance: Tackling a crisis for the health and wealth of nations. London: Review on Antimicrobial Resistance 2014.

[44] World Health Organization Global Strategy for Containment of Antimicrobial Resistance. Geneva 2001.

[45] O'Neill J. Tackling Drug-Resistant Infections Globally: Final Report and Recommendations the Review on Antimicrobial Resistance 2016.

[46] Tang KL, Caffrey NP, Nóbrega DB, et al. Restricting the use of antibiotics in food-producing animals and its associations with antibiotic resistance in food-producing animals and human beings: a systematic review and meta-analysis. Lancet Planet Health 2017; 1(8): e316-27.
[http://dx.doi.org/10.1016/S2542-5196(17)30141-9] [PMID: 29387833]

[47] Ejim L, Farha MA, Falconer SB, *et al.* Combinations of antibiotics and nonantibiotic drugs enhance antimicrobial efficacy. Nat Chem Biol 2011; 7(6): 348-50.
[http://dx.doi.org/10.1038/nchembio.559] [PMID: 21516114]

[48] Kalan L, Wright GD. Antibiotic adjuvants: multicomponent anti-infective strategies. Expert Rev Mol Med 2011; 13e5
[http://dx.doi.org/10.1017/S1462399410001766] [PMID: 21342612]

[49] Drawz SM, Bonomo RA. Three decades of beta-lactamase inhibitors. Clin Microbiol Rev 2010; 23(1): 160-201.
[http://dx.doi.org/10.1128/CMR.00037-09] [PMID: 20065329]

[50] Blanco P, Hernando-Amado S, Reales-Calderon JA, *et al.* Bacterial Multidrug Efflux Pumps: Much More Than Antibiotic Resistance Determinants. Microorganisms 2016; 4(1): 14.
[http://dx.doi.org/10.3390/microorganisms4010014] [PMID: 27681908]

[51] Zhou X, Jia F, Liu X, Wang Y. Total alkaloids of *Sophorea alopecuroides*-induced down-regulation of AcrAB-TolC efflux pump reverses susceptibility to ciprofloxacin in clinical multidrug resistant *Escherichia coli* isolates. Phytother Res 2012; 26(11): 1637-43.
[http://dx.doi.org/10.1002/ptr.4623] [PMID: 22371352]

[52] Molnár J, Engi H, Hohmann J, *et al.* Reversal of multidrug resitance by natural substances from plants. Curr Top Med Chem 2010; 10(17): 1757-68.
[http://dx.doi.org/10.2174/156802610792928103] [PMID: 20645919]

[53] Cushnie TP, Cushnie B, Lamb AJ. Alkaloids: an overview of their antibacterial, antibiotic-enhancing and antivirulence activities. Int J Antimicrob Agents 2014; 44(5): 377-86.
[http://dx.doi.org/10.1016/j.ijantimicag.2014.06.001] [PMID: 25130096]

[54] Lomovskaya O, Warren MS, Lee A, *et al.* Identification and characterization of inhibitors of multidrug resistance efflux pumps in *Pseudomonas aeruginosa*: novel agents for combination therapy. Antimicrob Agents Chemother 2001; 45(1): 105-16.
[http://dx.doi.org/10.1128/AAC.45.1.105-116.2001] [PMID: 11120952]

[55] Nakayama K, Ishida Y, Ohtsuka M, *et al.* MexAB-OprM-specific efflux pump inhibitors in *Pseudomonas aeruginosa*. Part 1: discovery and early strategies for lead optimization. Bioorg Med Chem Lett 2003; 13(23): 4201-4.
[http://dx.doi.org/10.1016/j.bmcl.2003.07.024] [PMID: 14623001]

[56] Banoee M, Seif S, Nazari ZE, *et al.* ZnO nanoparticles enhanced antibacterial activity of ciprofloxacin against *Staphylococcus aureus* and Escherichia coli. J Biomed Mater Res B Appl Biomater 2010; 93(2): 557-61.
[http://dx.doi.org/10.1002/jbm.b.31615] [PMID: 20225250]

[57] Yehia HM, Hassanein WA, Ibraheim SM. Studies on molecular characterizations of the outer membrane proteins, lipids profile, and exopolysaccharides of antibiotic resistant strain *Pseudomonas aeruginosa*. BioMed Res Int 2015; 2015651464
[http://dx.doi.org/10.1155/2015/651464] [PMID: 25710016]

[58] Vaara M, Vaara T. Polycations as outer membrane-disorganizing agents. Antimicrob Agents Chemother 1983; 24(1): 114-22.
[http://dx.doi.org/10.1128/AAC.24.1.114] [PMID: 6194743]

[59] Ghai I, Ghai S. Exploring bacterial outer membrane barrier to combat bad bugs. Infect Drug Resist 2017; 10: 261-73.
[http://dx.doi.org/10.2147/IDR.S144299] [PMID: 28919790]

[60] Candel FJ, Peñuelas M. Delafloxacin: design, development and potential place in therapy. Drug Des Devel Ther 2017; 11: 881-91.
[http://dx.doi.org/10.2147/DDDT.S106071] [PMID: 28356714]

[61] Petty LA, Henig O, Patel TS, Pogue JM, Kaye KS. Overview of meropenem-vaborbactam and newer

antimicrobial agents for the treatment of carbapenem-resistant *Enterobacteriaceae*. Infect Drug Resist 2018; 11: 1461-72.
[http://dx.doi.org/10.2147/IDR.S150447] [PMID: 30254477]

[62]  Eljaaly K, Alharbi A, Alshehri S, Ortwine JK, Pogue JM. Plazomicin: A Novel Aminoglycoside for the Treatment of Resistant Gram-Negative Bacterial Infections. Drugs 2019; 79(3): 243-69.
[http://dx.doi.org/10.1007/s40265-019-1054-3] [PMID: 30723876]

[63]  Sutcliffe JA, O'Brien W, Fyfe C, Grossman TH. Antibacterial activity of eravacycline (TP-434), a novel fluorocycline, against hospital and community pathogens. Antimicrob Agents Chemother 2013; 57(11): 5548-58.
[http://dx.doi.org/10.1128/AAC.01288-13] [PMID: 23979750]

[64]  Ito A, Nishikawa T, Matsumoto S, *et al.* Siderophore Cephalosporin Cefiderocol Utilizes Ferric Iron Transporter Systems for Antibacterial Activity against Pseudomonas aeruginosa. Antimicrob Agents Chemother 2016; 60(12): 7396-401.
[PMID: 27736756]

[65]  Draper MP, Weir S, Macone A, *et al.* Mechanism of action of the novel aminomethylcycline antibiotic omadacycline. Antimicrob Agents Chemother 2014; 58(3): 1279-83.
[http://dx.doi.org/10.1128/AAC.01066-13] [PMID: 24041885]

[66]  Schiebel J, Chang A, Lu H, Baxter MV, Tonge PJ, Kisker C. Staphylococcus aureus FabI: inhibition, substrate recognition, and potential implications for *in vivo* essentiality. Structure 2012; 20(5): 802-13.
[http://dx.doi.org/10.1016/j.str.2012.03.013] [PMID: 22579249]

[67]  Kaul M, Mark L, Zhang Y, *et al.* TXA709, an FtsZ- targeting benzamide prodrug with improved pharmacokinetics and enhanced *in vivo* efficacy against methicillin- resistant Staphylococcus aureus. Antimicrob Agents Chemother 2015; 59(8): 4845-55.
[http://dx.doi.org/10.1128/AAC.00708-15] [PMID: 26033735]

**CHAPTER 8**

# Analytical Advancement for Pharmaceuticals Quantification in Environmental Matrices

**Anushka Pandey[1], Manisha Bhateria[1] and Sheelendra Pratap Singh[1,*]**

[1] *CSIR-Indian Institute of Toxicology Research (CSIR-IITR), Lucknow, India*

**Abstract:** The pharmaceutical residues and their metabolites present in soil and water have been considered as active pollutants, posing various health risks to humans. Major sources from where pharmaceutical compounds enter the environment are hospitals, pharmaceutical industries, domestic wastes, and improper disposal of medicines. Metabolism of drugs in humans is sometimes incomplete, resulting in their excretion in either the unchanged form or in the form of metabolites. However, biodegradation of pharmaceutical compounds and/or their metabolites in the environment is not easy; therefore, their repeated addition to the environment makes them even more persistent. The pharmaceuticals, based on their physicochemical properties, bind to soil particles or enter the aquatic system. The most adverse effect of increasing the concentration of pharmaceuticals in environmental matrices is the development of resistance in certain bacteria against antibiotics, which is a serious health concern. Steroidal hormones can alter the steroidogenesis of aquatic and terrestrial life and cause endocrine disruption, leading to cognitive and brain development problems. The concentration of pharmaceutical residues in the environment is very low; therefore, highly sensitive instruments for their quantification are required like liquid chromatography coupled with mass spectroscopy (LS-MS/MS) and gas chromatography with mass spectroscopy (GS-MS). The techniques allow the identification of various analytes with improved detection limits. The pharmaceutical residues are considered lethal pollutants, even if present in ng/kg or ng/l, and can cause potential harm upon exposure. This chapter aims to review various analytical approaches for pharmaceutical residue analysis and recent advancements made in analytical techniques.

**Keywords:** Active pollutants, Analytical techniques, Biodegradation, Gas chromatography, Liquid chromatography, Mass spectroscopy, Pharmaceutical compounds.

---

* **Corresponding author Sheelendra Pratap Singh:** Analytical Chemistry/Pesticide Toxicology Laboratory CSIR - Indian Institute of Toxicology Research, Lucknow, India; E-mail:sheelendra@iitr.res.in

Tahmeena Khan, Abdul Rahman Khan, Saman Raza, Iqbal Azad and Alfred J. Lawrence (Eds.)
All rights reserved-© 2021 Bentham Science Publishers

## INTRODUCTION

In recent years, there has been a momentous increase in the use of pharmaceuticals in the field of medicine and personal care products to counter various health-related problems, and to improve the quality of life. The improper handling and disposal of these pharmaceuticals lead to an increase in the concentration of pharmaceutical residues in various environmental matrices [1]. Pharmaceutical residues in the environment are considered as one of the active pollutants, even in trace amounts, causing severe health-related issues to the wildlife and producing toxic effects to the ecosystem. These pharmaceutical compounds show the same behaviour as several harmful xenobiotics, that can accumulate in the environment, to produce a negative impact on living organisms [2]. Water and soil are most contaminated, as a large amount of pharmaceutical waste is dumped or disposed of inappropriately, into these matrices. The identification and determination of pharmaceutical residues are one of the major concerns to detect their contamination level in the environment [3].

There are several classes of pharmaceuticals that are extensively targeted for the development of analytical methods, to enhance their traceability in different environmental compartments. Pharmaceuticals are categorized into the following eight classes that include hormones, antibiotics, lipid regulators, nonsteroidal anti-inflammatory drugs, beta-blockers, antidepressants, anticonvulsants, and antineoplastics [4]. The use of these pharmaceuticals on a broad spectrum, for medication, has led to the contamination of the environment. Hormones like estrogens, at their polluting levels, are responsible for causing breast cancer in females and prostate cancer in males, as well as altering the physiology of fish and reproductive patterns of domestic and wild animals [5]. The increasing antibiotic pollution in the environment results in developing resistance in certain bacteria, against a particular dose of antibiotics, which is a serious health concern [6]. In a study, the European pharmaceutical review (EPR) found that around 10% of pharmaceutical compounds, that have been disposed into the environment, can cause damage to it. Amongst the compounds that are of major concern are hormones, painkillers, and antidepressants [7]. Due to the complex nature of the matrix and their low level of occurrence, rigorous quantification of pharmaceutical residues in the environment is an analytical challenge. There have been various methods for the quantification of pharmaceutical residues, even at their trace level, such as nanograms (ng) or picograms, in different environmental matrices. Presently, gas chromatography (GC) and liquid chromatography (LC), along with several steps of extraction, derivatization, clean-up method, and detection using mass spectroscopy (MS), are used to detect and quantify various pharmaceutical compounds as well as their metabolites, in different environmental matrices. For the analysis of pharmaceuticals, there is another method known as capillary electrophoresis (CE). CE is less complicated and cost-effective; however, it is less sensitive in comparison to GC and LC, having detection limits in micrograms. To enhance the analytical approaches towards the detection of pharmaceutical residues, various advancements have been made in the instrumentation and the sample preparation, derivatization, and clean up processes [8].

Before the instrumental analysis, there is an essential step of sample preparation that allows the removal of all the possible interferences and matrix effects. The sample preparation is tedious and time-consuming yet an important and compulsory step, for the determination of pharmaceutical residues or any other compounds present in the environment. The method of sample preparation involves the preservation, extraction, and clean-up procedures that help in processing an extract with a high concentration of analytes, that must be detected or analysed on a particular instrument. Several sample preparation techniques have been used for the determination of pharmaceutical compounds, of which the most used techniques are liquid-liquid extraction (LLE), dispersive liquid-liquid micro-extraction (DLLME), solid-phase extraction (SPE), solid-phase micro-extraction (SPME), and stir bar sorptive extraction (SBSE). Recently, with the emergence of new advancements in sample preparation techniques, several new techniques, such as pressurized liquid extraction (PLE) or ultrasonic extraction (USE) have been used, either individually or coupled with SPE clean-up, to prepare samples for solid matrices of the environment.

For the analysis of pharmaceuticals in environmental samples, both GC and LC are considered suitable methods. GC is applicable to analyse the non-polar and volatile compounds whereas LC helps to separate polar organic compounds. Mass spectrometry, coupled with LC or GC, has been used to detect numerous pharmaceutical compounds along with various detectors, such as fluorescence, UV (ultraviolet detector), PDA (Photo Diode array), FID (Flame ionization detector), ECD (electron capture detectors), *etc* [8].

## Analytical Methods for the Determination of Pharmaceutical Residues in the Environment

In the present day, advancements in analytical methods have made it possible to detect complex pharmaceutical compounds and their metabolites in the environment, even at low concentrations or trace levels. Chromatographic techniques such as GC and LC are used to analyse the environmental samples to detect the pharmaceutical residues. However, every analytical method involves various synergetic processes to determine the desired compound in a sample. After sample collection, the sample undergoes several consecutive steps in the laboratory, that are given below:

## Sample Preservation

Sample preservation aims to slow down the imminent chemical and biological changes that might occur after sample collection.

The stability of pharmaceutical residues in the sample can be affected by various factors, such as it may undergo degradation by the action of heat or light, or it may hydrolyse, due to the formation of a complex with other minerals present in the sample or adsorption on organic matter or due to cross-contamination during sample collection. All these factors are responsible for the loss of analytes in the sample. Hence, to reduce this loss, it is important to preserve the sample after its collection. The techniques that are used for the preservation of pharmaceuticals in the aquatic environment are [7].

### i. Filtration

In water samples, the pharmaceuticals are dissolved in the aqueous phase as well as adhered to the dispersed solid particles and other dissolved organic compounds. Therefore, sample filtration helps to separate the pharmaceuticals in the aqueous phase from the organic matter, to determine the residues of the aqueous phase in the water sample. The filtration should be done right after the sampling to prevent the pharmaceutical residues from being degraded or desorbed from organic matter. However, this also removes the adsorbed pharmaceuticals along with the organic matter, yet it helps to improve the analysis by reducing the interferences in the matrix. The selection of suitable filters can be made based on its sieving capacity for the selected pharmaceuticals. Filters that are used for filtration are glass fibre filters, cellulose acetate membrane, nylon membrane, cellulose acetate, and non-specified membrane filters [9 - 11].

### ii. Non-acidic Preservative Agent

Certain chemical substances are used to preserve the pharmaceuticals from bacterial growth at all temperatures, such as sodium azide, formaldehyde, and methanol. Sodium azide binds with the metal ion of the bacterial enzyme as well as to the metal ions present in the water sample, constraining the growth of bacteria. However, sodium azide is poisonous and is therefore considered toxic for humans when exposed while experimenting. On the other hand, formaldehyde in its aqueous solution produces a biocidal effect by introducing an alkyl radical or ion into the protein structure of micro-organisms. Being alcohol, methanol can inhibit bacterial growth, however, it has a very short half-life and therefore, undergoes decomposition readily. To inhibit the ability of certain pharmaceuticals to form chelates and precipitate out with metal ions, the addition of chelating agents like ethylenediaminetetraacetic acid (EDTA) is preferred. By forming a complex with metal ions, EDTA restricts the availability of metal ions to bind with the pharmaceutical compounds, to form a precipitate. A study on the action of EDTA was reported which states that its addition to the water sample, enhances the antibiotic recovery, as in the case of tetracyclines, macrolides, and fluoroquinolones, during solid-phase extraction. Various studies have been done to check the stability of pharmaceuticals upon the addition of preservatives, which have shown that they usually do not have any influence on the selected pharmaceuticals, although they can enhance the stability of hormones in the sample. It was seen that in the surface water samples, the addition of sodium azide at room temperature does not affect the stability of pharmaceuticals but increases the stability of hormones like progesterone and oestrogen. Formaldehyde and methanol have been reported to enhance the stability of estriol, which is otherwise unstable in the sample. Methanol is found suitable for the preservation of NSAIDs, caffeine trimethoprim, and sulfamethoxazole. However, it has been seen that methanol affects gemfibrozil, a lipid regulator, in a negative way. Furthermore, when these chemical substances are added to the sample as a preservative, for a longer period, they may react with certain pharmaceutical compounds or their metabolites and may lead to their degradation [7].

### *iii. Acidifying Agents*

These can also be called pH regulators. It has been seen that pH 5 to 9 is optimum for the growth of bacteria, hence, by altering the pH of the environmental samples, bacterial growth can be prevented. Hereupon it has been anticipated that the addition of certain acidic compounds can successfully hinder the biodegradation of pharmaceutical compounds in the environmental matrix, at the time of storage. Also, a decrease in the adsorption of pharmaceutical compounds on organic matter present in the sample can be seen at low pH. The main disadvantage of the acidifying agent is that it might cleave certain hormone pharmaceutical products that exist in conjugation, thereby resulting in an inaccurate result for their observed recoveries. The acidic compounds preferred are hydrochloric acid, sulphuric acid, and phosphate buffer, to check their impact on the pharmaceutical compounds present in the samples stored at 4°C. To protect the pharmaceutical compounds from getting degraded, pH 2 was found to be suitable. The application of HCl and $H_2SO_4$ was found useful in stabilizing the hormone pharmaceutical products, fluoxetine (antidepressant), and trimethoprim (antibiotic) in environmental samples [7].

## Sample Preparation

Sample preparation refers to all the modifications done in a sample to make it susceptible to get analysed over LC-MS/MS or GC. The process of sample preparation helps to reduce the interaction between the analyte and its matrices, which depends on their chemical and physical properties. Several techniques have been used for sample preparation. However, the purpose of every technique is similar, that is- to eliminate all the possible interference co-extractives, to enhance the concentration of an analyte in the sample, and to develop a rugged reproducible method that remains unaffected by the changes in the sample matrix [12]. The sample preparation technique involves processes like extraction and clean-up procedure. The extraction process refers to the isolation of analytes. The method involves the extraction of analytes of interest from the sample, into a solvent, to get analysed over LC-MS or GC. Commonly used extraction solvents are ethyl acetate, dichloromethane, acetone, and acetonitrile. The extraction solvent of choice must be compatible with the analyte and should not affect its stability. The clean-up process is aimed to remove the co-extractives from the samples without compromising the recoveries of target analytes. Many sample preparation techniques have been used in recent years with some advancement made, which includes a reduction in sample size with a limited amount of organic solvent, precise extraction, and enhanced ability to get automated. By applying an appropriate extraction method, it is possible to detect the pharmaceutical even at a very low concentration. After the clean-up procedure, the pharmaceutical samples are subjected to derivatization before their analysis on GC-MS. However, if the analysis must be done on LC-MS, then there is no need for derivatization of the sample [13].

### *i. Liquid-Liquid Extraction (LLE)*

As the name suggests, LLE is the extraction of analytes from one liquid phase to another, based on their solubility. Analytes present in liquid samples can be extracted directly by the partition of an immiscible solvent. LLE is governed by the equilibrium distribution/partition coefficient based on the relative solubility of an analyte in two immiscible phases. The difference in the solubilizing power (polarity) of two immiscible liquid phases allows the extraction of an analyte. Generally, a separating funnel is used to mix the two immiscible phases by shaking, and then the two layers of the liquids are separated. In several cases, the addition of salt to the sample solution can reduce the formation of an emulsion. For further analysis, the organic layer can be extracted. Multiple extractions may be applied to perform the complete extraction of an analyte into the required phase. Solvents such as hexane, diethyl ether, dichloromethane, ethyl acetate, and methyl tert-butyl ether (MTBE) are popular extraction solvents, as they are capable of extracting analytes with a wide range of polarity. Dichloromethane is generally used for the analysis of pesticide residues in an aqueous sample. Normally, hexane is better for the extraction of non-polar compounds while ethyl acetate and ether are preferably used for the extraction of semi-polar compounds. Even though LLE generally yields clean extracts, the method suffers from various setbacks, such as multiple extraction steps, a large volume of toxic solvents, chances of emulsion formation during stirring, and disposal problems of the post-extraction solvents. LLE proceeds by shaking the aqueous sample with the solvent, in a separating funnel. The mixture is shaken two to three times, followed by the separation of the lower layer of the solvent. In the case, where the analyte is partitioned in the upper layer of the solvent, careful separation of the upper layer is done. It has been observed that the LLE gave better recoveries by using a multi-step extraction procedure. However, extraction through the multi-step procedure is treated as undesirable as it requires a lot of time as well as labour. The LLE technique, being laborious, also does not give good recoveries for polar compounds until a reagent is added to the sample, which can facilitate the partitioning of analytes in the organic phase. Sometimes, an emulsion is formed, and it is too difficult to remove the solvent from the aqueous phase. The emulsion is removed or broken by any of the following methods- adding anhydrous sodium chloride, adding ethanol, sonicating the mixture or continuous LLE can be performed on the sample (but it is not performed for the thermodynamically unstable compounds due to its long extraction time). Further, to achieve the desired concentration of the analyte, the excess solvent is generally evaporated.

LLE cartridges have been used recently to avoid problems at the time of separating the two phases. LLE cartridges consist of a column filled with diatomaceous earth (a soft siliceous sedimentary rock) to which an aqueous sample is added. The diatomaceous earth acts as a sorbent for the adsorption of water together with the targeted analytes on its surface, which can be either buffered or unbuffered depending on the pH of the aqueous phase. On the addition of the organic phase to the column, the analytes undergo phase separation between the two solvents based on their solubility. As the organic phase is loosely held on to the column, therefore, it can be easily moved out of it as a separate solvent. The advantage of using LLE cartridges is that it has prevented the formation of emulsion while mixing the two solvents [14]. Fig. (**1**) illustrates the steps involved in a general LLE technique.

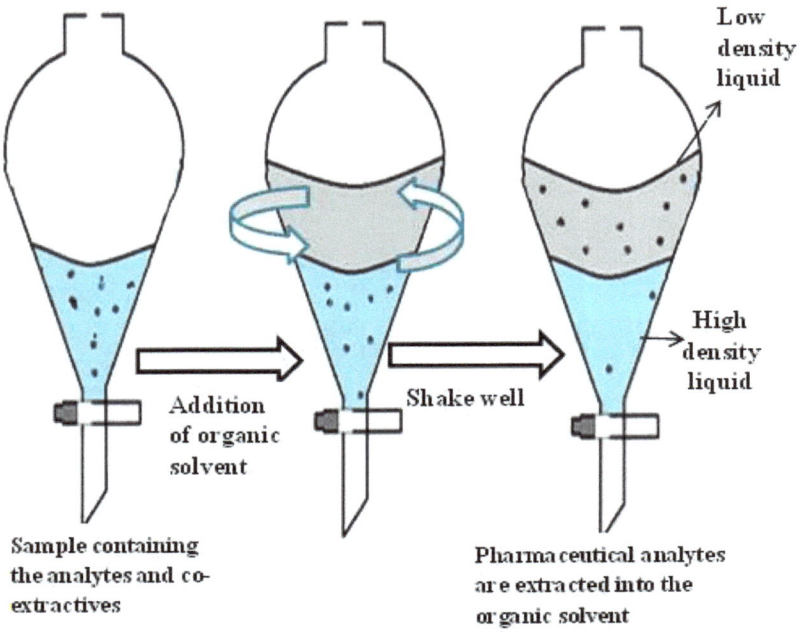

**Fig. (1).** Diagrammatic representation of Liquid- Liquid extraction (LLE).

Teresa I.A. Gouveia *et al.* collected water samples from the wastewater treatment plant situated in northern Portugal in May 2019, for the analysis of 14 cytostatic pharmaceutical compounds (Bicalutamide, capecitabine, cyclophosphamide, cyproterone, doxorubicin, etoposide, flutamide, ifosfamide, imatinib, megestrol, mycophenolate mofetil, mycophenolic acid, paclitaxel, and prednisone) by the application of LLE technique. Acetonitrile was added as an extraction solvent into the water sample in 1:1 ratio (20 ml: 20ml) and vortexed, followed by storage at -18 °C for 1 hour. The aqueous sample was solidified after freezing at -18 °C and the organic phase was then extracted from the mixture. Acetonitrile was added to the extract for washing purposes and was dried under nitrogen gas. The analytes were then reconstituted with methanol and analysed over LC – MS/MS, with detected concentration up to 1624 ng/l [15].

W. Arher *et al.* detected pharmaceutical compounds such as paracetamol, clofibric acid, penicillin V, naproxen, bezafibrate, carbamazepine, diclofenac, ibuprofen, and mefenamic acid, in several river-water samples, by combining HPLC or CE with mass spectroscopy. SPE was used as a sample preparation technique for the analysis of pharmaceuticals on HPLC-MS while for CE-MS, extraction of pharmaceuticals was done by LLE techniques followed by SPE clean-up procedure. For the extraction through the LLE technique, the sample pH was maintained at 2 (as it provides better recoveries for the analytes than at pH values 5.5 and 8.5) before the addition of 50g sodium sulphate. The sample was then extracted by using 25 ml of hexane: MTBE in a 1:1 ratio. Re-extraction of the organic phase was done by using 50 ml of 2 mM sodium hydroxide, followed by dilution with 250 ml water, at pH 2. Once the extraction was over, clean-up was done by using an SPE cartridge and reconstituted with 50 µl methanol carrier electrolyte, to get analysed over CE-MS [16].

## ii. Dispersive Liquid-liquid Microextraction (DLLME)

DLLME is a miniaturized form of liquid-liquid extraction that falls under the category of liquid phase micro-extraction. This extraction technique has become popular in recent years due to its environment-friendly properties and high enrichment factor. The amount of solvent used in this technique is very less. DLLME technique is based on the use of a ternary solvent in which a small amount of an organic extraction solvent (in µL) combined with a few mL of dispersive solvent is injected into an aqueous sample containing the targeted analytes. The addition of a sufficient extraction solvent along with a dispersive solvent into the sample causes the fine microdroplets to disperse into the solution giving a cloudy appearance to it. These microdroplets offer a large surface area which enables the partition of analytes into the extraction solvent instantaneously. After the partitioning of analytes, the solution is subjected to the centrifugation process that leads to the differentiation of the solution into two phases. The sedimented phase containing the analyte of interest is withdrawn for further analysis [17, 18]. The most important thing in the DLLME process is the selection of an appropriate extraction solvent, based on various requirements. The solvent should be immiscible in water and should form tiny droplets in it. It must have a high affinity for the analytes and its volume should be optimized with utmost care. Besides, the compatibility of the solvent should also be assessed with the desired instrument. Likewise, the optimization of disperser solvent type and volume is also very important. The disperser solvent is employed to allow the extracting solvent to partition uniformly in the aqueous sample, to achieve good extraction efficiency. Careful consideration should be given to the optimization of the ratio of extracting solvent to the disperser solvent. A significant influence of the disperser solvent type and volume has been observed on the volume of the sedimented phase. Various modifications in DLLME have been described by researchers to overcome the possible drawbacks of the process such as restriction in solvent selection (most of the available solvents are halogenated and hazardous), careful optimization of various parameters, and automation. DLLME offers several advantages over LLE such as it is rapid, economical, reliable, simple to operate, high preconcentration, and recovery factors [18]. In Fig. (2) the diagrammatic representation of the DLLME technique has been shown.

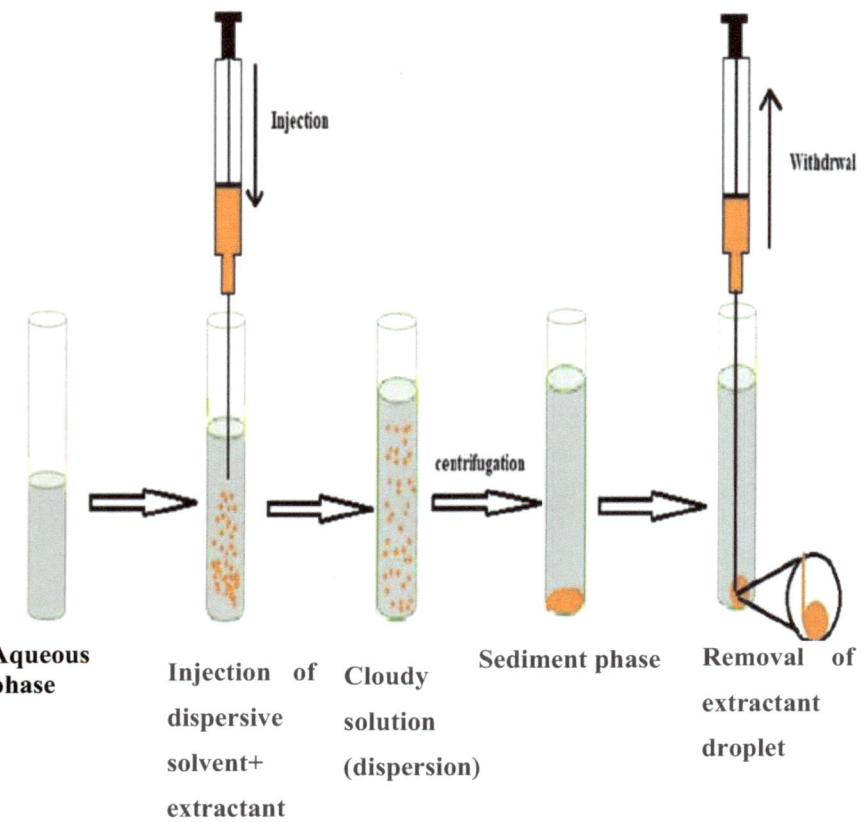

**Fig. (2).** Diagrammatic representation of dispersive liquid-liquid microextraction (DLLME).

Various modes of DLLME have been used to extract the organic and inorganic residues from various environmental samples such as ultrasound-assisted DLLME (UA– DLLME), low-density solvent DLLME (LDS–DLLME), ultrasound-assisted reverse micelle-DLLME (UA–RM DLLME), ionic liquid DLLME (IL–DLLME), and ultrasound-assisted ionic liquid DLLME (UA–IL DLLME) [17].

J.N. Sun *et al.* in their work isolated the two phenolic pharmaceutical compounds, 2-naphthol and 4-nitrophenol from the environmental water samples by using 1-hexyl-3-methylimidazolium hexafluorophosphate [$C_6$MIm][$PF_6$] as extraction solvent (65 µl) and hydrophilic ionic liquids 1-ethyl-3-methylimidazolium tetrafluoroborate ([EMIm][BF4] and 1-sulfobutyl-3-methylimidazolium trifluoromethane sulphate [BSO3HMIm][OTf]) as dispersive solvent (0.15 ml in the ratio of 4:1). Sediment phase was obtained by ultrasonication and centrifugation and extracted out from the aqueous followed by dilution with 60 μl methanol. The samples were then analysed by using HPLC instrument [19].

Guan J. *et al.* developed an analytical procedure to determine 12 pharmaceutical compounds in water samples from the city of Shenyang, China. They used the Ultrasound-assisted DLLME procedure to extract the analytes from the aqueous samples. The extraction solvent used in the procedure was dichloromethane (800 μl) and the dispersive solvent was methanol/ acetonitrile (1200 μl). After ultrasonication and centrifugation, the extractant was evaporated to dryness and made up with a combination of acetonitrile and methanol in a 1:1 ratio by volume. Finally, 5 μl of the sample was injected into the UPLC–MS/MS instrument for analysis [20].

Park and Myung used DLLME procedure to extract and concentrate the pharmaceutical residues in aquatic samples, followed by their LC-MS analysis. Seven NSAIDs that include indoprofen, ketoprofen, naproxen, diclofenac, ibuprofen, mefenamic acid, and tolfenamic acid were extracted by using chloroform as the extraction solvent and acetonitrile as the dispersive solvent, at pH 3.6, from water samples. The process of sample preparation involved the immediate addition of a mixture of 200 μl of extraction solvent and 1ml of dispersive solvent to 5 ml of the water sample. After centrifugation, 100 μl of the organic supernatant was extracted out and dried, using nitrogen gas, and was then reconstituted in methanol. The extract was then analysed with the LC-MS system, with detection and quantification limits ranging from 0.65-1.3 and 2.2-4.4μg/L, respectively [21].

### *iii. Solid – Phase Extraction (SPE)*

SPE is a technique that is used to extract the analyte, remove interfering compounds (clean–up process) and concentrate the analyte from large sample volumes, before proceeding for analysis. SPE has an upper hand over LLE as it improves the recovery yield quantitatively. SPE is fast, can be easily performed, and is automated. SPE technique works on the principle of partitioning, adsorption, and ion exchange. It consists of two phases- a solid adsorbent and a liquid matrix. Separation of the analyte occurs based on the difference between the binding property of the analyte and co-extractives. The difference in their binding properties enables the separation of the analyte of interest and the interfering compounds, into their respective phases. The adhesive force between the solid phase and the analyte needs to be higher than that between the analyte and the liquid matrix so that the analyte gets adsorbed on the solid sorbent and can be extracted out by an appropriate solvent.

Typically, SPE consists of four main steps (Fig. 3): condition, load, wash and elute. In the first step, the sorbent bed is conditioned with solvent to improve the reproducibility of the retention of the analyte and to reduce the concentration of any contaminants present. Next, the analyte is added to the conditioned sorbent bed and gets sorbed together with undesirable matrix components. In the third step, the column is washed with a weak solvent to remove the impurities or undesirable matrix components. Lastly, a sufficiently stronger solvent is employed for the elution of the analyte while leaving the unwanted constituents on the bed. The steps involved in the SPE technique have been explained in Fig. (3).

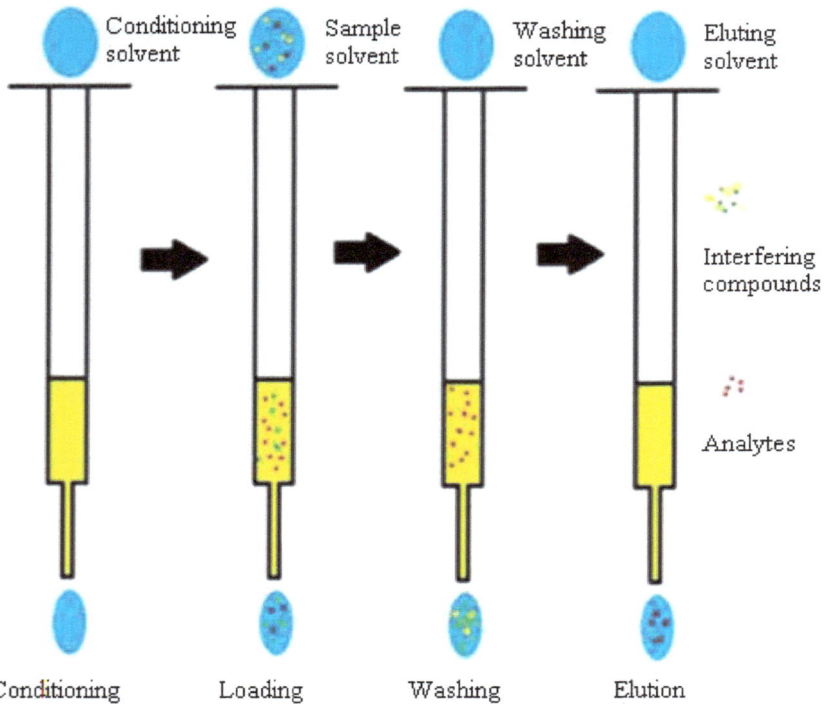

**Fig. (3).** Steps of Solid Phase Extraction (SPE).

The extent, to which the analyte can bind onto its surface, the specificity, and the potentiality of extraction, depends on the nature of the sorbent used in the SPE method. Therefore, the selection of a suitable sorbent is important and is determined by the binding ability of analytes at the sorbent's surface as well as the nature of the interaction of the sample matrix. Typically, there are two types of sorbents used in the SPE technique – silica sorbent and polymeric sorbent. The most widely used sorbents are C8 or C18 which are chemically linked with silica, macroporous styrene-divinylbenzene, St-DVB (a copolymer of styrene and divinylbenzene), immunosorbents (use natural antibodies for molecular recognition and are highly molecule specific), molecularly imprinted polymers, (MIPs, act as a selective adsorbent in SPE method) and restricted access materials (RAMs, they hold on the low molecular weight compound while excluding the macromolecules).

Silica sorbents are unstable at higher pH and consist of silanols groups to which tetracyclines undergo irreversible binding, however, the clean-up procedure for hormone samples like oestrogen is done by silica-gel along with C18 SPE concentration. Pre-concentration of pharmaceuticals with sufficient hydrophobic character can be done by using the reversed-phase material. Polar compounds do not show a good result with sorbents of silica origin and St-DVB, however, several sorbents of polymeric origin, offering large surface area for adsorption, have been recently developed. Presently, a co-polymer named polyvinylpyrrolidone (PVP) which is commercially available as Oasis HLB, has been widely used for the extraction of tetracyclines and sulfonamides from water samples. Strata-X, a synthetic polymer of divinylbenzene having piperidone groups, is used as a sorbent in SPE method to retain and extract the pharmaceutical compounds, such as sulfonamides, tetracyclines, fluoroquinolones, penicillin G, procaine, and trimethoprim [12, 22].

Immunosorbents are suitable for extracting, concentrating, and isolating the analytes from a large volume of sample from the aquatic environment [23]. RAM shows a good output for pharmaceuticals that are smaller in size; J. Chico *et al.* have suggested that for the analysis of tetracyclines in the water sample, the restricted access material has been used in the sample clean-up procedure [24]. MIP has specific sites that are suitable to bind with the functional group of the analyte and is highly molecule specific and gave a positive result with polar compounds, such as sulfaguanidine [25].

Y. P. Duan *et al.* synthesized MIP sorbent to use as SPE extracting phase while tracing acidic pharmaceuticals, such as ibuprofen, naproxen, ketoprofen, diclofenac, clofibric acid, fenoprofen, and mefenamic acid, in the samples selected from lake water, wastewater and sediments. They used the solid phase extraction technique to extract the pharmaceutical residues from water samples whereas to extract the analytes from sediments, the ultrasonic extraction technique was used [26].

A. Togola, H. Budzinski *et al.* developed a multi-residue analysis approach for the determination of 18 pharmaceutical compounds in the aquatic environment. Out of the 18 pharmaceutical compounds (anti-inflammatories, antidepressants, and hypolipidemic compound), 7 were basic and 11 were acidic. Extraction was done with SPE cartridges. MCX was selected for the purpose after comparing it with C18 and HLB that have shown that the MCX is more effective for extraction of both basic and acidic compounds and is also cost-effective. This was followed by derivatization and analysis through GC-MS. The extraction procedure involves- a) the pre-treatment of the sample *via* filtration through GFF fibre filters, b) conditioning of SPE cartridges with ethyl acetate and Mili-Q water (pH=2) before sample loading, c) loading of the sample at the flow rate of 12-15 ml min$^{-1}$, d) elution using 3 consecutive organic solvents (3ml ethyl acetate, 3ml ethyl acetate + acetone (50/50, v/v) and 3ml ethyl acetate + acetone + ammonium hydroxide (48/48/2, v/v/v/v) and lastly, e) drying to evaporate the solvent under nitrogen gas and then reconstituting with ethyl acetate [27].

## *iv. Solid-Phase Micro Extraction (SPME)*

SPME is an efficient sample preparation technique that has been recently developed for extracting and concentrating the analyte from various states of matter *i.e.*, solid, liquid, and gaseous sample matrix. In this method, phase separation of analytes takes place between the two phases *i.e.*, the extracting phase and the sample matrix, followed by a suitable desorption technique to get analysed over an instrument of interest, such as GC-MS or LC-MS/MS. The mode of action of SPME is analogous to that of the SPE method, with the only difference being the nature of sorbent used. A polymeric compound of organic origin, such as polyacrylate or Carbowax–divinylbenzene laminated over a fused silica fibre on a syringe, has been used as a sorbent or stationary phase in the SPME method. The method is aimed to create a state of equilibrium between the coated silica fibre (sorbent) and the sample matrix. When SPME is used in association with gas chromatography for the extraction and detection of the volatile or partially volatile organic compound present in the environmental sample, the analyte undergoes thermal desorption from the fibre's surface *via* GC injection port. For the analysis of less volatile compounds, SPME coupled with high-performance liquid chromatography (HPLC), and a suitable interface along with desorption of analyte to the solvent, is needed before its analysis. Furthermore, a newly introduced technique known as in-tube SPME, a combination of SPME with HPLC, uses silica fibre, in the form of a tubular capillary column, as the stationary phase. In-tube SPME is automated in nature, less time consuming and gives accurate and precise results as compared to other techniques. The SPME involves three modes of extraction through fibre: a) Direct extraction or immersion, which is known as DI-SPME, b) Headspace configuration, known as HS-SPME, and c) Membrane-protected system. DI-SPME involves the direct introduction of coated silica fibre into the sample for adsorption of analytes over its surface. HS-SPME is suitable for the extraction of analytes from a sample that contains high molecular weight interfering compounds. In this method, the fibre that is coated with polymeric substance is positioned in the air at the top of a sealed container, so that the volatile analytes get adsorbed at its surface, leaving the co-extractives behind. On the other hand, the membrane-protected SPME is suitable for those samples that consist of both non-volatile and interfering compounds with large molecular weight. However, the SPME method usually lacks the characteristic of producing an appropriate result upon making some discretionary changes with the experimental conditions. The silica fibre cannot be used for a long period and the continuous use of the needle damages it as well [12, 28]. Diagrammatic representation of the SPME technique has been shown in Fig. (4).

M. Moeder et al. explained the SPME method for the extraction of polar pharmaceutical compounds (ibuprofen, paracetamol, phenazone, carbamazepine, and xenoestrogens) from the aquatic environment and their determination through GC-MS, with subsequent derivatization, and compared it with the solid phase extraction. Ibuprofen was one of the principal pharmaceuticals that were detected in the ground and river water sample of Germany. The ideal time for the determination of these pharmaceutical compounds was observed to be 30 minutes and their limit of detection (LOD) value ranged from 0.2 to 50 µg/l. Hence, SPME is less time-consuming and detects pharmaceuticals at trace levels [29].

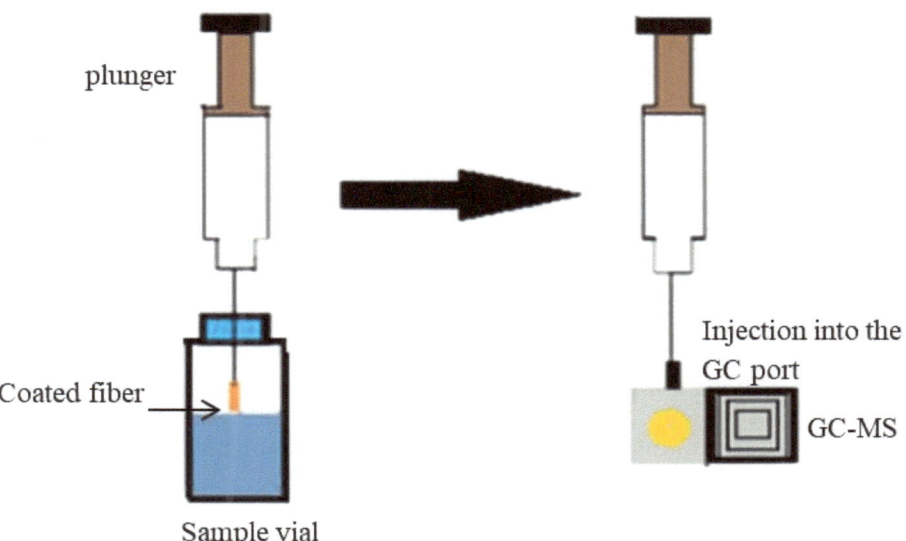

**Fig. (4).** Diagrammatic representation of Solid Phase Micro Extraction (SPME).

N. Unceta *et al*. collected water samples from wastewater treatment plants to determine the concentration of pharmaceutical residues. The extraction of the pharmaceutical analytes from the sample was done by employing the SPME method. PDMS/DVB fibre was used as an adsorbent to extract out the pharmaceuticals. Acetonitrile and methanol were the organic solvents used for the desorption of analytes from the fibre. To maintain the ionic strength of the sample, sodium chloride and sodium sulphate were compared, and it was found that sodium chloride had less interference with the cleaning procedure of the fibre. The final analysis was done with the LC-MS system equipped with a C18 reversed-phase column and the mobile phase used was 20mM ammonium formate and acetonitrile [30].

## v. Stir- bar Sorptive Extraction (SBSE)

The SBSE is an approach for extracting and concentrating the desired analytes, which are volatile or semi-volatile, from environmental matrices. The SBSE has the same mechanism as that of SPME, however, the sorbent used in SBSE is a stir bar coated with a polymeric substance like polydimethylsiloxane (PDMS), in place of coated silica fibre. The method provides a large surface area for adsorption of analytes and an increased amount of extracting phase in comparison with SPME. Stir bars, that are used for the extraction, consist of three vital elements- a) a magnetic stirring rod that is needed for shifting the circular motion of the stirring rod into the sample, b) a thin glass cover that protects the polymer coating from getting decomposed by the catalytic action of the metals present on the rod, and c) a covering of polymeric sorbent onto which the analyte get adsorbed. The extraction procedure involves placing the polymer-coated stirring rod into the liquid sample and blending it in a continuous motion. The time duration for the extraction of analytes from a sample depends on various factors, such as the amount of sample, speed with which the rod is moving, and the size of the stirring bar. The analyte gets partitioned in between the sorbent and the sample matrix, depending on its ability to bind at the surface of the extracting phase; then the stir bar is taken out from the sample and is allowed to dry. Once the drying is done, desorption of analytes occurs, either thermally by injection port of GC or by using an appropriate liquid to get analysed over LC [31, 32].

The SBSE is environment friendly as there is no use of any harmful solvent. It also allows the detection of trace amounts of biologically active compounds present in the environmental samples, with better LOD values compared to SPME [31].

N. Gilart *et al*. explained the use of new polymeric compounds, other than PDMS, as a sorbent to perform SBSE, such as polyacrylate (PA) with a portion of polyethyleneglycol (PEG), and ethylene glycol (EG) modified polysiloxane. It was observed that in comparison to PDMS (non-polar structure), pharmaceutical compounds with a high polarity showed better extraction with PA and EG coatings on stir bar, due to their polar structure [33]. A.R.M. Silva *et al.* demonstrated the extraction of six acidic pharmaceutical compounds that are categorized as NSAIDs and lipid regulators (*o*-acetylsalicylic acid, ibuprofen, diclofenac sodium, naproxen, mefenamic acid, and gemfibrozil) from water sample by using the SBSE method. Two polymeric substances- PDMS and polyurethane (PU), were used as sorbents and among them, PU showed better recovery results for the targeted pharmaceutical compounds [32].

D. Bratkowska *et al.* synthesized a copolymer of methacrylic acid and divinylbenzene, which was used as a polymeric phase to extract the polar pharmaceutical compounds from samples of the aquatic environment, through SBSE. The extract thus obtained was analysed over LC-MS/MS [33].

Other sample preparation techniques that are used for extraction purpose are liquid-phase microextraction (LPME), supercritical fluid extraction (SFE), pressurized liquid extraction (PLE), matrix solid-phase dispersion (MSDP), dispersive solid-phase extraction (DSPE), ultrasonic extraction (USE) and microwave-assisted solvent extraction (MASE) [12].

LPME is the miniaturization of LLE and is performed by applying a membrane that acts as an interface between the sample matrix and the extraction solvent, separates the two phases, and prevents them from getting mixed. LPME has several advantages over LLE, such as the usage of organic solvent is less, making it environment-friendly, it is simple, cost-effective, and highly selective, and clear extracts are obtained. When a hollow porous fibre of polypropylene is used as a membrane to separate the two phases, the technique is known as hollow fibre liquid phase micro-extraction (HF-LPME). Other liquid-liquid extraction techniques that use membranes are - a) supported liquid membrane extraction that is an aqueous-organic-aqueous system, containing a hydrophobic polymeric membrane in between the aqueous layer, b) microporous membrane LLE, which is an aqueous-organic system separated by a polymeric membrane [12].

SFE is another sample preparation technique in which supercritical liquid is used for the extraction phase. Supercritical fluids that are used as extracting solvent are- carbon dioxide (most widely used), nitrous oxide (harmful due to its oxidizing nature), xenon, ethane, propane, n-pentane, fluoroform, sulphur hexafluoride, and water [12].

The PLE is a technique that combines pressure and temperature with the organic solvent, to attain a faster and systematic extraction procedure for extracting the analyte of interest from the sample [13]. M.P. Schlusener *et al*. combined the PLE technique with SPE clean-up procedure for the determination of antibiotics (macrolides erythromycin, roxithromycin, oleandomycin, tylosin, ivermectin, salinomycin and monensin, and tiamulin) in soil samples, using LC-MS/MS analysis [34].

MSPD is an approach used to prepare and extract analytes from the sample, which is solid, semi-solid, or viscous. In this method, the sample is blended with a solid-support of the same bonded-phase, in the ratio of 1:4 that acts as an abrasive to disrupt the sample, as well as to bind the solvent that aids in achieving the sample disruption. Its ability to perform the quantitative and selective extraction made it a more feasible sample preparation technique. The columns used in MSPD allow the extraction of various polar compounds. The dispersive solid-phase extraction (DSPE) is the same as MSPD, except that in DSPE, the solid sorbent phase is added to a portion of the liquid sample. Thus, this reduces the amount of sorbent used as well as the sample size [12].

The pharmaceutical analytes in a solid sample can be effectively extracted out by using ultrasonic extraction (USE). This method involves the use of ultrasound waves to extract the analytes from the sample matrix. Separation of extracted analytes is done by either centrifugation of the sample or by vacuum filtration. The use of ultrasound waves enhances the effectiveness of extraction. Similarly, in MASE, the mixture of an extraction solvent and sample matrix is subjected to heat by using microwave radiation, under a specified temperature and pressure atmosphere. However, the method is suitable for non-volatile pharmaceutical analytes only [12].

## Chromatographic Techniques for Pharmaceuticals Analysis

With recent advancements, HPLC and GC, in combination with mass spectroscopy, have proved efficient to detect the pharmaceutical residues in environmental samples, both quantitatively and qualitatively. These analytical techniques are considered excellent in the field of pharmaceutical analysis because of their high sensitivity, better resolution, and less time taken [35].

## Analysis of Pharmaceutical Compounds by Gas Chromatography (GC)

GC is a chromatographic technique that is used to identify and separate the chemical compounds in the various sample matrices. Gas chromatography is best suited for volatile or thermally stable compounds. The instrument comprises of three essential parts:

1. Injector- Its function is to inject the sample into the instrument.
2. Column- The column is packed with the stationary phase which can be a polymeric compound. The analytes get isolated as per their ability to combine with the stationary phase and the mobile phase (which is a carrier gas in the case of gas chromatography). Partitioning of the analyte takes place between the stationary phase and the mobile phase.
3. Detector- It detects the structure and concentration of the targeted analyte.

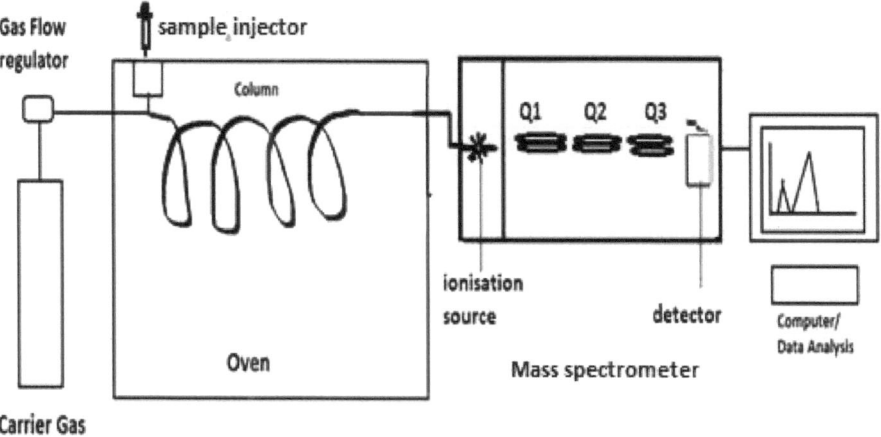

**Fig. (5).** Diagrammatic representation of GC- MS/MS System.

The inert gases such as nitrogen, hydrogen, argon, or helium are used as the carrier gas or the mobile phase in GC. The mobile phase is responsible for the elution of components from the column to the detector. According to the nature of the analyte, a variety of columns are used that include an open tubular column, also known as a capillary column, and a packed column that consists of polymeric substance as a stationary phase, such as diatomaceous earth [36]. The time at which the compound enters the detector with the mobile phase, under a definite pressure and temperature, is specific for a particular compound and is known as the retention time. The various types of detectors that are used in gas chromatography, along with their detector gas, are summarised in Table **1** [37, 38]. The components of the GC-MS/MS system have been shown in Fig (**5**).

The principle of GC is that the greater the ability of the stationary phase to bind with the targeted analyte, the more is the retention of the compound in the column, leading to longer elution and detection. GC-MS technique allows determining the structure of the compound by combining the retention time and elution pattern for a compound in a sample, along with its mass spectra. In GC-

MS once the compound is isolated from its matrix by gas chromatography, it reaches the ionization chamber, along with the mobile phase. Inside the ionization chamber, the compound gets ionized on bombardment with fast-moving electrons which produces various fragment ions of that compound. These ions are then subjected to mass analysis and are detected by the electron multiplier, which is then processed and a mass spectrum for a particular compound is produced. The mass spectroscopy produces a mass to charge ratio (m/e) for each ion which helps to determine the molecular weight of the compound [36].

**Table 1. Detectors used in GC along with their limit of detection.**

| S.No. | Detector | Detector Gas | Nature of Detected Compound |
|---|---|---|---|
| 1 | Flame ionisation detector (FID) | Hydrogen & air | Organic compounds, except formaldehyde and formic acid |
| 2 | Thermal conductivity detector (TCD) | Not require | All compounds, except the carrier gas |
| 3 | Barrier discharge ionization detector (BID) | Helium | All compounds, except helium and neon |
| 4 | Electron capture detectors (ECD) | Nitrogen | Halogenated and metallic organic compound |
| 5 | Flame Thermionic Detectors (FTD) | Hydrogen & air | Organic compounds containing nitrogen and phosphorus, Inorganic compounds containing phosphorus |
| 6 | Flame Photometric Detectors (FPD) | Hydrogen & air | Organo-tin compounds, Inorganic sulphur and phosphorus compound |
| 7 | Sulphur Chemiluminescence Detectors (SCD) | Hydrogen & oxygen | Inorganic and organic sulphur compounds |

ECD, FTD, FPD and SCD are considered highly selective and sensitive detectors. However, the combination of mass spectroscopy with gas chromatography is found to be more potent in comparison with other detectors [38].

Jolanta D Cebska *et al.* explained the analysis of several pharmaceutical compounds, such as chloramphenicol, diazepam, diclofenac, hydroxyibuprofen, ibuprofen, carbamazepine, carboxyibuprofen, acetylsalicylic acid, salicylic acid, metoprolol, naproxen, and primidone in water samples, by GC–MS technique. The method involves the pretreatment of the sample by filtration and acidification, followed by enrichment with the SPE method using C-18 column and methanol, and ether as the eluting solvent. The analysis of the pharmaceutical compounds involves the step of derivatization with $CH_2N_2$ (diazomethane), TMSH (trimethylsulfonium hydroxide) and MSTFA (N-Methyl-N-(trimethylsilyl) trifluoroacetamide) [13].

## Analysis of Pharmaceutical Compounds by Liquid Chromatography (LC)

LC or HPLC, coupled with mass spectroscopy, is another analytical approach that is used for the separation, identification, and quantification of chemical compounds from a variety of samples. In recent years, it has emerged as an important technique for the quantification of pharmaceutical residues in environmental samples. HPLC consists of a) a column with the stationary phase, b) a reservoir that stores the mobile phase, c) a pump that assists with the movement of the mobile phase into the column, d) an injection port to inject the sample into the mobile phase, and e) a detector which is responsible to analyse detect the analyte. The column is made up of stainless steel packed with the stationary phase. The mobile phase used in LC is a mixture of organic and aqueous solvents, based on the nature of the compound to be analysed. The injection of the sample takes place through an injection port into the mobile phase which is supplied by a high-pressure pump. Then the sample, along with the mobile phase, enters the column containing the stationary phase and gets separated onto the column. The separation mechanism depends on the nature of the stationary phase used and the physicochemical properties of the analyte. It can work on the mechanism of adsorption, partition, ion pair, ion exchange or size exclusion/gel permeation [38].

Detectors that are used in combination with liquid chromatography for the analysis of pharmaceutical compounds are- UV-VIS detector, photodiode array detector (PDA), fluorescence detector (FLD), conductometric and colourimetric detector, mass detector, and evaporative light scattering detector (ELSD). For the analysis of pharmaceuticals in surface and wastewater, diode array UV absorbance and fluorescence detection are cost-effective. However, the detection through mass-spectroscopy becomes more reliable with its combination with the analyser, such as orbitrap analyser, quadrupole time-of-flight (QTOF) and High-Resolution Mass Spectrometry (HRMS). Orbitrap analyser provides better resolution with high accuracy, however, its scanning speed decreases at a resolution above 100,000 due to which it cannot be coupled with the ultra HPLC. Orbitrap analyser is found useful for the analysis of nitrosamines in wastewater and the identification and quantification of targeted analytes. The screening of antibiotic compounds is done by using the QTOF. HRMS can do the screening of a variety of organic pollutants in the environment, along with their degradation products, offering a high sensitivity and accuracy for targeted and non-targeted compounds [39]. Tandem mass spectroscopy is a more sensitive technique that provides high resolution due to multiple reaction monitoring [22]. In comparison to gas chromatography, liquid chromatography has been proved to be an efficient method for the analysis of non-volatile and thermally labile pharmaceutical compounds. Another advantage of liquid chromatography over gas chromatography is that it does not require the step of derivatization of pharmaceutical compounds, making it less time-consuming. A diagrammatic representation of the LC-MS/MS system is shown in Fig. (6).

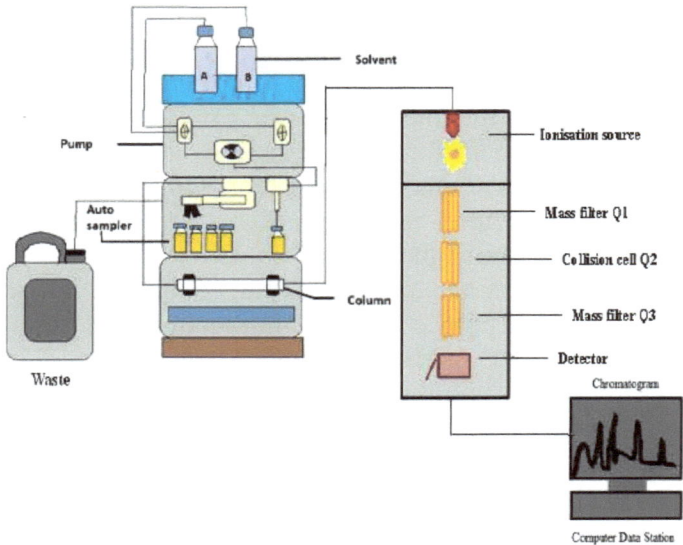

**Fig. (6).** Diagrammatic representation of the LC-MS/MS system.

J. D Cebska *et al.* described the application of HPLC-MS for determining the analytes of pharmaceutical origin, in water samples. The pharmaceutical compounds analysed by employing the LC-MS method were naproxen, ibuprofen, diclofenac, diazepam, carbamazepine, and antibiotics (penicillins, tetracyclines, sulphonamides, macrolide antibiotics, norfloxacin and ciprofloxacin). Sample pre-treatment was done by filtration and acidification followed by enrichment, using the SPE method with a C-18 column and methanol and ether as the eluting solvent, and then analysis over HPLC-ESI-MS [13].

## Methods Used for the Analysis of Pharmaceuticals in Different Environmental Analysis

Previously, various studies have been done to develop an appropriate analytical procedure to determine the nature and concentration of pharmaceutical compounds, in environmental matrices, by employing suitable sample preparation methods and a compatible analytical mode of detection. Several sample preparation methods have been discussed for determining the pharmaceutical residues in different environmental matrices, such as river water, wastewater, soil, sewage, sludge, *etc.*, by optimizing the analytical conditions for their analysis either by LC or GC techniques, in combination with mass spectroscopy, for their better quantification. In the sample preparation techniques, such as SPE, LLE or DLLME, the selection of the sorbent and the extraction solvent plays an important role as it helps to retain the analytes based on their interaction and binding ability to the analytes. Optimizing the conditions such as pH, ionic strength of the sample, temperature or storage conditions also have an impact on the retention of pharmaceutical analytes. In a study with 148 pharmaceutical compounds, it has been observed that most of the compounds have shown effective extraction under acidic conditions while certain benzodiazepines, antibiotics, and NSAIDs require basic conditions for better recovery [39]. Temperature also plays a key role in the effective extraction of analytes. Analytes can degrade thermally at high temperature while at low temperatures they can lose their extraction efficiency; therefore, maintaining an optimum temperature is important as well. LC-MS/MS or GC-MS are highly selective and sensitive analytical instruments that are widely used for the analysis of pharmaceutical residues in the environment. Various parameters in LC and GC which must be optimized to generate a better recovery result for targeted analytes. For GC, an inert carrier gas is responsible for carrying the analyte into the column packed with a stationary phase where, depending on the binding affinity of the stationary phase with the analyte, the separation takes place. However, GC has proved useful for the analysis of volatile and non-polar compounds while for the polar, non-volatile, and thermally stable compounds with high molecular weight, LC is the better option. During the analysis through LC, the properties of the column used, and the nature of the mobile phase are the key factors for the elution of analytes. For example, Atlantis T3 C-18 column (100×2.1 mm, 3 µm) was used for the analysis of 148 pharmaceutical compounds in sewage matrices, however, it had a negative impact on the sensitivity and the peak area of morphine and ecgonine methyl ester (EME), in comparison to the pentafluorophenyl (PFP) column, whereas the PFP column may not be suitable for the lipophilic compounds. For the compounds such as penicillins, beta-blockers, steroids, opioids and opiates, cocaine compounds, antidepressants, anaesthetics, benzodiazepines, amphetamine compounds, and antiepileptics, the use of methanol as the organic component of the mobile phase has shown better utility than acetonitrile [39]. In Table **2**, various sample preparation methods have been summarized along with the suitable chromatographic technique for the analysis of pharmaceutical residues in the environment.

From Table **2** it can be concluded that the SPE is the most widely used extraction and clean-up technique, in combination with LC-MS/MS or GC-MS. However, with recent advancements, the micro-extraction techniques been evolved, such as SPME, SBSE, and LPME that are more cost-effective, less time-consuming, and environment-friendly as compared to LLE and SPE. PLE and USE are extraction techniques that are effective in extracting pharmaceutical residues from solid samples, such as soil, sediments, and sewage sludge.

Table 2. Analytical conditions for determination of pharmaceutical compounds in different environmental matrices.

| S. No. | Pharmaceutical Compounds/ Class of the Pharmaceutical Compound | Environmental Matrices | Sample Preparation Method | Analytical Conditions | Reference |
|---|---|---|---|---|---|
| 1 | Diclofenac, ibuprofen, ketoprofen, indomethacine, naproxen, fenoprofen, clofibric acid, bezafibrate, gemfibrozil, etofibrate, fenofibrate, fenofibric acid, carbamazepine, pentoxifylline, and diazepam | Ground water | SPE sorbent: octadecyl carbon chain-bonded silica eluting solvent: acetone | Instrument: GC/MS Column: DB 35 Mobile phase: Helium | [37] |
| 2 | Sulphanilamide, Cotinine, Acetaminophen, Sulfadiazine, 1,7Dimethylxanthine, Sulfathiazole, Codeine, Sulfamerazine, Lincomycin, Caffeine, Sulfamethizole, Trimethoprim, Thiabendazole, Sulfamethazine, Cefotaxime, Carbadox, Ormetoprim, Norfloxacin, Sulfachloropyridazine Ofloxacin, Ciprofloxacin, Sulfamethoxazole, Lomefloxacin, Enrofloxacin, Sarafloxacin, Clinafloxacin, Digoxigenin, Oxolinic acid, Sulfadimethoxine, Sulfamethoxazole, Diphenhydramine, Penicillin G Azithromycin, Flumeqine, Ampicillin, Diltiazem Carbamazepine, Penicillin V, Erythromycin, Tylosin, Oxacillin, Dehydronifedipine, Digoxin, Fluoxetine, Cloxacillin, Virginiamycin, Clarithromycin, Erythromycin anhydrate, Roxithromycin, Miconazole, Norgestimate | Drinking water, Wastewater, Groundwater | SPE Sorbent: Oasis HLB Eluting solvent: methanol | Instrument: LC-MS/MS Column: C18 Mobile phase: A = 0.3% Formic Acid and 0.1% ammonium formate in distilled water; B = 1:1 acetonitrile: methanol | [37] |

(Table 2) cont.....

| S. No. | Pharmaceutical Compounds/ Class of the Pharmaceutical Compound | Environmental Matrices | Sample Preparation Method | Analytical Conditions | Reference |
|---|---|---|---|---|---|
| 3 | Carbamazepine-10-11-epoxide 10-dihydro-carbamazepine 2-OH-carbamazepine Iminoquinone Acridone Nordiazepam Oxazepam Temazepam | Hospital wastewater | SPE Sorbent: C18 Eluting solvent: methanol | Instrument: LC-MS/MS Column: C18 Mobile phase: A= water + 1.0% formic acid B= methanol + 1.0% formic acid | [40] |
| 4 | Antibiotics, analgesic and/or anti-inflammatory drugs, antiepileptics, benzodiazepines, antipsychotics, and antidepressants, and illicit drugs, including opiates, opioids, cocaine, amphetamines, cannabinoids, and their metabolites | Sewage sludge | USE Extraction solvent: Methanol + water with ultrasonic extraction. | Instrument: LC–MS/MS Column: C18 Mobile phase: For positive detection: A= water + 0.1% formic acid, B= methanol For negative mode: A= water + 1milimolar L-1 ammonium formate, B= Methanol + acetonitrile at constant proportion | [39] |
| 5 | Fluoroquinolones | Wastewater | SPE Sorbent: mixed-phase cation exchange cartridge Eluting solvent: 5% ammonia solution in 15% methanol neutralised with 85% phosphoric acid | Instrument: LC – FLD Column: RP-AmideC16 Mobile phase: A= Phosphoric acid, B= Acetonitrile | [41] |
| 6 | Estrogens | Surface water | IT-SPME Sorbent: fused silica with polyetheretherketone tubing | Instrument: HPLC – MS/MS Column: C8 Mobile phase: 0.01% ammonia/acetonitrile | [42] |
| 7 | carbamazepine, diclofenac, fluoxetine, propranolol, sulfamethazine | Soil and Crops | Extraction – USE Extraction solvent: methanol followed by ultra-sonication Clean-up- SPE Sorbent: Oasis HLB Eluting solvent: methanol | Instrument: LC-MS/MS Column: C18 Mobile phase: A= 0.1% formic acid B= Acetonitrile | [43] |

(Table 2) cont.....

| S. No. | Pharmaceutical Compounds/ Class of the Pharmaceutical Compound | Environmental Matrices | Sample Preparation Method | Analytical Conditions | Reference |
|---|---|---|---|---|---|
| 8 | 14 steroid compounds including natural and synthetic reproductive hormones | Surface water and sediments | LLE Extraction solvent: dichloromethane | Instrument: GC-MS By using capillary column | [44] |
| 9 | Venlafaxine, fluvoxamine, fluoxetine, citalopram and sertraline | Water | SPME Sorbent: PDMS – DVB fibre in a headspace vial | GC-MS | [45] |
| 10 | Clofibric acid, ibuprofen, naproxen, ketoprofen, diclofenac, and triclosan estrone | Soil | Extraction: USE Extraction solvent: combination of acetone with methanol, ethyl acetate, and dichloromethane or methanol with ethyl acetate and dichloromethane or ethyl acetate and dichloromethane followed with ultra-sonication. Clean up: SPE Sorbent: C18 Eluting solvent: ethyl acetate. | Instrument: GC – MS Column: HP – 5MS Mobile phase: Helium | [46] |
| 11 | Clofibric acid, diclofenac, fenoprofen, gemfibrozil, ibuprofen, 2-hydrox--ibuprofen, indomethacin, ketoprofen, naproxen, Clarithromycin, Erythromycin, Oleandomycin, Roxithromycin, Sulfadiazine, Sulfamethazine, Sulfamethoxazole, Sulfapyridine, Trimethoprim | River sediments | Extraction: USE Extraction solvent: methanol, acetone and ethyl acetate. Clean-up: SPE Sorbent- Oasis MCX Eluting solvent: acetone (acidic pharmaceuticals) methanol (antibiotics) | Instrument: LC-MS/MS Column: C18 Mobile phase: For acidic pharmaceuticals – A= acetonitrile B= water acidified with acetic acid For antibiotics – A=ammonia solution acidified with acetic acid + acetonitrile B= 200ml eluent A + 800ml acetonitrile | [47] |
| 12 | Tetracyclines | Groundwater | SPE Sorbent: C18 and Oasis HLB Eluting solvent: methanol | Instrument: LC-MS/MS Mobile phase: A= water + 5% formic acid + acetonitrile B= methanol | [48] |

(Table 2) cont.....

| S. No. | Pharmaceutical Compounds/ Class of the Pharmaceutical Compound | Environmental Matrices | Sample Preparation Method | Analytical Conditions | Reference |
|---|---|---|---|---|---|
| 13 | Oxytetracycline, chlortetracycline, sulfadiazine, erythromycin, tylosin | Soil | Extraction: PLE Using extraction buffer- 50% methanol + 50% 0.2M citric acid Clean up: SPE Sorbent: strong anion exchange and oasis HLB Eluting solvent: methanol | Instrument: LC-MS/MS Column: C18 | [49] |
| 14 | Sulphonamide antibiotics | Wastewater | Extraction: SPME Sorbent- polydimethylsiloxane (PDMS), carbowax/divinylbenzene (CW/DVB), carbowax-templated resin (CW/TPR), polyacrylate (PA), and PDMS/DVB. Desorbing solvent- methanol Clean up: SPE cartridges followed with methanol + 5% ammonium hydroxide as the eluting solvent. | Instrument: LC-MS/MS Column: C18 Mobile phase: A= ammonium acetate + 0.1% formic acid B= ammonium acetate in 2:1 acetonitrile: Methanol | [50] |
| 15 | Paracetamol, naproxen, diclofenac, antipyrine, propranolol, carbamazepine | River water and waste water | SBSE Sorbent: poly(VPD-c--DVB) Desorbing solvent: methanol | Instrument: LC-MS/MS Column: C18 Mobile phase: water + acetonitrile adjusted to (pH)= 3 with formic acid | [51] |
| 16 | Acidic pharmaceuticals | Wastewater | HF-LPME Extraction phase: tubular polypropylene membranes loaded with 1- octanol (organic solvent) | Instrument: LC – MS/MS Column: phenyl –.hexyl column Mobile phase: A= 20% Methanol B= 95% methanol Both containing 10 mM Tributylamine and 0.5% acetic acid | [52] |

*(Table 2) cont.....*

| S. No. | Pharmaceutical Compounds/ Class of the Pharmaceutical Compound | Environ mental Matrices | Sample Preparation Method | Analytical Conditions | Reference |
|---|---|---|---|---|---|
| 17 | Clotrimazole | River water and sewage effluent | DLLME Extraction solvent: trichloroethylene Dispersive solvent: ethanol | Instruments: LC-MS/MS Column: C18 Mobile phase: A= 5 mM Ammonium formate + water B = Methanol | [53] |

## CONCLUDING REMARKS

Pharmaceutical compounds in the environment are considered as potential pollutants causing harm to life on earth. Therefore, it is important to analyse the concentration of pharmaceuticals in different environmental samples to eliminate or minimize their harmful effect on the environment and living beings. With recent advancements in the analytical field, it has become possible to detect trace amounts of pharmaceuticals in different compartments of the environment. Chromatographic techniques, such as gas chromatography and liquid chromatography, coupled with mass spectroscopy and various sample preparation techniques are effective for pharmaceutical residue analysis in environmental matrices. No single technique is superior and hence the choice of an appropriate technique for analysis and sample preparation depends on the nature of the sample and the analytes that are to be analysed.

## CONSENT FOR PUBLICATION

Not Applicable.

## CONFLICT OF INTEREST

The author confirms that this chapter contents have no conflict of interest.

## ACKNOWLEDGEMENT

Declared none.

## REFERENCES

[1] Heberer T. Occurrence, fate, and removal of pharmaceutical residues in the aquatic environment: a review of recent research data. Toxicol Lett 2002; 131(1-2): 5-17.
[http://dx.doi.org/10.1016/S0378-4274(02)00041-3] [PMID: 11988354]

[2] Hernando MD, Mezcua M, Fernández-Alba AR, Barceló D. Environmental risk assessment of pharmaceutical residues in wastewater effluents, surface waters and sediments. Talanta 2006; 69(2): 334-42.

[http://dx.doi.org/10.1016/j.talanta.2005.09.037] [PMID: 18970571]

[3] Peake B, Braund R, Tong A, Tremblay L. Regulatory practices to control the discharge of pharmaceuticals into the environment 2016.
[http://dx.doi.org/10.1016/B978-1-907568-25-1.00007-4]

[4] Kachhawaha AS, Nagarnaik PM, Labhasetwar P, Banerjee K. A Review of Recently Developed LC-MS/MS Methods for the Analysis of Pharmaceuticals and Personal Care Products in Water. J AOAC Int 2020; 103(1): 9-22.
[http://dx.doi.org/10.5740/jaoacint.19-0209] [PMID: 31526434]

[5] Adeel M, Song X, Wang Y, Francis D, Yang Y. Environmental impact of estrogens on human, animal and plant life: A critical review. Environ Int 2017; 99: 107-19.
[http://dx.doi.org/10.1016/j.envint.2016.12.010] [PMID: 28040262]

[6] Kraemer SA, Ramachandran A, Perron GG. Antibiotic Pollution in the Environment: From Microbial Ecology to Public Policy. Microorganisms 2019; 7(6)E180
[http://dx.doi.org/10.3390/microorganisms7060180] [PMID: 31234491]

[7] Mompelat S, Jaffrezic A, Jardé E, Le Bot B. Storage of natural water samples and preservation techniques for pharmaceutical quantification. Talanta 2013; 109: 31-45.
[http://dx.doi.org/10.1016/j.talanta.2013.01.042] [PMID: 23618138]

[8] Fatta-Kassinos D, Meric S, Nikolaou A. Pharmaceutical residues in environmental waters and wastewater: current state of knowledge and future research. Anal Bioanal Chem 2011; 399(1): 251-75.
[http://dx.doi.org/10.1007/s00216-010-4300-9] [PMID: 21063687]

[9] Mompelat S, Thomas O, Le Bot B. Contamination levels of human pharmaceutical compounds in French surface and drinking water. J Environ Monit 2011; 13(10): 2929-39.
[http://dx.doi.org/10.1039/c1em10335k] [PMID: 21912785]

[10] Shaaban H, Górecki T. Optimization and validation of a fast ultrahigh-pressure liquid chromatographic method for simultaneous determination of selected sulphonamides in water samples using a fully porous sub-2 μm column at elevated temperature. J Sep Sci 2012; 35(2): 216-24.
[http://dx.doi.org/10.1002/jssc.201100754] [PMID: 22162242]

[11] García-Galán MJ, Garrido T, Fraile J, Ginebreda A, Díaz-Cruz MS, Barceló D. Simultaneous occurrence of nitrates and sulfonamide antibiotics in two ground water bodies of Catalonia (Spain). J Hdrol 2010; 383(1-2): 93-101.
[http://dx.doi.org/10.1016/j.jhydrol.2009.06.042]

[12] Pavlović DM, Babić S, Horvat AJM, Kaštelan-Macan M. Sample preparation in analysis of pharmaceuticals. TrAC 2007; 26(11): 1062-75.

[13] Dębska J, Kot-Wasik A, Namieśnik J. Fate and Analysis of Pharmaceutical Residues in the Aquatic Environment. Crit Rev Anal Chem 2004; 34(1): 51-67.
[http://dx.doi.org/10.1080/10408340490273753]

[14] Rossi DT, Miller KG. Sample preparation methods for the analysis of pharmaceutical materials.Sep Sci Technol. Academic Press 2004; Vol. 5: pp. 165-201.
[http://dx.doi.org/10.1016/S0149-6395(03)80010-7]

[15] Gouveia TIA, Silva AMT, Ribeiro AR, Alves A, Santos MSF. Liquid-liquid extraction as a simple tool to quickly quantify fourteen cytostatics in urban wastewaters and access their impact in aquatic biota. Sci Total Environ 2020; 740139995
[http://dx.doi.org/10.1016/j.scitotenv.2020.139995] [PMID: 32559532]

[16] Ahrer W, Scherwenk E, Buchberger W. Determination of drug residues in water by the combination of liquid chromatography or capillary electrophoresis with electrospray mass spectrometry. J Chromatogr A 2001; 910(1): 69-78.
[http://dx.doi.org/10.1016/S0021-9673(00)01187-0] [PMID: 11263577]

[17] Quigley A, Cummins W, Connolly D. Dispersive Liquid-Liquid Microextraction in the Analysis of

Milk and Dairy Products. Rev J Chem 2016; 20164040165

[18] Padrón ME, Afonso-Olivares C, Sosa-Ferrera Z, Santana-Rodríguez JJ. Microextraction techniques coupled to liquid chromatography with mass spectrometry for the determination of organic micropollutants in environmental water samples. Molecules 2014; 19(7): 10320-49.
[http://dx.doi.org/10.3390/molecules190710320] [PMID: 25033059]

[19] Sun J-N, Chen J, Shi Y-P. Multiple functional ionic liquids based dispersive liquid-liquid microextraction combined with high performance chromatography for the determination of phenolic compounds in water samples. Talanta 2014; 125: 329-35.
[http://dx.doi.org/10.1016/j.talanta.2014.03.013] [PMID: 24840452]

[20] Guan J, Zhang C, Wang Y, Guo Y, Huang P, Zhao L. Simultaneous determination of 12 pharmaceuticals in water samples by ultrasound-assisted dispersive liquid-liquid microextraction coupled with ultra-high performance liquid chromatography with tandem mass spectrometry. Anal Bioanal Chem 2016; 408(28): 8099-109.
[http://dx.doi.org/10.1007/s00216-016-9913-1] [PMID: 27614980]

[21] Park SY, Myung SW. Simultaneous Determination of Nonsteroidal Anti□Inflammatory Drugs in Aqueous Samples Using Dispersive Liquid–Liquid Microextraction and HPLC Analysis. B Korean Chem. Soc 2015; 36: 2901-6.

[22] Buchberger WW. Current approaches to trace analysis of pharmaceuticals and personal care products in the environment. J Chromatogr A 2011; 1218(4): 603-18.
[http://dx.doi.org/10.1016/j.chroma.2010.10.040] [PMID: 21067760]

[23] Pichon V, Bouzige M, Miège C, Hennion M-C. Immunosorbents: natural molecular recognition materials for sample preparation of complex environmental matrices. TrAC 1999; 18: 219-35.
[http://dx.doi.org/10.1016/S0165-9936(98)00120-4]

[24] Chico J, Meca S, Companyó R, Prat MD, Granados M. Restricted access materials for sample clean-up in the analysis of trace levels of tetracyclines by liquid chromatography. Application to food and environmental analysis. J Chromatogr A 2008; 1181(1-2): 1-8.
[http://dx.doi.org/10.1016/j.chroma.2007.12.033] [PMID: 18190920]

[25] Yi LX, Fang R, Chen GH. Molecularly imprinted solid-phase extraction in the analysis of agrochemicals. J Chromatogr Sci 2013; 51(7): 608-18.
[http://dx.doi.org/10.1093/chromsci/bmt024] [PMID: 23537564]

[26] Duan Y-P, Dai C-M, Zhang Y-L, Ling-Chen . Selective trace enrichment of acidic pharmaceuticals in real water and sediment samples based on solid-phase extraction using multi-templates molecularly imprinted polymers. Anal Chim Acta 2013; 758: 93-100.
[http://dx.doi.org/10.1016/j.aca.2012.11.010] [PMID: 23245900]

[27] Togola A, Budzinski H. Multi-residue analysis of pharmaceutical compounds in aqueous samples. J Chromatogr A 2008; 1177(1): 150-8.
[http://dx.doi.org/10.1016/j.chroma.2007.10.105] [PMID: 18054788]

[28] https://www.chromedia.org/chromedia?waxtrapp=npuhcHsHiemBpdmBlIEcCKJ&subNav=abffyDsHiemBpdmBlIEcCtBDF

[29] Moeder M, Schrader S, Winkler M, Popp P. Solid-phase microextraction-gas chromatography-mass spectrometry of biologically active substances in water samples. J Chromatogr A 2000; 873(1): 95-106.
[http://dx.doi.org/10.1016/S0021-9673(99)01256-X] [PMID: 10757288]

[30] Unceta N, Sampedro MC, Abu Bakar NK, Gómez-Caballero A, Goicolea MA, Barrio RJ. Multi-residue analysis of pharmaceutical compounds in wastewaters by dual solid-phase microextraction coupled to liquid chromatography electrospray ionization ion trap mass spectrometry. J Chromatogr A 2010; 1217(20): 3392-9.
[http://dx.doi.org/10.1016/j.chroma.2010.03.008] [PMID: 20362295]

[31] Sánchez-Rojas F, Bosch-Ojeda C, Cano-Pavón JM. A Review of Stir Bar Sorptive Extraction. Chromatographia 2009; 69: 79-94.
[http://dx.doi.org/10.1365/s10337-008-0687-2]

[32] Gilart N, Miralles N, Marcé RM, Borrull F, Fontanals N. Novel coatings for stir bar sorptive extraction to determine pharmaceuticals and personal care products in environmental waters by liquid chromatography and tandem mass spectrometry. Anal Chim Acta 2013; 774: 51-60.
[http://dx.doi.org/10.1016/j.aca.2013.03.010] [PMID: 23567116]

[33] Bratkowska D, Fontanals N, Cormack PAG, Borrull F, Marcé RM. Preparation of a polar monolithic stir bar based on methacrylic acid and divinylbenzene for the sorptive extraction of polar pharmaceuticals from complex water samples. J Chromatogr A 2012; 1225: 1-7.
[http://dx.doi.org/10.1016/j.chroma.2011.12.064] [PMID: 22226555]

[34] Schlüsener MP, Spiteller M, Bester K. Determination of antibiotics from soil by pressurized liquid extraction and liquid chromatography-tandem mass spectrometry. J Chromatogr A 2003; 1003(1-2): 21-8.
[http://dx.doi.org/10.1016/S0021-9673(03)00737-4] [PMID: 12899294]

[35] Pathuri R, Muthukumaran M, Krishnamoorthy B, Nishat A. A review on analytical method development and validation of pharmaceutical technology. Curr Pharma Res 2013; 3: 855.
[http://dx.doi.org/10.33786/JCPR.2013.v03i02.009]

[36] Mani D, Kalpana MS, Patil DJ, Dayal AM. Organic Matter in Gas Shales: Origin, Evolution, and Characterization.Shale Gas. Elsevier 2017; pp. 25-54.
[http://dx.doi.org/10.1016/B978-0-12-809573-7.00003-2]

[37] Caban M, Kumirska J, Białk-Bielińska A, Stepnowski P. Analytical Techniques for Determining Pharmaceutical Residues in Drinking Water - State of Art and Future Prospects. Curr Anal Chem 2015; 12: 1-11.

[38] Niessen WM. Liquid chromatography-mass spectrometry. CRC press 2006.
[http://dx.doi.org/10.1201/9781420014549]

[39] Gago-Ferrero P, Borova V, Dasenaki ME, Thomaidis NS. Simultaneous determination of 148 pharmaceuticals and illicit drugs in sewage sludge based on ultrasound-assisted extraction and liquid chromatography-tandem mass spectrometry. Anal Bioanal Chem 2015; 407(15): 4287-97.
[http://dx.doi.org/10.1007/s00216-015-8540-6] [PMID: 25716466]

[40] de Almeida CA, Oliveira MS, Mallmann CA, Martins AF. Determination of the psychoactive drugs carbamazepine and diazepam in hospital effluent and identification of their metabolites. Environ Sci Pollut Res Int 2015; 22(21): 17192-201.
[http://dx.doi.org/10.1007/s11356-015-4948-y] [PMID: 26139407]

[41] Golet EM, Alder AC, Hartmann A, Ternes TA, Giger W. Trace determination of fluoroquinolone antibacterial agents in urban wastewater by solid-phase extraction and liquid chromatography with fluorescence detection. Anal Chem 2001; 73(15): 3632-8.
[http://dx.doi.org/10.1021/ac0015265] [PMID: 11510827]

[42] Mitani K, Fujioka M, Kataoka H. Fully automated analysis of estrogens in environmental waters by in-tube solid-phase microextraction coupled with liquid chromatography-tandem mass spectrometry. J Chromatogr A 2005; 1081(2): 218-24.
[http://dx.doi.org/10.1016/j.chroma.2005.05.058] [PMID: 16038212]

[43] Carter LJ, Harris E, Williams M, Ryan JJ, Kookana RS, Boxall AB. Fate and uptake of pharmaceuticals in soil-plant systems. J Agric Food Chem 2014; 62(4): 816-25.
[http://dx.doi.org/10.1021/jf404282y] [PMID: 24405013]

[44] Kolpin DW, Furlong ET, Meyer MT, et al. Pharmaceuticals, hormones, and other organic wastewater contaminants in U.S. streams, 1999-2000: a national reconnaissance. Environ Sci Technol 2002; 36(6): 1202-11.

[http://dx.doi.org/10.1021/es011055j] [PMID: 11944670]

[45] Lamas JP, Salgado-Petinal C, García-Jares C, Llompart M, Cela R, Gómez M. Solid-phase microextraction-gas chromatography-mass spectrometry for the analysis of selective serotonin reuptake inhibitors in environmental water. J Chromatogr A 2004; 1046(1-2): 241-7.
[http://dx.doi.org/10.1016/j.chroma.2004.06.099] [PMID: 15387194]

[46] Xu J, Wu L, Chen W, Chang AC. Simultaneous determination of pharmaceuticals, endocrine disrupting compounds and hormone in soils by gas chromatography-mass spectrometry. J Chromatogr A 2008; 1202(2): 189-95.
[http://dx.doi.org/10.1016/j.chroma.2008.07.001] [PMID: 18639882]

[47] Löffler D, Ternes TA. Determination of acidic pharmaceuticals, antibiotics and ivermectin in river sediment using liquid chromatography-tandem mass spectrometry. J Chromatogr A 2003; 1021(1-2): 133-44.
[http://dx.doi.org/10.1016/j.chroma.2003.08.089] [PMID: 14735982]

[48] Zhu J, Snow DD, Cassada DA, Monson SJ, Spalding RF. Analysis of oxytetracycline, tetracycline, and chlortetracycline in water using solid-phase extraction and liquid chromatography-tandem mass spectrometry. J Chromatogr A 2001; 928(2): 177-86.
[http://dx.doi.org/10.1016/S0021-9673(01)01139-6] [PMID: 11587336]

[49] Jacobsen AM, Halling-Sørensen B, Ingerslev F, Hansen SH. Simultaneous extraction of tetracycline, macrolide and sulfonamide antibiotics from agricultural soils using pressurised liquid extraction, followed by solid-phase extraction and liquid chromatography-tandem mass spectrometry. J Chromatogr A 2004; 1038(1-2): 157-70.
[http://dx.doi.org/10.1016/j.chroma.2004.03.034] [PMID: 15233531]

[50] Balakrishnan VK, Terry KA, Toito J. Determination of sulfonamide antibiotics in wastewater: a comparison of solid phase microextraction and solid phase extraction methods. J Chromatogr A 2006; 1131(1-2): 1-10.
[http://dx.doi.org/10.1016/j.chroma.2006.07.011] [PMID: 16879830]

[51] Bratkowska D, Marcé RM, Cormack PAG, Borrull F, Fontanals N. Development and application of a polar coating for stir bar sorptive extraction of emerging pollutants from environmental water samples. Anal Chim Acta 2011; 706(1): 135-42.
[http://dx.doi.org/10.1016/j.aca.2011.08.028] [PMID: 21995920]

[52] Quintana JB, Rodil R, Reemtsma T. Suitability of hollow fibre liquid-phase microextraction for the determination of acidic pharmaceuticals in wastewater by liquid chromatography-electrospray tandem mass spectrometry without matrix effects. J Chromatogr A 2004; 1061(1): 19-26.
[http://dx.doi.org/10.1016/j.chroma.2004.10.090] [PMID: 15633740]

[53] Zgoła-Grześkowiak A, Grześkowiak T. Application of dispersive liquid-liquid microextraction followed by HPLC-MS/MS for the trace determination of clotrimazole in environmental water samples. J Sep Sci 2013; 36(15): 2514-21.
[http://dx.doi.org/10.1002/jssc.201300271] [PMID: 23720393]

# CHAPTER 9

# Use of Bioisosteric Functional Group Replacements or Modifications for Improved Environmental Health

**Nidhi Singh**[1,*] and **Jaya Pandey**[1]

[1] *Amity University, Lucknow, India*

**Abstract:** Bioisosteres are chemical substituents, groups, atoms, or moieties that have similar physical and chemical properties, producing analogous biological effects but with greater impact and potency. Bioisostere replacement is an impactful concept in medicinal chemistry. Bioisostere replacement is used for attenuation of toxicity, enhancement of the activity of the lead compound, or alterations in pharmacokinetics and toxicity of the lead. This chapter deals with the degradation or minimization of ecotoxic waste through bioisostere replacement. The chapter details bioisosteric replacements for the degradation of eco-hazardous wastes in two ways, *i.e.*, direct way and indirect way. The direct way involves bioisosteric changes in insecticides, which directly affects the environment, while the indirect way involves bioisosteric modifications in drug molecules to increase their bioavailability and half-life period so that maximum drug is consumed within the body, providing better efficacy against the disease and release of a minimum amount of waste into the environment. These modifications prove to be eco-friendly. Some important bioisosteric groups used for replacement are -fluoro, -deutero, -nitro, -t-butyl, and others. This chapter gives an insight into the plausible alterations with improved functional groups in bioisosterism to improve the eco-detrimental effects of compounds or drugs.

**Keywords:** Bioisosteres, Drugs, Ecofriendly, Half-life period, Insecticides, Medicinal chemistry, Metabolites.

## INTRODUCTION

The interactions between molecular interfaces of two or more biomolecules, like receptor (DNAs, RNAs, proteins, peptides, and polysaccharides) and ligand (physiologically active substances) play a major role in molecular recognition, which is important for metabolic events of life action [1]. The specific interaction

---

[*] **Corresponding author Nidhi Singh:** Amity School of Applied Sciences, Amity University, Lucknow, India; E-mail: nidhi.singh23081993@gmail.com

Tahmeena Khan, Abdul Rahman Khan, Saman Raza, Iqbal Azad and Alfred J. Lawrence (Eds.)
All rights reserved-© 2021 Bentham Science Publishers

of binding site and substrate is an important concern for drug targeted action in medicinal chemistry which are often comparable to the 'Lock-and-Key model'. These interactions are attributed to various intermolecular forces, such as hydrogen bonding, ionic bonding, van der Waals force, and dipole interaction, generated due to specific functional groups, atoms, or chemical moieties. Improvement in drug efficacy, *in vivo* stability of drugs, their oral absorption, membrane permeability, and ADME (absorption, distribution, metabolism, excretion) properties can be achieved by bioisosterism. The modification of drug candidates by their corresponding bioisosteres is the best alternative for drug discovery investigations [2, 3]. In the present scenario, a large amount of work is being done for environmental sustainability and thus, greener alternatives are being explored, whether in terms of synthesis or degradation. Bioisosterism *via* computer-aided drug design is one such alternative that serves the purpose of environmentally safe drugs in terms of synthesis as well as degradation [4]. In this chapter, we have discussed bioisostere replacement in chemical compounds with improved functional groups to enhance their degradation characteristics or to reduce their toxicity, for safer and lesser hazardous environment waste. The chapter details bioisosteric replacements for the degradation of eco-hazardous wastes in two ways, *i.e.* direct way and indirect way. The direct way involves bioisosteric changes in agricultural insecticides and pesticides entering the metabolism of insects and pests, while the indirect way involves bioisosteric modifications in drug molecules to increase their bioavailability, half-life period, and *in vivo* stability, so that maximum drug is consumed within the body, providing better efficacy against the disease and release of a minimum amount of waste into the environment.

## BIOISOSTERISM - DIRECT EFFECT ON ENVIRONMENT

Bioisosteric modifications directly affecting the environment include changes or modifications in chemical compounds like insecticides and pesticides, in terms of bioisostere replacement. Synthetic insecticides and pesticides play a major role in the integrated pest management system, limiting pests that are harmful to crop yields. However, long-term usage of conventional synthetic insecticides and pesticides is posing a serious threat to the environment, creating eco-biological problems. Therefore, to overcome this grave threat to the environment, there is a pressing need to search for novel compounds (insecticides/pesticides) which are potent, follow a modified mode of action, bear eco-friendly properties, such as easy biodegradability to non-toxic residues and minimize or cause no metabolic disturbances in human biological systems [5].

## BIOISOSTERIC MODIFICATIONS FOR ANTHRANILIC DIAMIDES

Most of the insecticidal bioisosteric modifications were performed on the established class of compounds; anthranilic diamides. Anthranilic diamides are portrayed through three chemical moieties: A) aromatic bridge amide moiety, B) N-pyridyl-pyrazole moiety and C) terminal aliphatic amide moiety. Bioisosteric modifications were made on these three groups, especially amide functional groups, to produce novel compounds with better efficiency and lesser toxicity [6, 7]. Major work done on this class of insecticides, to reduce eco-detrimental effects of insecticides, has been discussed in detail in this chapter.

## BIOISOSTERIC MODIFICATIONS AT AROMATIC BRIDGED AMIDE FUNCTIONAL GROUP

Wang *et al.* worked on bioisosteric replacements in anthranilic diamides, yielding novel pyridyl pyrazole acid derivatives as a potent alternate insecticide. Anthranilic diamides are an important and potent class of conventional insecticides. The ryanodine receptor is considered to be an important receptor class for insecticidal compounds. Chlorantraniliprole and Cyantraniliprole (Fig. **1**) are important and established anthranilic diamide class of insecticides. These compounds exhibit good insecticidal activity on a broad range of insects such as Lepidopterans, Coleopterans, Dipterans, and Isopterans. Wang and his co-workers explored the bioisosteric replacement of the amide functional group or moiety for improved, novel insecticidal compounds.

**Fig. (1).** Structures of anthranilic diamide class of insecticides.

Bioisosteric changes were made in these two anthranilic diamides to yield potent molecules with greater efficacy. The changes made in the molecule are represented through schemes. The modifications were mainly concerned with benzene moiety or amide terminal functional group or amide bridge-group [8].

Scheme **1** details the bioisosteric replacement alterations at benzene moiety and amide bridge group. The changes of group and moieties are detailed in Scheme **1**. Bioisosteric replacements in these anthranilic diamides (chlorantraniliprole and cyantraniliprole) yield 1-[2-(3-Chloro-pyridin-2-yl)-2*H*-pyrazole-3-carbony-]-3-phenyl-thiourea derivatives. The efficiencies of the novel compounds and established compounds were compared against larvicidal forms. Chlorantraniliprole showed 100% larvicidal activity at 5 mg/L, 10 mg/L, 25 mg/L, 50 mg/L and 100 mg/L, while, compound 1-[5-Bromo-2-(3-chloro-p-ridin-2-yl)-2*H*-pyrazole-3-carbonyl]-3-(2-bromo-4,6—difluoro-phenyl)-thiourea (1c) and 1-[5-Bromo-2-(3-chloro-pyridin-2-yl)-2*H*-pyrazole-3-carbonyl]-3-(-,4,6-trichloro-phenyl)-thiourea (1d) witnessed 100% larvicidal activity at 50mg/L and 100 mg/L, against Oriental Armyworm (*Mythimna separata walker*) and the rest of the compounds displayed lower efficiency. The acyl thiourea modifications made in these compounds have an upper hand in the fact that they have low toxicity for non-target organisms, thus showing the lesser hazardous influence on the environment and more targeted action on the specific organism.

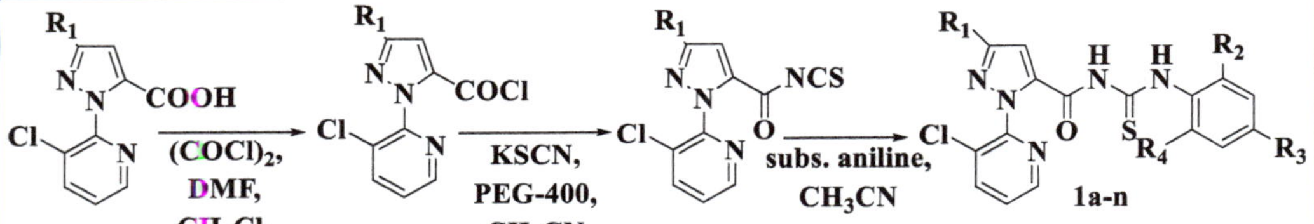


**1a**: $R_1$=Cl, $R_2$=H, $R_3$=$i$-$C_3F_7$, $R_4$=$CH_3$;
**1b**: $R_1$=Br, $R_2$=H, $R_3$=$i$-$C_3F_7$, $R_4$=$CH_3$;
**1c**: $R_1$=Br, $R_2$=F, $R_3$=F, $R_4$=Br;
**1d**: $R_1$=Br, $R_2$=Cl, $R_3$=Cl, $R_4$=Cl;
**1e**: $R_1$=Cl, $R_2$=Br, $R_3$=$CF_3O$, $R_4$=Br;
**1f**: $R_1$=Br, $R_2$=Br, $R_3$=$CF_3O$, $R_4$=Br;
**1g**: $R_1$=Br, $R_2$=Cl, $R_3$=H, $R_4$=$NO_2$;
**1h**: $R_1$=Br, $R_2$=H, $R_3$=$OCH_3$, $R_4$=$NO_2$;
**1i**: $R_1$=Br, $R_2$=H, $R_3$=F, $R_4$=H;
**1j**: $R_1$=Br, $R_2$=H, $R_3$=Br, $R_4$=$CH_3$;
**1k**: $R_1$=Br, $R_2$=H, $R_3$=I, $R_4$=$CH_3$;
**1l**: $R_1$=Br, $R_2$=$CH_3$, $R_3$=$CH_3$, $R_4$=$CH_3$;
**1m**: $R_1$=Br, $R_2$=$CH_3$, $R_3$=H, $R_4$=$C_2H_5$;
**1n**: $R_1$=Br, $R_2$=$C_2H_5$, $R_3$=H, $R_4$=$C_2H_5$.

**Scheme 1.** Synthesis of novel 1-[2-(3-Chloro-pyridin-2-yl)-2*H*-pyrazole-3-carbonyl]-3-phenyl-thiourea derivatives.

Scheme **2** also details bioisosteric replacement alterations at benzene moiety and amide bridge group. The changes of groups and moieties are detailed in the scheme. The efficiencies of the novel substituted 2-(3-Chloro-pyridin-2-yl)-2*H*-pyrazole-3-carbonyl]-thioureido-3-methyl benzamide derivatives and established compounds were compared against various larvicidal forms. Chlorantraniliprole showed 100% larvicidal activity at 0.5 mg/L, 1 mg/L, 2.5 mg/L, 5 mg/L, 10 mg/L, 25 mg/L, 50 mg/L and 100 mg/L, while, all the compounds (2a-2g) of this series witnessed 100% larvicidal activity at 10 mg/L, 25 mg/L, 50 mg/L and 100 mg/L, against Oriental Armyworm (*Mythimna separata walker*) and surprisingly 5-chloro-2-{3—chloro -pyridin-2-yl)-5- (2,2,2- trifluoro-ethoxy) -2H-pyrazole --carbonyl] -thioureido}-3 ,N-dimethyl-benzamide (2a) displayed 100% larvicidal activity at a concentration as low as 0.5mg/L indicating quite high efficiency, and the carbonyl thiourea modification results in low toxicity. Chlorantraniliprole showed 100% larvicidal activity at 1.25 mg/L, 2.5 mg/L and 25 mg/L, while, 5-Bromo-2-{3-[2-(3 -chloro-pyridin-2-yl) -5-(2,2,2-trifluoro- ethoxy)-2H-pyrazoe-3- carbonyl]-thioureido}-N-cyclopropyl-3-methyl-benzamide (2g) witnessed 100% larvicidal activity at 1.25 mg/L, 2.5 mg/L and 25 mg/L, against Diamondback moth (*Plutella xylostella L.*), while 5-chloro-2-{3—chloro-pyrdin-2-yl)-5-(2,2,2-trifluoro-ethoxy)-2H-pyrazole-3- carbonyl]-thioureido} -3,--dimethyl -benzamide (2a) displayed 100% larvicidal activity at 0.125 mg/L, 1.25 mg/L, 2.5 mg/L and 25 mg/L; and N-tert-Butyl-5-chloro-2-{3-[2--3-chloro-pyridin-2-yl)-5-(2,2,2-trifluoro-ethoxy)-2H-pyrazole-3-carbonyl] -thiou reido}-3-methyl-benzamide (2e) showed 100% larvicidal activity at 0.1mg/L, 0.125 mg/L, 1.25 mg/L, 2.5 mg/L and 25 mg/L and surprisingly 5-Chloro-2-{3-[2-(3-chloro-pyridin-2-yl)-5-(2,2,2-trifluoro-ethoxy)-2H-pyrazole-3-carbonyl]- thioureido}- N-isopropyl-3-methyl-benzamide (2c) displayed 100% larvicidal activity at 0.01mg/L, 0.1mg/L, 0.125 mg/L, 1.25 mg/L, 2.5 mg/L and 25 mg/L indicating higher efficiency than the standard compound (chlorantraniliprole) against Diamondback moth (*Plutella xylostella L.*), and the carbonyl thiourea modification resulted in low toxicity, thus, displaying a perfect combination of higher efficacy and lower toxicity against larval forms of Diamondback moth. Chlorantranilipore displayed 80% larvicidal activity at 1 mg/L and 10 mg/L against Beet Armyworm (*Laphygma exigua hübner*), while, from this series N-tert-Butyl-5 -chloro-2-{3-[2- (3-chloro-pyridin -2-yl)-5-(2,2, 2-trifluoro-ethoxy)-2H-pyrazole- 3-carbonyl]- thioureido}- 3- methyl-benzamide (2e) showed 80% larvicidal activity at 10 mg/L. Both the compounds did not show much effect for this larval form, although activity of the synthesized compound was found to be comparable to the marketed insecticide. Besides this, chlorantraniliprole displayed 100% larvicidal activity at 25 mg/L against Corn Borer (*O. nubilalis*), while, 5-chloro-2-{3—chloro-pyridin-2-yl)---(2,2,2-trifluoro-ethoxy)-2H-pyrazole -3-carbonyl]-thioureido}-3,N- dimethyl-benzamide (2a), 5-Chloro-2-{3-[2-(3-chloro- pyridin-2-yl)-5-(2,2, 2-trifluoro-ethoxy)-2H -pyrazole-3 -carbonyl]- thioureido}-N-isopropyl-3-methyl-benzamide (2c) and N-tert-Butyl-5-chloro-2-{3-[2-(3-chloro- pyridin-2-yl)-5-(2,2,2-trifuoro-ethoxy) -2H-pyrazole-3-carbonyl]- thioureido}-3-methyl-benzamide (2e) also showed 100% larvicidal activity at 25 mg/L, showing justifiably relatable values for established and novel compounds with a superseding feature of low toxicity, in the novel compounds [8].

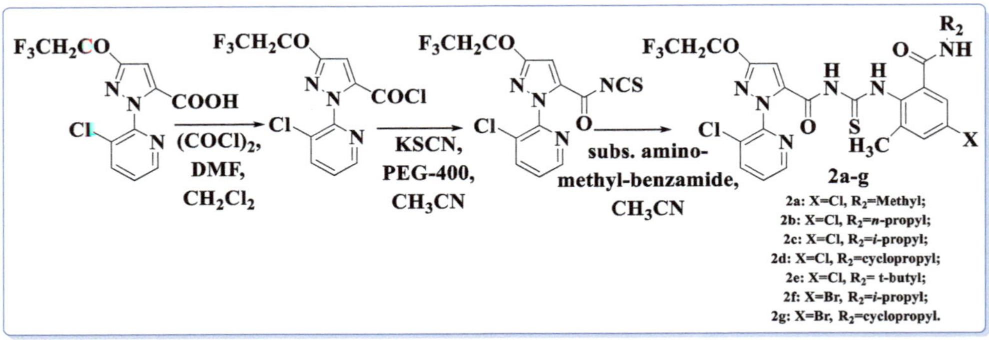

**Scheme 2.** Synthesis of novel substituted 2-(3-Chloro-pyridin-2-yl)-2H-pyrazole-3-carbonyl]-thioure-do-3-methyl benzamide derivatives.

To summarize, it can be stated that the synthesized novel compounds, with alterations at bridge amide group by carbonyl urea and alterations for terminal trifluoroethoxy group, have shown considerable activity at low concentrations (low toxicity) against various larval forms of agricultural pests. The higher efficiency and lower toxicity of these compounds prove that these novel insecticides formed by bioisostere replacements are an achievement for preventing environmental toxicity, as their low dosage is potent to eliminate the larva, and their target-specific approach of action prevents its effects in non-targeted objects.

## BIOISOSTERIC MODIFICATIONS AT ALIPHATIC AMIDE FUNCTIONAL GROUPS

Liu *et al.* worked on bioisosteric modifications at terminal aliphatic amide functional groups. Recent works on thiadiazole and oxazoline displayed good results for larvicidal activities, so bioisosteric modification at aliphatic amide functional group with 1,3,4-oxadiazole group was considered for greater efficiency of compounds. The other modifications such as thienyl, pyridyl and pyrazinyl moieties attached to the oxadiazole moiety were responsible for enhancing insecticidal activity [9]. Modifications were also made at the bridge amide group, but they did not yield any significant results in terms of larvicidal activity.

Scheme **3** gives the synthetic route of Pyridinyl-pyrazolo-substituted-[1,-,4-oxadiazolyl]-phenyl-amide derivatives, and also demonstrates introduction of oxadiazole moiety at the terminal aliphatic amide position. These bioisosteric modifications gave some good results for low toxicity compounds. 3-(3-Chloo-pyridin-2-yl)-4-methyl-pyrazole-1-carboxylic acid {4-bromo-2- methyl-6--

5-(2H- 1-thiophen-1-yl)- [1, 3, 4]oxadiazol-2- yl]-phenyl}- amide (3a) and 3-(-
-Chloro- pyridin-2-yl)-4-methyl- pyrazole-1-carboxylic acid [4-bromo-2-methyl -
6-(5-phenyl- [1, 3, 4]oxadiazol- 2-yl)-phenyl]- amide (3b) showed 100%
larvicidal activity at 100µg/ml, while 3-(3-Chloro-pyridin- 2-yl)-4-methyl -
pyrazole-1-carboxylic acid [4-bromo-2-methyl-6-(5-pyrazin-2-yl- [1, 3,
4]oxadiazol-2-yl)- phenyl]-amide (3d), 3-(3-Chloro-pyridin-2-yl)-4- methyl-
pyrazole-1-carboxylic acid [4-chloro-2-methyl-6-(5-pyridin-4-yl- [1, 3,
4]oxadiazol-2-yl)-phenyl]-amide (3g) and 3-(3-Chloro-pyridin-2 -yl)-4-methyl-
pyrazole-1- carboxylic acid [4-chloro-2-methyl-6-(5-pyrazin-2-yl- [1, 3,
4]oxadiazol-2-yl)-phenyl]-amide (3h) displayed 100% larvicidal activity only at
minimum concentration of 40µg/ml. This indicated the importance of these
bioisosteric functional group replacements for efficient and low toxicity
insecticides [10].

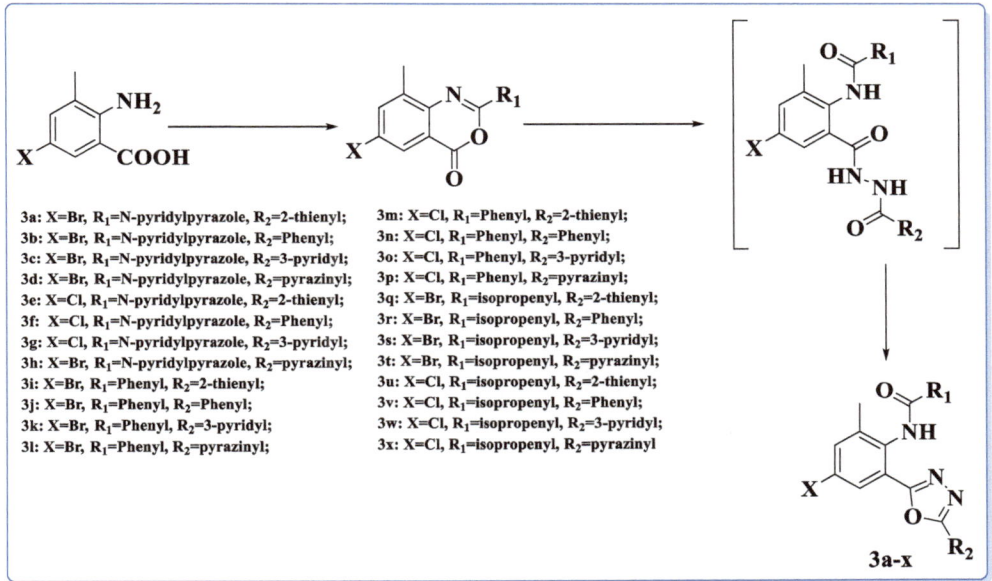

**Scheme 3.** Synthesis of novel substituted Pyridinyl-pyrazolo-substituted-[1,3,4-oxadiazolyl]-phenyl-amide derivatives.

Liu *et al.* further expanded their work on bioisosteric modifications at terminal aliphatic amide groups. As the aliphatic amide was converted into 1,3,4-oxadiazine ring, further modifications were made in groups associated with it. The bioactivity of these novel compounds was tested against Diamondback moth (*Plutella xylostella)*, for its consideration as an insecticide.

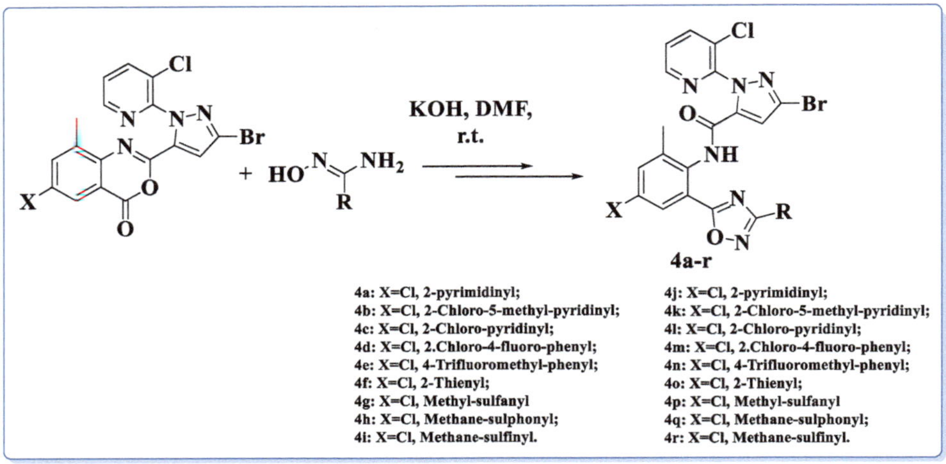

**Scheme 4.** Synthesis of novel substituted Pyridinyl-pyrazolo-substituted-[1,3,4-oxadiazolyl] derivatives.

All the compounds of the Scheme **4** possessed 100% larvicidal activity against the Diamondback moth at 100 mg/L concentration. Compounds 4a, 4b, 4j, and 4k showed 100% larvicidal activity at 10 mg/L, while, compounds 4g, 4h, 4i, and 4p displayed 100% larvicidal activity at 5 mg/L concentration and compound 4r exhibited 100% larvicidal activity at a concentration as low as 2.5 mg/L. Thus, some of the novel compounds were found to exhibit low toxicity, owing to the lesser concentration potency of these molecules [10].

## BIOISOSTERIC MODIFICATIONS FOR ORGANOCHLORINES

Martins *et al.* worked on the bioisosteric replacements of organochlorine compounds. They suggested fluorine as a bioisosteric replacement of chlorine yielding organofluorine compounds instead of organochlorine compounds (Fig. **2**). The basis of this work was the hazardous nature of organochlorine insecticides. They were found to be harmful to human and animal health as well as the environment. Organochlorine compounds have a lipophilic character, which augments the persistence of these compounds in soil and water, and due to the relative ease of C-Cl bond cleavage, the compounds are likely to be absorbed through the plants and enter the food chain through it [11]. Due to these hazardous effects, these insecticides were banned in some countries and these novel organofluorine compounds can be a suitable alternative for them, as they have lower lipophilicity and reduced steric effects with relatively strong C-F bond so that the compound does not enter the food chain [12, 13].

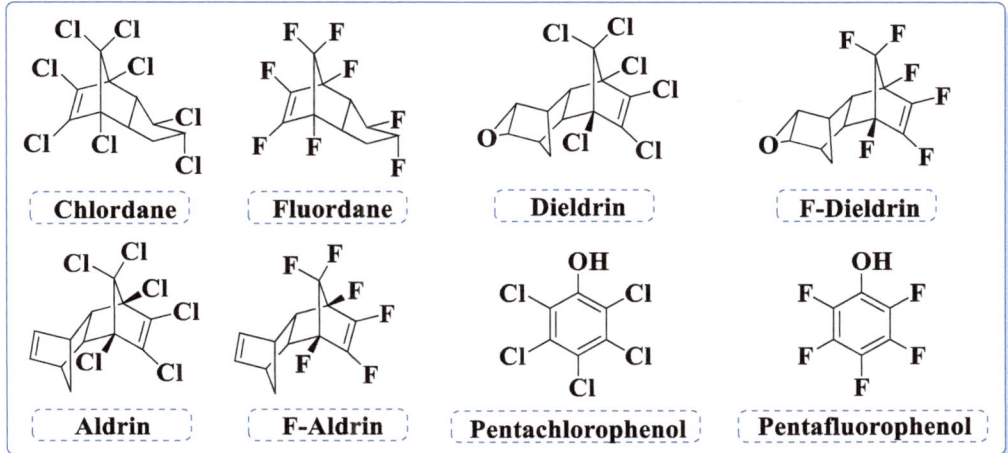

**Fig. (2).** Structures of organochlorine insecticides and their bioisosterically replaced organofluorine compounds.

The chlorine group in organochlorine insecticides like chlordane, dieldrin, aldrin, and pentachlorophenol was bioisosterically replaced by fluorine to form fluordane, F-dieldrin, F-aldrin, and pentafluorophenol. Fig. **1** represents the structures of both the established insecticide and the bioisosterically replaced insecticide. Log $LC_{50}$ of the proposed molecule was calculated against rainbow fish *(Poicilia reticulata)*, predicting the lethal concentration for the compounds at which they killed 50% of the organisms [14].

The log $LC_{50}$ values for chlordane, dieldrin, aldrin, and pentachlorophenol were established as -2.8, -2.6, -2.4, and -0.3, respectively, while, it was found to be 0.8, 0.1, 0.1 and 2.2 for fluordane, F-dieldrin, F-aldrin, and pentafluorophenol, respectively. The lower log $LC_{50}$ values indicate greater bioactivity, showing higher bioactivities for organochlorine compounds, but their hazardous effects overshadow this feature, while from amongst bioisosterically replaced organofluorine compounds, F-dieldrin and F-aldrin were found to possess good bioactivities. The bioisosteric switch of chlorine with fluorine in organochlorine compounds may be beneficial in all ways, either in case of reduced side effects on human health or in case of minimized damaging effects on the environment [14].

## BIOISOSTERISM - INDIRECT EFFECT ON ENVIRONMENT

Bioisosteric modifications in drugs exert an indirect effect on environmental health. Besides treating human health, the metabolic waste of drugs also affects the health of the environment. Therefore, bioisosteric modifications in drugs are required so that the metabolized waste of drug molecules effused into the environment is less toxic, more degraded or less susceptible to degradation under environmental conditions. Thus, this is the indirect effect of bioisosteric modifications of medicines or drugs on the environment.

## BIOISOSTERIC MODIFICATIONS FOR DIARYLPYRIMIDINE DERIVATIVES

Kang *et al*. worked on bioisosteric replacements in diarylpyrimidines functioning as non-nucleoside reverse transcriptase inhibitors (NNRTIs). In this work, benzenesulphonamide functional group of the lead compound was bioisosterically replaced with a fluorine group to reduce activity for the hERG ion channel and improve the half-life period of drug candidates. The increased half-life periods of the drug candidates indirectly affect the release of toxic waste into the environment, since the longer the half-life, the more metabolically stable will be the compound, thereby increasing the positive effects of the drug in humans and the negative effects on the environment [15, 16].

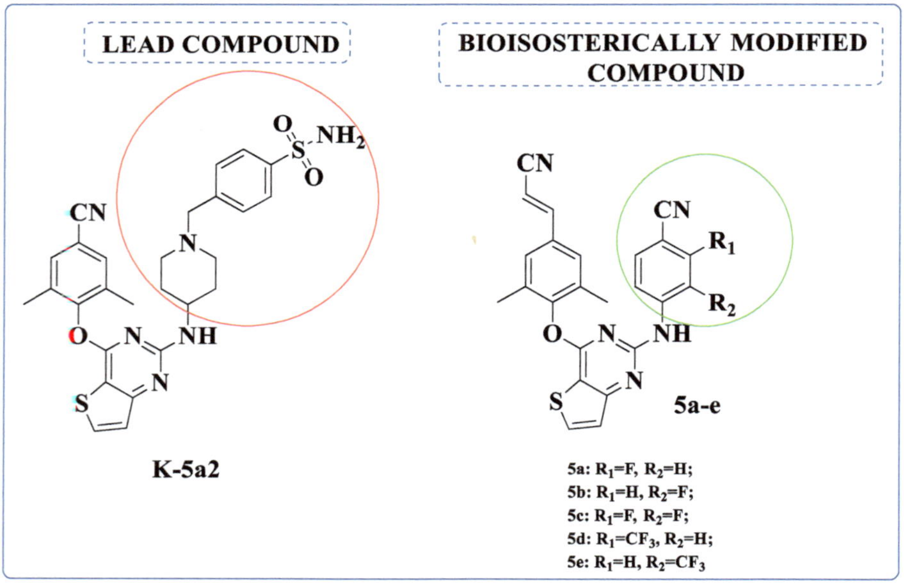

**Scheme 5.** Structures of diarylpyrimidine NNRTIs and their bioisosterically modified fluorine derivatives.

The lead compound selected for this work was K-5a2. The benzenesulphonamide group of this lead was replaced with fluorine (Scheme **5)**. For NNRTIs treatment of Human Immunodeficiency Virus (HIV), binding hERG channels can be a cause of one of the major side-effects, tachyarrhythmia. The flexible piperidine-linked benzenesulphonamide motif, containing a tertiary amine group and a hydrogen acceptor (present in K-5a2), displayed the most favourable characteristics for hERG inhibitory activities. To overcome this concern, a series of fluorine substituted diarylpyrimidine derivatives were synthesized *via* molecular hybridization and bioisosterism strategies. Some modifications were also made in the central core and terminal phenyl cyanide group of the compound for enhancing the bioactivity of the compound.

The synthesized derivatives were tested for their bioactivity and half-life periods. Out of the entire series, Compound 5b was found to be the most active inhibitor, possessed broad-spectrum activity ($EC_{50}$=3.60-21.5nM), and lower cytotoxicity ($CC_{50}$=155μM), and reduced hERG activity ($IC_{50}$>30 μM). Crystallographic studies also demonstrated good binding for the same. The half-life period of compound 5b, when compared with the lead compound (K-5a2) was found to be 80.1 ($T_{1/2}$ min.) against 2.4 ($T_{1/2}$ min.) for K-5a2, which is much higher than the lead compound's half-life, indicating metabolically stable compounds having higher efficacy and low toxicity towards the environment [17].

## BIOISOSTERIC MODIFICATIONS FOR CARBOHYDRATES

Carbohydrates are related to the treatment of several diseases, including cancer. Besides, they also play an important role in immune regulation and suppression. Owing to their high significance, they form an important class of targets for therapeutic development. But native carbohydrates possess certain drawbacks like possessing lower binding affinities, majorly due to weak interactions, susceptibility to hydrolysis in both acidic pH of gut or endogenous glycosidases, lower oral bioavailability due to passive permeation through intestinal enterocyte layer, owing to its high polar surface area. Besides this, poor binding kinetics and rapid renal removal of the compounds from the body are also major barriers for carbohydrates as drugs, both for health and the environment [18, 19]. As rapid renal removal suggests more toxic waste (less degraded) is released into the environment.

To overcome these innate poor properties of carbohydrates, synthetic modifications through glycomimetics and bioisosterism were done, to improve their therapeutic potential. These modifications render improved binding affinity and selectivity for carbohydrates. Bioisosteric modifications are incorporated in carbohydrates to improve binding affinity by removing or exchanging hydroxyl groups or introducing hydrophobic fragments to reduce polar surface area. For improvement of pharmacokinetic properties like better oral bioavailability and longer half-life periods, bioisosteric functional group replacement of *O*-glycosidic atom of carbohydrate with more stable atoms as *C-, S-, N-etc.* is encouraged [20, 21]. Thus, bioisosteric modifications can potentially improve both the binding affinity, like enhanced metal chelation and pharmacokinetic parameters for improved oral bioavailability and longer serum half-life.

A summary of bioisosteric replacements in carbohydrates and their basis for replacement is depicted in Table 1.

Table 1. Summary representing isosteric replacements in carbohydrates and their effects.

| Functional Group | Bioisosteric Replacement | Rationale |
|---|---|---|
| O-Glycosidic linkage | N-,C-,S-,Se-Glycosides | Susceptibility to *in vivo* chemical/enzymatic hydrolysis is reduced. |
| Endocyclic O atom | Imino-, thio-, carbasugars, phostones, phostines | Enhanced stability, reduced polar surface area, improved serum half-life, iminosugars mimic charged oxocarbenium transition state. |
| -OH | Deoxygenation | Reduced polar surface area, increased hydrophobic contacts with proteins improving passive permeation |
| -OH | Deoxyfluorination | Reduced polar surface area and increased H-bond acceptor ability. |
| -OH | Methyl etherification | Reduced polar surface area and improved binding affinity |
| -OH | SH/ SeH substitution | Reduced polar surface area, enhanced atom polarizability, enhanced *pi*-interactions and enhanced affinity |
| -H | Fluorination | Longer serum half-life and chemically inert after release into the environment |
| NHAc | C-,N- derivatives | Enhanced metal chelation and improved bioconjugation |
| $CO_2^-$ | Amide, sulphonate, phosphonate | Reduced polar surface area and enhanced charged protein interactions |

These bioisosteric modifications in carbohydrates lead to the improved therapeutic potential of carbohydrates, with enhanced pharmacokinetic and pharmacodynamic properties. Some of these bioisosteric replacements are depicted structurally through Fig. 3.

These bioisosteric replacements in carbohydrates increase their serum half-life period and binding affinity, thereby leading to better metabolic degradation and slow removal from the body. This factor indirectly affects the environment as a more degraded chemical release in the environment means a less hazardous pollutant. Bioisosteric replacements like fluorine modifications render the molecule to be chemically inert, which even on release into the environment do not pose any ecological threat [22].

**Fig. (3).** Bioisosteric replacements in Carbohydrates.

## SOME IMPORTANT EXAMPLES OF BIOISOSTERIC FUNCTIONAL GROUP MODIFICATIONS FOR IMPROVED ENVIRONMENT

### 1. Ivacaftor

Ivacaftor or Kalydeco is used for the treatment of cystic fibrosis in the cystic fibrosis transmembrane conductance regulator (CFTR) gene. It is extensively metabolized into hydroxy and carboxylate metabolites in the human body, which are rapidly cleared or removed from the body [23]. However, the deuterated analogue of Ivacaftor CTP-656 exhibits reduced metabolism or is metabolically stable, with enhanced half-life and reduced oral clearance of the compound. These improved features of the drug, achieved through functional group bioisosteric replacement, not only affect the health and human biology but also indirectly affect the environment. These metabolically stable drugs, with an enhanced half-life, are gradually released into the environment, providing enough time for degradation of the harmful and metabolized chemical waste [24]. The comparative structures of both, Ivacaftor and its deuterated analogue are displayed in Fig. (4).

**Fig. (4).** Structure of Ivacaftor and its bioisosterically replaced Deuterated analogue (CTP-656).

### 2. Tetrabenazine

Tetrabenazine is a drug used for the treatment of Huntington's disease. It primarily acts as a reversible high-affinity inhibitor of monoamine uptake [25]. The drawback of this drug usage was that it had low bioavailability and a reduced half-life period. To overcome this drawback, its deuterated analogue, deutetrabenazine, was synthesized. Deutetrabenazine was found to exhibit a half-life period twice as that of tetrabenazine [24], thus allowing administration of reduced doses, thereby reducing peak concentration adverse effects on both human health as well as environment. The comparative structures of Tetrabenazine and Deutetrabenazine are displayed in Fig. (5).

Fig. (5). Structure of Tetrabenazine and bioisosterically replaced Deutetrabenazine.

## 3. JNJ-38877605

JNJ-38877605 was a drug used as an antitumor agent, but its usage was terminated due to its *in vivo* renal toxicity, as a result of its aldehyde oxidase (AO) metabolites [26]. The bioisosteric replacement of terminal hydrogen of quinoline moiety with deuterium suggested the reduced appearance of aldehyde oxidase metabolites, thereby decreasing toxicity of the compound [24]. The decreased toxicity positively impacted human health as well as atmospheric health, after releasing from the body into the environment. The comparative structures of JNJ-38877605 and its deuterated analogue are displayed in Fig. (6).

Fig. (6). Structure of JNJ-38877605 and its bioisosterically replaced deuterated analogue.

## 4. SCH-48461

SCH-48461 is a potential cholesterol inhibitor, but the compound underwent cleavage to produce some unwanted metabolites [27]. Besides, the compound had low plasma concentrations indicating rapid removal from the body, relating to lower efficacy in the human body and higher risk to the environment. This drawback was overcome through bioisosteric modification of the compound with fluorine and hydroxyl functional groups. The bioisosterically modified compound,

Ezetimibe, possessed a stronger C-F bond, reducing the risk of cleavage to produce unwanted metabolites and was found to be multiple times more efficient than SCH-48461, indicating enhanced and prolonged plasma concentrations [24], thereby increasing efficacy as drug and reducing toxicity as a pollutant, after removal from the body. The comparative structures of SCH-48461 and its bioisosterically modified analogue are displayed in Fig. (**7**).

**Fig. (7).** Structure of SCH-48461 and its bioisosterically modified analogue-Ezetimibe. 7

## 5. Etofenprox

Etofenprox is a pyrethroid ether insecticide. The drawback associated with this compound was that the C-H bond was not quite strong and underwent cleavage or degradation in soil, producing some products which could be toxic to the environment. Also, the compound was found to be susceptible to oxidative metabolism, generating harmful free radicals in the environment [28]. Thus, etofenprox was bioisosterically modified with fluorine and silicon groups to improve its toxicological profile. The bioisosterically modified compound possessed a strong C-F bond, making the compound equally effective on pests but less toxic to the environment, as the C-F bond makes the compound more inert in soil and does not break down easily to release free radicals [24]. The comparative structures of Etofenprox and its bioisosterically modified analogue are displayed in Fig. (**8**).

## 6. Trifluoromethyl Ketone

Trifluoromethyl ketones are a class of drug compounds that exhibits high lipophilicity and low metabolic activity [29]. Their bioisosterically modified analogues overcame these problems and possessed reduced lipophilicity, increased lipophilic efficiency, and enhanced metabolic stability, indicating improved half-life periods, and the release of less toxic waste into the environment [24]. The comparative structures of Trifluoromethyl ketone and its bioisosterically modified analogue, trifluoromethyloxetane, are displayed in Fig. (**9**).

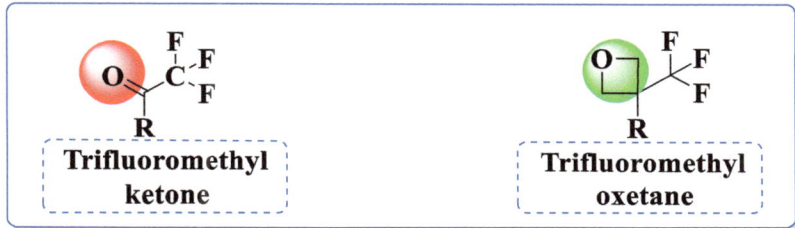

**Fig. (8).** Structure of Etofenprox and its bioisosterically modified analogue-Silafluofen.

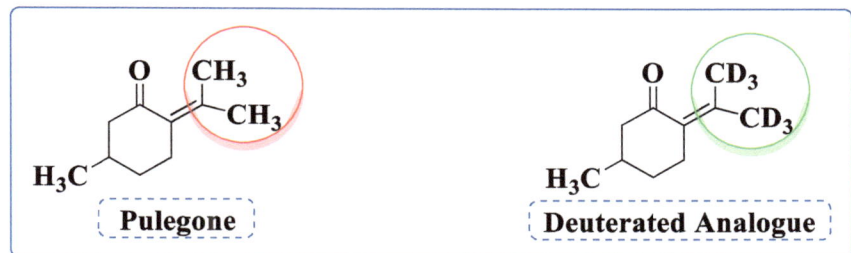

**Fig. (9).** Structure of Trifluoromethyl ketone and its bioisosterically modified analogue-Trifluoromethyl oxetane.

**Fig. (10).** Structure of Pulegone and its Deuterated analogue.

## 7. Pulegone

Pulegone is used for the treatment of seizures and aromatherapy. But the compound was found to be highly hepatotoxic *in vivo* and if used in conjugation with other oils [30], its hepatotoxic effect is reduced, as it is excreted easily, but then it raises toxicity risks to the environment. As a solution to this problem, perdeuteration of hydrogens of allylic methyl groups of pulegone was performed as a bioisosteric modification. This bioisosteric replacement by deuterium reduced the extent of CYP-450 mediated allylic oxidation, thereby reducing the toxicity of the compound manifolds, making it a safer alternative for humans as well as the environment. The structures of Pulegone and its bioisosterically modified deuterated analogue are displayed in Fig. (**10**).

## 8. Efavirenz

Efavirenz is a class of non-nucleoside reverse transcriptase inhibitors (NNRTIs), used for the treatment of the Human Immunodeficiency Virus (HIV) [31]. Its metabolism follows a pathway that affords a glutathione-derived conjugate, responsible for the nephrotoxicity of the compound. To subside this toxic effect, the hydrogen at the labile propargylic site was bioisosterically replaced by deuterium, which reduced the formation of cyclopropylcarbinol, responsible for the production of glutathione-derived conjugate [24], thus leading to reduced toxicity for humans as well as the environment. The comparative structures of Efavirenz and its bioisosterically modified deuterated analogue are displayed in Fig. (**11**).

**Fig. (11).** Structure of Efavirenz and its Deuterated analogue.

## 9. Iloprost

Iloprost is an orally bioavailable drug, used for the treatment of pulmonary arterial hypertension (PAH), scleroderma, and other diseases involving blood vessels [32]. The compound faced the drawback of a short half-life period ($T_{1/2}$=20-30mins.) thereby reducing drug serum levels before it reaches the efficacious dose. To overcome this, a bioisosteric modification was made in the compound, while it is in one of its metabolic pathways, which involves the replacement of β-carbon with a heteroatom. The bioisosterically modified compound was called Cicaprost and it exhibited approximately 2-3 times higher half-life than Iloprost [24], thereby increasing efficiency towards treatment and decreasing the hazardous risk for the environment. The comparative structures of Iloprost and its bioisosterically modified analogue, Cicaprost, are displayed in Fig. (**12**).

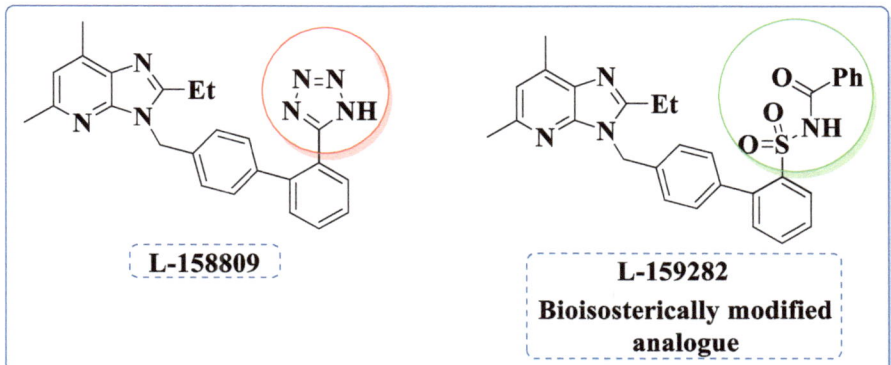

**Fig. (12).** Structure of Iloprost and its Bioisosterically modified analogue, Cicaprost.

## 10. L-158809

L-158809 is a potent Angiotensin antagonist. However, its clinical trials suggested a short half-life and rapid clearance of the administered drug [33], thus leading to lowered efficacy in the body. Since the release was quick, a potentially toxic amount of drug was released into the atmosphere, damaging the environment also. The cause of this rapid clearance of the drug was found to be glucuronidation of tetrazole moiety. Therefore, to suppress this potential drawback of the compound, the tetrazole moiety was bioisosterically replaced with acylsulfonamide, thereby preserving its potency and extending its efficacy and half-life, due to resistance of acylsulfonamide for glucuronidation [24]. Thus, this modification proved beneficial for both human health and the environment. The comparative structures of L-158809 and its bioisosterically modified analogue are displayed in Fig. (**13**).

**Fig. (13).** Structure of L-158809 and its Bioisosterically modified analogue, L-159282.

## CONCLUDING REMARKS

To conclude, the chapter focuses on the application of bioisosterism in chemical compounds for improving their toxicological profile, for reduced environmental toxicity. It details a more eco-friendly approach for chemicals and drugs, to act more effectively on targets and less destructively on the environment. Various novel bioisosterically modified molecules and various bioisosterically replaced lead molecules have been discussed in this chapter, with their modifications, and the effect of these bioisosteric altercations on the compounds. These alterations suggest that bioisosteric replacement of compounds with improved functional groups affects their degradation characteristics and their toxicological profile, thereby improving the chemical, ecological, and biomedicinal worth of the compounds.

## CONSENT FOR PUBLICATION

Not Applicable.

## CONFLICT OF INTEREST

The author confirms that this chapter contents have no conflict of interest.

## ACKNOWLEDGEMENT

Declared none.

## REFERENCES

[1] Thornber C. Isosterism And Molecular Modification in Drug Design. Chem Soc Rev 1979; 8(4): 563.
[http://dx.doi.org/10.1039/cs9790800563]

[2] Lima LM, Barreiro EJ. Bioisosterism: a useful strategy for molecular modification and drug design. Curr Med Chem 2005; 12(1): 23-49.
[http://dx.doi.org/10.2174/0929867053363540] [PMID: 15638729]

[3] Patani GA, LaVoie EJ. Bioisosterism: A Rational Approach in Drug Design. Chem Rev 1996; 96(8): 3147-76.
[http://dx.doi.org/10.1021/cr950066q] [PMID: 11848856]

[4] Hamada Y, Kiso Y. The application of bioisosteres in drug design for novel drug discovery: focusing on acid protease inhibitors. Expert Opin Drug Discov 2012; 7(10): 903-22.
[http://dx.doi.org/10.1517/17460441.2012.712513] [PMID: 22873630]

[5] Sechser B, Reber B. Using A Sequential Testing Scheme Under Laboratory and Field Conditions with The Bumble Bee Bombus Terrestris To Evaluate the Safety of Different Groups of Insecticides. InEcotoxicology 1998; pp. 166-74.
[http://dx.doi.org/10.1007/978-1-4615-5791-3_17]

[6] Lahm GP, Stevenson TM, Selby TP, *et al.* Rynaxypyr: a new insecticidal anthranilic diamide that acts as a potent and selective ryanodine receptor activator. Bioorg Med Chem Lett 2007; 17(22): 6274-9.
[http://dx.doi.org/10.1016/j.bmcl.2007.09.012] [PMID: 17884492]

[7] Hughes K, Lahm G, Selby T, Stevenson T. Cyanoanthranilamide insecticides. WO2004067528 2004.

[8] Wang BL, Zhu HW, Ma Y, *et al.* Synthesis, insecticidal activities, and SAR studies of novel pyridylpyrazole acid derivatives based on amide bridge modification of anthranilic diamide insecticides. J Agric Food Chem 2013; 61(23): 5483-93.
[http://dx.doi.org/10.1021/jf4012467] [PMID: 23687975]

[9] Zhang X, Li Y, Ma J, *et al.* Synthesis and insecticidal evaluation of novel anthranilic diamides containing N-substitued nitrophenylpyrazole. Bioorg Med Chem 2014; 22(1): 186-93.
[http://dx.doi.org/10.1016/j.bmc.2013.11.038] [PMID: 24326275]

[10] Liu Q, Chen K, Wang Q, *et al.* Synthesis, Insecticidal Activity, Structure–Activity Relationship (SAR) And Density Functional Theory (DFT) Of Novel Anthranilic Diamides Analogs Containing 1,3,4-Oxadiazole Rings. RSC Advances 2014; 4(98): 55445-51.
[http://dx.doi.org/10.1039/C4RA06356B]

[11] Puiu D, Popescu M, Niculescu M, Pascu L, Galaon T, Postolache C. Mobility of Some High Persistent Organochlorine Compounds from Soil to Mentha Piperita. Revista de Chimie 2019; 70(1): 278-82.
[http://dx.doi.org/10.37358/RC.19.1.6899]

[12] Fujiwara T, O'Hagan D. Successful Fluorine-Containing Herbicide Agrochemicals. J Fluor Chem 2014; 167: 16-29.
[http://dx.doi.org/10.1016/j.jfluchem.2014.06.014]

[13] Murphy M, Loi E, Kwok K, Lam P. Ecotoxicology of organofluorous compounds. Heidelberg: Springer 2011; pp. 339-63.

[14] Martins FA, Daré JK, Freitas MP. Theoretical study of fluorinated bioisosteres of organochlorine compounds as effective and eco-friendly pesticides. Ecotoxicol Environ Saf 2020; 199110679
[http://dx.doi.org/10.1016/j.ecoenv.2020.110679] [PMID: 32402896]

[15] Kang D, Fang Z, Huang B, *et al.* Structure-Based Optimization of Thiophene[3,2-d]pyrimidine Derivatives as Potent HIV-1 Non-nucleoside Reverse Transcriptase Inhibitors with Improved Potency against Resistance-Associated Variants. J Med Chem 2017; 60(10): 4424-43.
[http://dx.doi.org/10.1021/acs.jmedchem.7b00332] [PMID: 28481112]

[16] Meanwell NA. Fluorine and Fluorinated Motifs in the Design and Application of Bioisosteres for Drug Design. J Med Chem 2018; 61(14): 5822-80.
[http://dx.doi.org/10.1021/acs.jmedchem.7b01788] [PMID: 29400967]

[17] Kang D, Ruiz FX, Feng D, *et al.* Discovery and Characterization of Fluorine-Substituted Diarylpyrimidine Derivatives as Novel HIV-1 NNRTIs with Highly Improved Resistance Profiles and Low Activity for the hERG Ion Channel. J Med Chem 2020; 63(3): 1298-312.
[http://dx.doi.org/10.1021/acs.jmedchem.9b01769] [PMID: 31935327]

[18] Dugger SA, Platt A, Goldstein DB. Drug development in the era of precision medicine. Nat Rev Drug Discov 2018; 17(3): 183-96.
[http://dx.doi.org/10.1038/nrd.2017.226] [PMID: 29217837]

[19] Veber DF, Johnson SR, Cheng HY, Smith BR, Ward KW, Kopple KD. Molecular properties that influence the oral bioavailability of drug candidates. J Med Chem 2002; 45(12): 2615-23.
[http://dx.doi.org/10.1021/jm020017n] [PMID: 12036371]

[20] Hevey R, Ling CC. Recent advances in developing synthetic carbohydrate-based vaccines for cancer immunotherapies. Future Med Chem 2012; 4(4): 545-84.
[http://dx.doi.org/10.4155/fmc.11.193] [PMID: 22416779]

[21] Biffinger JC, Kim HW, DiMagno SG. The polar hydrophobicity of fluorinated compounds. ChemBioChem 2004; 5(5): 622-7.
[http://dx.doi.org/10.1002/cbic.200300910] [PMID: 15122633]

[22] Hevey R. Bioisosteres of Carbohydrate Functional Groups in Glycomimetic Design. Biomimetics (Basel) 2019; 4(3): 53.
[http://dx.doi.org/10.3390/biomimetics4030053] [PMID: 31357673]

[23] McPhail GL, Clancy JP. Ivacaftor: the first therapy acting on the primary cause of cystic fibrosis. Drugs Today (Barc) 2013; 49(4): 253-60.
[http://dx.doi.org/10.1358/dot.2013.49.4.1940984] [PMID: 23616952]

[24] Swair C. Bioisosteric Replacements https://www.cambridgemedchemconsulting.com/resources/bioisoteres

[25] Frank S. Tetrabenazine: the first approved drug for the treatment of chorea in US patients with Huntington disease. Neuropsychiatr Dis Treat 2010; 6: 657-65.
[http://dx.doi.org/10.2147/NDT.S6430] [PMID: 20957126]

[26] Lolkema MP, Bohets HH, Arkenau HT, *et al.* The c-Met Tyrosine Kinase Inhibitor JNJ-38877605 Causes Renal Toxicity through Species-Specific Insoluble Metabolite Formation. Clin Cancer Res 2015; 21(10): 2297-304.
[http://dx.doi.org/10.1158/1078-0432.CCR-14-3258] [PMID: 25745036]

[27] Sybertz E, Davis H, Salisbury B, Burrier R, Clader J, Burnett D. SCH 48461, A Novel Inhibitor of Cholesterol Absorption. Atherosclerosis 1994; 109(1-2): 89.
[http://dx.doi.org/10.1016/0021-9150(94)93372-3]

[28] Tucker NS, Kaufman PE, Weeks EN. Identification of permethrin and etofenprox cross-tolerance in Rhipicephalus sanguineus sensu lato (Acari: Ixodidae). Pest Manag Sci 2019; 75(10): 2794-801.
[http://dx.doi.org/10.1002/ps.5391] [PMID: 30809952]

[29] Kelly CB, Mercadante MA, Leadbeater NE. Trifluoromethyl ketones: properties, preparation, and application. Chem Commun (Camb) 2013; 49(95): 11133-48.
[http://dx.doi.org/10.1039/c3cc46266h] [PMID: 24162741]

[30] Tisserand R, Young R. Essential of oil safety e-book A guide for health care professionals. Elsevier Health Sciences 2013.

[31] Maggiolo F. Efavirenz. Expert Opin Pharmacother 2007; 8(8): 1137-45.
[http://dx.doi.org/10.1517/14656566.8.8.1137] [PMID: 17516877]

[32] Olschewski H, Simonneau G, Galiè N, *et al.* Aerosolized Iloprost Randomized Study Group. Inhaled iloprost for severe pulmonary hypertension. N Engl J Med 2002; 347(5): 322-9.
[http://dx.doi.org/10.1056/NEJMoa020204] [PMID: 12151469]

[33] Lund DD, Brooks RM, Faraci FM, Heistad DD. Role of angiotensin II in endothelial dysfunction induced by lipopolysaccharide in mice. Am J Physiol Heart Circ Physiol 2007; 293(6): H3726-31.
[http://dx.doi.org/10.1152/ajpheart.01116.2007] [PMID: 17965276]

# CHAPTER 10

# Gold and Silver Nanoparticle Synthesis by *Pyrus* and *Eurya*: Environment-Friendly Therapeutic Agents

**Dhara Shukla**[1] and **Padma S. Vankar**[2,*]

[1] *Shree R. International, Chakeri II, Kanpur, India*
[2] *Bombay Textile Research Association, Mumbai, India*

**Abstract:** Two new metal-containing biosources *i.e. Pyruspaschia* fruits and *Eurya acuminate* leaves were used in the preparation of gold and silver nanoparticles. *Pyruspashia* has many medicinal uses as it is used in gastrointestinal disorders, fever, and headache, hysteria, and epilepsy. The fruits are sedative, febrifuge, and laxative. *Eurya acuminate* leaves are used as a treatment for cholera, diarrhoea, and other stomach diseases. The leaves are applied as a poultice on skin eruptions. These bio-sources are metal chelators used for binding natural dye to textile. These both metal-bearing plant parts were first time used to produce nanoparticles which further can be used therapeutically based on their size. This approach can add results to an environment-friendly medicinal agent. The nano-particles so generated were characterized by UV-Visible spectroscopy, FESEM (field emission scanning electron microscopy), TEM (transmission electron microscopy), and AFM (atomic force microscopy) techniques. The particles were found to be crystalline and both Au and Ag nanoparticles were pure and their mother liquor did not have significant sedimentation as impurities. FT-IR (Fourier transformed infrared spectroscopy) analysis authenticates the role of phytochemicals in this work. The synthesis of silver and gold nanoparticles using the above biological resources suggests an eco-friendly/green possibility, in comparison to many available methods based on chemical or physical techniques. Their application as therapeutic agents in various diseases and cancerous growth is of great prospect.

**Keywords:** AFM, *Eurya acuminate* leaves, FESEM, FT-IR, gold and silver nanoparticles, *Pyruspaschia* fruits, TEM.

---

\* **Corresponding author Padma S. Vankar:** Bombay Textile Research Association, Mumbai, India; E-mail: padma.vankar@gmail.com

Tahmeena Khan, Abdul Rahman Khan, Saman Raza, Iqbal Azad and Alfred J. Lawrence (Eds.)
All rights reserved-© 2021 Bentham Science Publishers

# INTRODUCTION

The current field of synthetic and medicinal chemistry has now gained a significant impetus in terms of research into the preparation of small and sub small level molecules beneficial for the human race. Nanotechnology in this respect has gained much attention and found a respectable place in the therapeutic arena as well. It involves the synthesis and maintenance of such nano/small molecules to almost any structure. These methods are used today to produce a wide variety of useful chemicals such as pharmaceuticals or commercial polymers [1]. Nanoparticles present an exciting and potential tool to operate at a cellular level as almost all physical reactions in the human body involving enzymes and biochemical/biomolecules manoeuvres at the nano-scale. Successful management of these procedures with manufactured nano-scale molecules gives rise to a new and very powerful arena for future therapeutics, especially where deadly/lethal diseases are involved. In this regard, metal nanoparticles (NPs) are of great importance due to their specific role, determined by their size, shape, composition, and crystallinity [2]. The green/eco-friendly chemistry process involving the production of specially design nanoparticles will eventually rise as an attractive technology to chemists, researchers, and industrialists for innovative chemistry research and applications involving medicinal/therapeutic demands [3]. It also connects to another branch of chemistry *i.e.*, environmental chemistry. This green chemistry is fundamental for the improvement of sustainable chemistry involving such small biomolecules [4]. Biosynthesis of metal nanoparticles using biological/green sources will produce and sustain methods/procedures which allow the procurement of uncommon shapes in nanoparticles such as nano-triangles and prisms [5]. The physical [6] and chemical processes [7] are the classical general methods used for the fabrication of nanoparticles, but due to the presence of non-environment friendly effects, these methods are not ecologically compatible [8] and can lead to unwanted results in the biomedical applications [9]. These shortcomings can be met or overcome by a microbe-mediated and plant-mediated biological process and this bio-route appeals to a larger perspective of eco-friendliness and biocompatibility [10]. A rapidly growing area of nanoscience and nanotechnology by green route has been widely recognized and gained quite an attention in recent years [11]. Thus, the controlled synthesis of metal nanoparticles with well-defined shapes and sizes is among one of the most fascinating aspects of nanoparticle research. In recent years, compared to bulk metals or metal ions, metallic nanoparticles, including silver (Ag), gold (Au), platinum (Pt), and palladium (Pd), have been extensively studied because of their unique properties, particularly the effect of quantum size and large surface area. Besides, metallic nanoparticles are compatible with the biological system, and therefore, they have been used for drug delivery, diagnostic imaging, labelling, and biosensors [12].

Therefore, a clean/green or environment-friendly method of synthesizing gold and silver nanoparticles that exhibit biological functions is demonstrated herein. It satisfies the need for developing environmentally friendly and sustainable methods for the synthesis of nanoparticles. There is a current drive to incorporate all the eco-friendly methodologies in designing environmentally benign tools and procedures. The utilization of various plant resources for the biosynthesis of metallic nanoparticles is called green nanotechnology, and it does not utilize any harmful chemical protocols. In this respect, a method of preparation of gold nanoparticles by the reduction of Auric chloride by aqueous extract of *Mirabilis jalapa* flowers has been reported [13].

The following study reports the plant-mediated synthesis of gold and silver nanoparticles using the plant leaf extract of *Eurya acuminate* and dried fruit extract of *Pyruspaschia*, in which plant extract are reductants. These nanoparticles were characterized by ultraviolet-visible spectroscopy, FTIR, scanning electron microscopy, AFM, and TEM.

The plant part of the study was *Pyruspaschia* fruit (Fig. **1**) which was used in the dried state. *Pyruspaschia* or the wild pear tree grows commonly at altitudes of between 700 and 2000 meters. *Pyruspaschia* is native to the Himalayas, west China and Myanmar, and Afghanistan [14]. The fruits of *Pyrus paschia* are used by local tribal people for eating. *Pyruspaschia* fruit contains about 6.8% sugars, 3.7% protein, 1% ash, and 0.4% pectin [15]. It also contains a low content of Vitamin C. *Pyruspaschia* also contains minerals, such as potassium, magnesium, phosphorous, calcium, and iron [16]. It is assumed that the presence of a little amount of Vitamin C is responsible for its reducing capability.

**Fig. (1).** Fruits of *Pyrus paschia*.

*Eurya acuminate* (Fig. **2**) is a shrub and 130 species of this plant are available which are found in North East India and many other countries of Southeast Asia and Nepal, Sri Lanka. The leaves of this plant are medicinal in nature and local inhabitants use it as a remedy for diarrhoea [17].

**Fig. (2).** Leaves of *Eurya acuminate*.

## MATERIAL AND METHODS

### Material Collection

The biomaterial *i.e.*, *Pyruspaschia* fruits and *Eurya acuminate* leaves were directly procured from Arunachal Pradesh. They were dried and then grounded and stored to produce gold (Au) and silver (Ag) nanoparticles. Inorganic salts to prepare Au and Ag nanoparticles (Chloroauric acid ($HAuCl_4$) and silver nitrate ($AgNO_3$) were purchased from HiMedia Chemicals and used as received.

### Instrumentation

Many different techniques were used to characterize nanoparticles. For UV-Visible spectroscopy, a diluted aqueous sample was used, and UV-Visible spectra were obtained from Heλios α Thermo Electron Corporation Spectrophotometer. Fourier transform–infra-red spectra of bio extract and gold, silver nanoparticles were recorded on Bruker Vertex 70 machine. TEM data were collected on TECNAI $G^2$software run FEI TECNAI 02. Molecular Imaging Agilent Machine was used to get AFM data and pictures were collected on PicoScan software. A diluted aqueous sample was loaded on glass coverslips and dried in an oven at ambient temperature for AFM measurements. Cantilevers μ Masch (Cu-Au) with Tip curvature less than 10 nm were used in the Molecular Imaging probe. The sample was loaded on tabs and dried a little for SEM analysis, whereas for TEM analysis, the sample was prepared by placing drops of the nanoparticle mixture over carbon-coated copper grids and allowing the water to disperse to get dried.

## Preparation of Bio-Extract

20 gm dried and ground leaves and fruits of both *Eurya acuminate* and *Pyruspaschia* were soaked in 100 ml boiling distilled water for 5-10 minutes and filtered through Whatman filter paper no. 42. The bio extract of dried fruit and leaves were always used fresh for generating gold and silver nanoparticles.

## RESULTS AND DISCUSSION

The appearance of brownish-red colour in solution started after some time as bio extract was added to $10^{-3}$ M aqueous Auric chloride ($HAuCl_4$) solution, indicating the formation of gold nanoparticles. The UV-Vis absorption spectrum of the above brownish solution showed the peak at 560nm (Fig. **3**) which corresponds to the distinctive surface Plasmon resonance (SPR) band of gold nanoparticles at this very wavelength. It can be supposed that phytochemicals of different nature play a significant role in reducing salt to get nanoparticles. The FTIR spectra of both the plant material indicated the presence of carbonyl keto and alcoholic groups as the components of biomolecules in a reaction mixture having both nanoparticles as a common constituent. After reaction with gold ions, however, an improvement in the signal analogous to a carbonyl group has been observed. *Pyrus* gave a very good yield of thin, comparatively flat, single-crystalline gold nano triangles as can be suggested by further analysis. The nano triangles seem to grow by a process in which fast reduction of inorganic salt is involved by biomolecules present in *Pyrus* extract along with Copper ion. Along with the triangles, other non-spherical particles were also detected, including nanodisks, nano hexagons. It can be assumed that the course of formation of nanoparticles involves amelioration of bio-extract with salt then bioreduction and then synthesis of nanostructures of various shapes depending upon the variables involved in the complete process, particularly the other metal present in the plant extract.

FT-IR of AgNP of *Pyrus* extract showed sharp carbonyl peaks. Well-defined and controlled nanostructures having macroscopic surfaces are sturdy procedures of self-assembly in which thin films of gold nanoparticles take place [18]. The formation of triangular disc type nanoparticles is very prominent in the SEM of *Pyrus*AuNP as shown in Fig. (**4**).

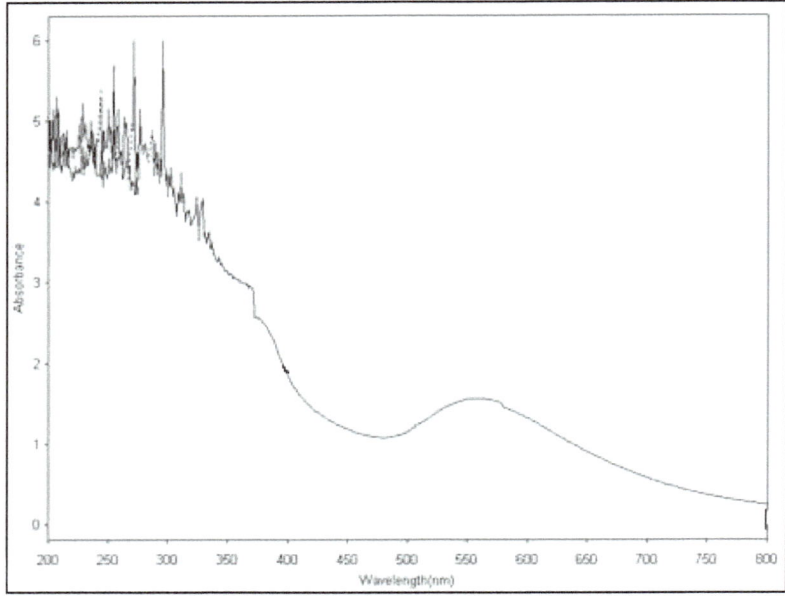

**Fig. (3).** UV-Visible spectra of *Pyrus* gold nanoparticles (Au NP).

**Fig. (4).** SEM of *Pyrus* gold nanoparticles.

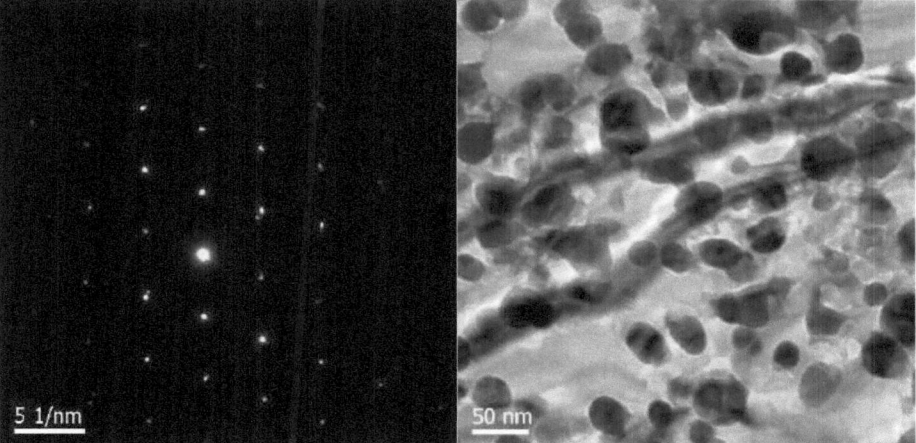

**Fig. (5).** TEM and SADP of *Pyrus* gold nanoparticles.

*Pyruspaschia* was found superior in reducing gold and silver nanoparticles in an aqueous system, although silver nanoparticle formation took a little more time to generate and stabilize due to the presence of copper in it. Their shape and size were well characterized. The shape was mostly triangular in the case of gold and spherical in the case of silver nanoparticles. These particles have shown a size range of 30 nm which is quite good for this kind of nanoparticles generation process. SEM of *Pyrus* AgNP clearly shows very uniform particle formation. TEM of *Pyrus* AuNP also shows very uniform particle formation (Fig. **5**) as shown by SADP (selected area diffraction pattern). The AFM of *Pyrus* AuNP's showed the presence of triangular gold nanoparticles abundantly. It has been observed through this analysis that spherical nanoparticles are a little big than triangular nanoparticles. Shapes and quantity of gold particles got confirmed as triangular gold nanoparticles along with some other shapes.

## Morphological Identification of Gold and Silver Nanoparticles Produced by *Eurya acuminate* Leaves

*Eurya acuminate* leaves proved to be a good bio source for the production of gold and silver nanoparticles. In a bio source, it can be assumed that temperature and occurrence of carbonyl compounds play an important role in reducing and paving way for a particular shape as a triangle in nanoparticle formation. Smaller nanoplates form triangular structures with sharp edges can be observed in this case (Fig. **6, 7, 8**). A stable (111) facet as clear from the analysis was confirmed in nano triangles. Here it can be assumed that the presence of a substantial amount of aluminium in the bio extract does not govern the size or the shape and multi-shaped nanoparticles of gold. However, the same extract when used in the preparation of silver nanoparticles behaves somewhat differently and controls the size and shape of silver nanoparticles. This change can be attributed to the difference in electrochemical potentials of Al *versus* Au and Ag cations.

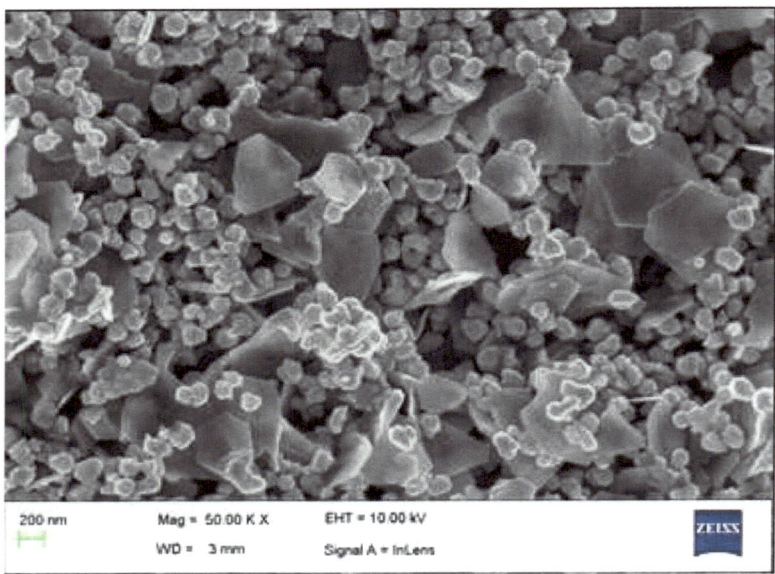

**Fig. (6).** SEM of *Eurya* gold nanoparticles.

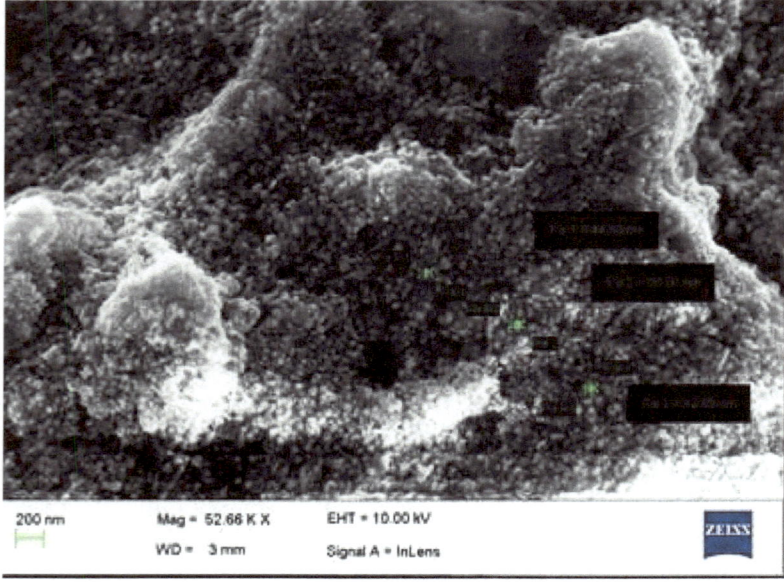

**Fig. (7).** SEM of *Eurya* silver nanoparticles.

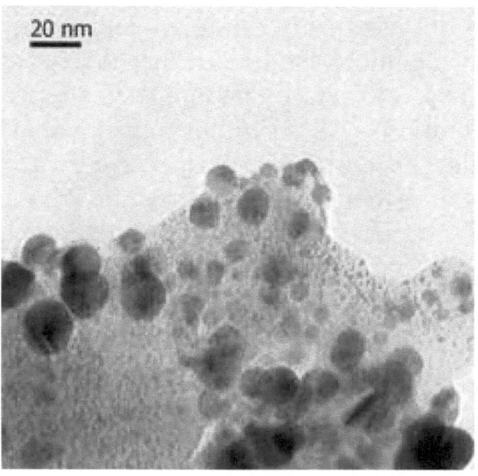

**Fig. (8).** TEM of *Eurya* silver nanoparticles.

## Effect of the Presence of Metal Ions on Nanoparticle formation

The UV-Visible absorption spectrum of the gold and silver nanoparticles showed a slight shift in the maximum absorption peak, its requisite wavelength, as the observed wavelength of the surface plasmon absorption band of nanoparticles when prepared by biotic reductant [19]. The size and polydispersity of the particles as well as the occurrence of any electropositive metal ion on the surface of the particles depend on the position and width of the plasmon band in the UV-Visible absorption spectrum of the gold and silver nanoparticles [20]. In the case of both *Eurya* and *Pyrus* respectively, mono-dispersed silver nanoparticles under 50 nm in size with no adsorbed substances (*i.e.* $Al^{3+}$ or $Cu^{2+}$ ions) have a narrow surface plasmon absorption peak at a wavelength of 390 to 450 nm that generally varies with uniformity and smaller particle size. The observed decrease in the size of the silver nanoparticles and their uniform distribution can certainly be attributed to the presence of copper, due to being more electropositive. It is observed often that colloidal dispersions of nanometal particles reveal absorption bands in the UV-Visible region, due to collective excitations of the free electrons (surface plasmon band) [21]. The width and position of the plasmon modes are influenced by the shape, volume, and surface/interface effects for scattered particles. Interactions between clusters and the surrounding environment, due to the presence of stabilizers (may be phytochemical of plant species containing metal) or due to the existence of cluster size or shape distributions, made the understanding of the optical behaviour difficult in many clusters or particle system [22].

Both the plant material *viz., Pyruspaschia* fruit and *Eurya acuminate* leaves were used as bio-mordant in the textile dyeing industry too. Production of nanoparticles from them is likely to be less costly, simpler, and require less energy and raw materials than existing chemical methods. The development of an eco-friendly procedure for the synthesis of metallic nanoparticles represents a significant step in the field of nanotechnology. This synthesis of nanoparticles particularly nano triangles will provide exceptional prospects towards the blueprint and development of fascinated 'green' gold nano triangles that can be used extensively in various biomedical purposes. Large structural anisotropy of the triangular nanocrystals should have a considerable outcome on the optical properties of the trials, including light absorption, scattering, and surface-enhanced resonance Raman spectroscopy [23]. Thus these nano triangles can very well adapt as therapeutic agents or as biomarkers in cell assays due to these properties.

## Biomedical or Therapeutic Applications Involving Gold and Silver NPs

There are several types of natural products which have a potent remedial effect, but their feeble solubility and bioavailability have inhibited their usage. The use of such plant or plant parts for synthesizing/producing nanoparticles can attribute their medicinal properties in the finish products *i.e.* NPs.

Examinations of *Pyruspaschia* fruits in terms of the presence of phytochemicals and pharmacological components revealed the presence of secondary metabolites like alkaloids, flavonoids, steroids, and tannins, lupeol, β-sitosterol, βsitosterol—β-D-glucoside [24] and possesses antimicrobial activity against *Klebsiella pneumonia, Shigellaflexneri*a nd *Eschericia coli* [25].

Phytochemical investigations of various *Eurya* species also reported various compounds like anthocyanins, ellagic acid, caffeine, flavone, flavonols, β---glucopyranoside, euryanoside, chrysoeriol *etc*. The compounds which were identified in leaf methanolic extract were phytol and β-sitosterol by extensive spectroscopic studies [26]. *Pyruspaschia* extract was found to be good in reducing gold and silver nanoparticles in an aqueous system, although silver nanoparticle formation took little more time to generate and stabilize may be due to the presence of copper in it. The shape was mostly triangular in the case of gold and spherical in the case of silver nanoparticles. These particles have shown a size range of 20-100 nm which is quite good for this kind of nanoparticles generation process. The rapid improvement of environmental chemistry has raised the possibility of using the nanoparticles which were premeditated and they specifically interact within biological environments for the treatment of diseases. Nanoparticles interacting with cells and the extracellular environment can activate a sequence of biological effects. These effects principally depend on the dynamic physicochemical features of nanoparticles, which govern the biocompatibility and efficacy of the intended outcomes [27].

Nanotechnology involves the use of techniques and tools from various disciplines, together with biology, chemistry, and medical science [28]. That is why it is considered to be a prominent branch of environmental chemistry. The fundamental plan behind nanotechnology science could be the structural property of a molecule that is typically not offered by individual molecules and bulk solids [29].

Biosynthesized or biogenic or environment-friendly metallic nanoparticles, particularly silver and gold nanoparticles (AgNPs and AuNPs, respectively), have been increasingly used because of their advantages, including high stability and loading capacity; moreover, these nanoparticles are synthesized using a green and cost-effective method [30].

AgNPs are used as coating materials for medical devices because of their exceptional anti-microbial properties. AgNPs guard the outer and inner surfaces of the devices and assist the nonstop release of silver ions to induce antibacterial activity [31]. Also, the synthesized AgNPs are dispersed in various types of vehicles as nano-carriers [32] to achieve an enhanced antibacterial effect. The antimicrobial activity of silver nanoparticles derived from lemon leaves showed enhancement in activity due to the synergistic effect of silver and essential oil components of lemon leaves [33]. AgNPs are used as excellent healing wound dressings because they accelerate re-epithelialization and increase the bacterial clearance from infected wounds [34]. These effects of AgNPs are because of a decrease in the activity of local matrix metalloproteinase (MMP) and an increase in the apoptosis of neutrophils within the wound [35]. AgNPs have been developed as vehicles for various drugs [36] and biomolecules, *e.g.*, oligodeoxynucleotides [37] and interleukin-10 [38], for treating cancers and inflammatory diseases, respectively.

Gold nanoparticles are one of the widely used particles as they have many therapeutic applications, such as drug delivery system for many diseases like cancer, cardiovascular diseases, *diabetes mellitus,etc*. biosensors, and environmental applications of dye degradation, bioremediation of toxic chemicals present in the environment (soil and atmosphere). Gold nanoparticles synthesis by the green route has become the latest development, because of the bioavailability of sources like plants or microorganisms, and it also reduces the utilization of toxic chemicals [39]. AuNPs allow *in-vitro* detection and act as a diagnostic agent for diseases such as cancer by readily conjugating with biomarkers such as oligonucleotides or antibodies to detect the target biomolecules [40]. Gold nanoparticles have also found applications in the manufacturing of biosensors due to their electron transferring ability and adsorptive capacity [41].

Surface functionalization is one of the most favourable properties of Au NPs in the biomedical domain. The surface of Au NPs can be functionalized with various biomolecules, such as DNA, peptides, and antibodies. There are two kinds of interactions. One is noncovalent interactions, the other is covalent interactions. Noncovalent modifications take place through electrostatic interactions, hydrophobic entrapment, and van der Waals forces [42].

In a study, results confirmed that gold nano triangles showed the highest cellular uptake, followed by gold nanorods and gold nanostars. Gold nanoparticle uptake was induced *via* various endocytosis mechanisms, dependent on the shape. All three shapes utilized the clathrin-mediated endocytic pathway. Gold nanorod uptake was also dependent on caveolae/lipid raft-mediated endocytosis and gold nanotriangle uptake was strongly associated with cytoskeletal rearrangement, as well as the dynamic pathway [43].

Drug delivery is a fascinating arena of research that has attracted the attention of researchers. It can be defined as a process for the release of biologically active medicine at a certain speed and a specific location. At present, it is vital to ameliorating specific drug delivery modes for clinical application. Nanomaterials offer immense probabilities for multiple, locus-specific drug delivery to the disease locus as their diminutive size can effectively penetrate across obstacles through small capillaries into individual cells. Specifically, Au NPs have revealed a great capacity for use as drug delivery platforms. Au NPs can deliver multiple drug molecules, recombinant proteins, vaccines, or nucleotides into their targets and can control drug release *via* biological stimuli [42].

Thus nanoparticles have unique properties of exhibiting larger surface area to volume ratio, size, shape like triangles, spherical or rod, *etc*. due to which they are being used in the various fields of diagnostic biological probes, optoelectronics, display instruments, catalysis, fabricating biological sensors, diagnosis or monitoring diseases like cancer cells, drug discovery, detecting environmental toxic metals or reagents and in therapeutic applications [44].

The nano-sized anticancer agents have been standardized against a variety of human cancer cells like prostate, colon, lung, cardio, and breast cancer [45].

## CONCLUDING REMARKS

Biomedical applications require new improved nanomaterials, and this could be attained by utilizing newer approaches in nanotechnology. Many unique properties are exhibited by nanoparticles whose size ranges from 1-100 nm, which cannot be observed in micron-sized or bulk materials. The reason for the nanoparticles to be a potential candidate for biomedical applications is their high surface activity due to large surface area, strong antioxidant activity, and more importantly their biocompatibility. These properties can be very well manoeuvred at the molecular level. Gold nanoparticles, carbon nanotube, quantum dot, grapheme, and ferro-ferric oxide are commonly used for biomedical applications. The most abundantly used is gold nanoparticle (AuNP) due to its unique optical, electronic, sensing, and biochemical properties. The size and shape of AuNPs make them good materials for optical and electrical applications [46]. These nanoparticles can be further functionalized by other biomolecules like drugs, genes, and targeting ligands [47], making them negatively charged which can act as drug carriers. Another added advantage with AuNPs is that they are nontoxic and biocompatible [48]. AuNPs have the presence of surface plasmon resonance (SPR) [49] along with distinct surface effect and macroscopic quantum.

In the case of treating infections or as drug carriers, the outer membrane does not allow larger molecular mass penetration [50]. Thus the usage of nanoparticles enhances their ability to improve hydrophilicity and the higher capacity to penetrate through the cell membrane [51]. Studies indicate that the positive charge on the silver ion plays a role in the antimicrobial activity by applying electrostatic attractions between the negatively charged cell membrane and the positively charged nanoparticles [52].

The ill effects on the human body and environment that are commonly associated with chemically synthesized metallic nanoparticles can be avoided by the green development of nanoparticles [53].

Hence the development of cost-effective and environmentally friendly methods to synthesize metallic nanoparticles without the use of any toxic chemicals is required in today's time [54]. Metallic nanoparticles can be made in bulk quantities that are free from contaminants, easily biosynthesized using plant extracts, and these nanoparticles have interestingly better-defined morphology and size. It is believed that there are good opportunities for developing industrial-scale production, where NPs have important applications [55]. Thus, the use of green synthesis of NPs/nanomaterials is an emerging and exciting area of nanotechnology and may have a substantial influence on further improvements in nanoscience. The green synthesis of nanomaterial has a great future in times to come.

## LIST OF ABBREVIATIONS

**NP**    Nanoparticles

**Ag NP** Silver Nanoparticles

**Au NP** Gold Nanoparticles

## CONSENT FOR PUBLICATION

Not Applicable.

## CONFLICT OF INTEREST

The author confirms that this chapter contents have no conflict of interest.

## ACKNOWLEDGEMENT

Declared none.

## REFERENCES

[1]   Hussain Z, Tahir S, Mahmood K. ALIa, A.; Arshad, M.; Ikram, S.; Nabi, M. A. U.; Ashfaq, A.; Rehman, U.; Uddassir, Y., Synthesis and Characterization of Silver Nanoparticles With Epoxy Resin Composites. Dig J Nanomater Biostruct 2020; 15(3): 873-83.

[2]   aSun Y, Xia Y. Shape-controlled synthesis of gold and silver nanoparticles. Science 2002; 298(5601): 2176-9.
[http://dx.doi.org/10.1126/science.1077229] [PMID: 12481134] bHormozi-Nezhad MR, Karami P, Robatjazi H. A simple shape-controlled synthesis of gold nanoparticles using nonionic surfactants. RSC Advances 2013; 3(21): 7726-32.
[http://dx.doi.org/10.1039/c3ra40280k]

[3]   Mostafa HED. Introductory Chapter: Principles of Green Chemistry. IntechOpen 2018.

[4]   Blum C, Bunke D, Hungsberg M, *et al.* The concept of sustainable chemistry: Key drivers for the transition towards sustainable development. Sustainable Chemistry and Pharmacy 2017; p. 5.

[5]   Verma VC, Singh SK, Solanki R, Prakash S. Biofabrication of Anisotropic Gold Nanotriangles Using Extract of Endophytic Aspergillus clavatus as a Dual Functional Reductant and Stabilizer. Nanoscale Res Lett 2011; 6(1): 16.
[http://dx.doi.org/10.1186/1556-276X-6-261] [PMID: 27502640]

[6]   Xu G-n, Qiao X-l, Qiu X-l, Chen J-g. Preparation and characterization of stable monodisperse silver nanoparticles *via* photoreduction. Colloids Surf A Physicochem Eng Asp 2008; 320(1–3): 222-6.
[http://dx.doi.org/10.1016/j.colsurfa.2008.01.056]

[7]   Wang H, Qiao X, Chen J, Ding S. Preparation of silver nanoparticles by chemical reduction method. Colloids Surf A Physicochem Eng Asp 2005; 256(2–3): 111-5.
[http://dx.doi.org/10.1016/j.colsurfa.2004.12.058]

[8]   Dubey SP, Lahtinen M, Sillanpää M. Tansy fruit mediated greener synthesis of silver and gold nanoparticles. Process Biochem 2010; 45(7): 1065-71.
[http://dx.doi.org/10.1016/j.procbio.2010.03.024]

[9]   Bar H, Bhui DK, Sahoo GP, Sarkar P, De SP, Misra A. Green synthesis of silver nanoparticles using latex of Jatropha curcas. Colloids Surf A Physicochem Eng Asp 2009; 339(1–3): 134-9.
[http://dx.doi.org/10.1016/j.colsurfa.2009.02.008]

[10]  Krumov N, Perner-Nochta I, Oder S, Gotcheva V, Angelov A, Posten C. Production of Inorganic Nanoparticles by Microorganisms. Chem Eng Technol 2009; 32(7): 1026-35.
[http://dx.doi.org/10.1002/ceat.200900046]

[11]  aKim J, Rheem Y, Yoo B, *et al.* Peptide-mediated shape- and size-tunable synthesis of gold nanostructures. Acta Biomater 2010; 6(7): 2681-9.
[http://dx.doi.org/10.1016/j.actbio.2010.01.019] [PMID: 20083240] bVerma VC, Kharwar RN, Gange AC. Biosynthesis of antimicrobial silver nanoparticles by the endophytic fungus Aspergillus clavatus.

Nanomedicine (Lond) 2010; 5(1): 33-40.
[http://dx.doi.org/10.2217/nnm.09.77] [PMID: 20025462] cHuang P, Kong Y, Li Z, Gao F, Cui D. Copper selenide nanosnakes: bovine serum albumin-assisted room temperature controllable synthesis and characterization. Nanoscale Res Lett 2010; 5(6): 949-56.
[http://dx.doi.org/10.1007/s11671-010-9587-0] [PMID: 20672034]

[12] Noruzi M, Zare D, Davoodi D. A rapid biosynthesis route for the preparation of gold nanoparticles by aqueous extract of cypress leaves at room temperature. Spectrochim Acta A Mol Biomol Spectrosc 2012; 94: 84-8.
[http://dx.doi.org/10.1016/j.saa.2012.03.041] [PMID: 22522293]

[13] Vankar PS, Bajpai D. Preparation of gold nanoparticles from Mirabilis jalapa flowers. Indian J Biochem Biophys 2010; 47(3): 157-60.
[PMID: 20653286]

[14] Tsering J, Gogoi BJ, Tag H. Ethnobotany and phytochemical analysis of Pyrus pashia leaves. Int J Pharm Sci Res 2012; 3(8): 2721.

[15] Kumari A, Dhaliwal Y. Formulation of kainth (Pyrus serotina) based functional food products and changes during storage. Int Res J Nat Appl Sci 2016; 3(11): 59-70.

[16] Tewari L M, Tewari G, Chopra N, Tewari A, Pandey N C, Kumar M. Phytochemical Screening And Antioxidant Potential Of Some Selected Wild Edible Plants Of Nainital District, Uttarakhand.

[17] Das G, Patra JK, Singdevsachan SK, Gouda S, Shin H-S. Diversity of traditional and fermented foods of the Seven Sister states of India and their nutritional and nutraceutical potential: a review. Front Life Sci 2016; 9(4): 292-312.
[http://dx.doi.org/10.1080/21553769.2016.1249032]

[18] Gao B, Arya G, Tao AR. Self-orienting nanocubes for the assembly of plasmonic nanojunctions. Nat Nanotechnol 2012; 7(7): 433-7.
[http://dx.doi.org/10.1038/nnano.2012.83] [PMID: 22683842]

[19] Turkevich J, Stevenson PC, Hillier J. A study of the nucleation and growth processes in the synthesis of colloidal gold. Discuss Faraday Soc 1951; 11(0): 55-75.
[http://dx.doi.org/10.1039/df9511100055]

[20] aMaillard M, Huang P, Brus L. Silver Nanodisk Growth by Surface Plasmon Enhanced Photoreduction of Adsorbed. Nano Lett 2003; 3(11): 1611-5. [Ag+].
[http://dx.doi.org/10.1021/nl034666d] bHenglein A. Colloidal Silver Nanoparticles: Photochemical Preparation and Interaction with O2, CCl4, and Some Metal Ions. Chem Mater 1998; 10(1): 444-50.
[http://dx.doi.org/10.1021/cm970613j]

[21] Wilcoxon J, Williamson R, Baughman R. Optical properties of gold colloids formed in inverse micelles. J Chem Phys 1993; 98(12): 9933-50.
[http://dx.doi.org/10.1063/1.464320]

[22] aRodriguez-Sanchez L, Blanco M, Lopez-Quintela M. Electrochemical synthesis of silver nanoparticles. J Phys Chem B 2000; 104(41): 9683-8.
[http://dx.doi.org/10.1021/jp001761r] Kreibig U, Vollmer M. Theoretical considerations.Optical Properties of Metal Clusters. Springer 1995; pp. 13-201.
[http://dx.doi.org/10.1007/978-3-662-09109-8_2] cGötz T, Hoheisel W, Vollmer M, Träger F. Characterization of large supported metal clusters by optical spectroscopy. Z Phys D At Mol Clust 1995; 33(2): 133-41.
[http://dx.doi.org/10.1007/BF01437432]

[23] Ashley MJ, Bourgeois MR, Murthy RR, *et al.* Shape and size control of substrate-grown gold nanoparticles for surface-enhanced raman spectroscopy detection of chemical analytes. J Phys Chem C 2018; 122(4): 2307-14.
[http://dx.doi.org/10.1021/acs.jpcc.7b11440]

[24] Khandelwal R, Paliwal S, Chauhan R, Siddiqui AA. Phytochemical screening of hexane soluble

fraction of *Pyrus pashia* fruits. Orient J Chem 2008; 24(2): 773-4.

[25] Saklani S, Chandra S. *In vitro* antimicrobial activity nutritional profile of medicinal plant of Garhwal, Himalaya. Int J Pharm Sci Res 2012; 3: 268-72.

[26] Faisal T, Ahsan M, Choudhury JA, Azam AZ. Phytochemical and Biological Investigations of Eurya acuminata (Theaceae). Dhaka University Journal of Pharmaceutical Sciences 2016; 15(2): 151-4.
[http://dx.doi.org/10.3329/dujps.v15i2.30928]

[27] Naahidi S, Jafari M, Edalat F, Raymond K, Khademhosseini A, Chen P. Biocompatibility of engineered nanoparticles for drug delivery. J Control Release 2013; 166(2): 182-94.
[http://dx.doi.org/10.1016/j.jconrel.2012.12.013] [PMID: 23262199]

[28] Fakruddin M, Hossain Z, Afroz H. Prospects and applications of nanobiotechnology: a medical perspective. J Nanobiotechnology 2012; 10(1): 31.
[http://dx.doi.org/10.1186/1477-3155-10-31] [PMID: 22817658]

[29] Pitkethly MJ. Nanomaterials–the driving force. Mater Today 2004; 7(12): 20-9.
[http://dx.doi.org/10.1016/S1369-7021(04)00627-3]

[30] Katas H, Moden N Z, Lim C S, *et al*. Biosynthesis and potential applications of silver and gold nanoparticles and their chitosan-based nanocomposites in nanomedicine 2018.
[http://dx.doi.org/10.1155/2018/4290705]

[31] Wilcox M, Kite P, Dobbins B. Antimicrobial intravascular catheters--which surface to coat? J Hosp Infect 1998; 38(4): 322-4.
[http://dx.doi.org/10.1016/S0195-6701(98)90084-6] [PMID: 9602982]

[32] Sharma S. Enhanced antibacterial efficacy of silver nanoparticles immobilized in a chitosan nanocarrier. Int J Biol Macromol 2017; 104(Pt B): 1740-5.
[http://dx.doi.org/10.1016/j.ijbiomac.2017.07.043] [PMID: 28736042]

[33] Vankar PS, Shukla D. Biosynthesis of silver nanoparticles using lemon leaves extract and its application for antimicrobial finish on fabric. Appl Nanosci 2012; 2(2): 163-8.
[http://dx.doi.org/10.1007/s13204-011-0051-y]

[34] Ge L, Li Q, Wang M, Ouyang J, Li X, Xing MM. Nanosilver particles in medical applications: synthesis, performance, and toxicity. Int J Nanomedicine 2014; 9: 2399-407.
[PMID: 24876773]

[35] Kirsner R, Orsted H, Wright JB. The role of silver in wound healing part 3 matrix metalloproteinases in normal and impaired wound healing: a potential role of nanocrystalline silver. Wounds 2001; 13(3) (Suppl. C).

[36] Benyettou F, Rezgui R, Ravaux F, *et al.* Synthesis of silver nanoparticles for the dual delivery of doxorubicin and alendronate to cancer cells. J Mater Chem B Mater Biol Med 2015; 3(36): 7237-45.
[http://dx.doi.org/10.1039/C5TB00994D] [PMID: 32262831]

[37] Brown PK, Qureshi AT, Moll AN, Hayes DJ, Monroe WT. Silver nanoscale antisense drug delivery system for photoactivated gene silencing. ACS Nano 2013; 7(4): 2948-59.
[http://dx.doi.org/10.1021/nn304868y] [PMID: 23473419]

[38] Baganizi DR, Nyairo E, Duncan SA, Singh SR, Dennis VA. Interleukin-10 conjugation to carboxylated PVP-coated silver nanoparticles for improved stability and therapeutic efficacy. Nanomaterials (Basel) 2017; 7(7): 165.
[http://dx.doi.org/10.3390/nano7070165] [PMID: 28671603]

[39] Menon S. S, R.; S, V. K., A review on biogenic synthesis of gold nanoparticles, characterization, and its applications. Resource-Efficient Technologies 2017; 3(4): 516-27.
[http://dx.doi.org/10.1016/j.reffit.2017.08.002]

[40] Huo Q, Colon J, Cordero A, *et al.* A facile nanoparticle immunoassay for cancer biomarker discovery. J Nanobiotechnology 2011; 9(1): 20.

[http://dx.doi.org/10.1186/1477-3155-9-20] [PMID: 21605409]

[41] Menon S, Rajeshkumar S, Kumar V. A review on biogenic synthesis of gold nanoparticles, characterization, and its applications. Resource-Efficient Technologies 2017; 3(4): 516-27.
[http://dx.doi.org/10.1016/j.reffit.2017.08.002]

[42] Kong F-Y, Zhang J-W, Li R-F, Wang Z-X, Wang W-J, Wang W. Unique roles of gold nanoparticles in drug delivery, targeting and imaging applications. Molecules 2017; 22(9): 1445.
[http://dx.doi.org/10.3390/molecules22091445] [PMID: 28858253]

[43] Xie X, Liao J, Shao X, Li Q, Lin Y. The effect of shape on cellular uptake of gold nanoparticles in the forms of stars, rods, and triangles. Sci Rep 2017; 7(1): 3827.
[http://dx.doi.org/10.1038/s41598-017-04229-z] [PMID: 28630477]

[44] aDavis TA, Volesky B, Mucci A. A review of the biochemistry of heavy metal biosorption by brown algae. Water Res 2003; 37(18): 4311-30.
[http://dx.doi.org/10.1016/S0043-1354(03)00293-8] [PMID: 14511701] bAnnamalai A, Babu ST, Jose NA, Sudha D, Lyza CV. Biosynthesis and characterization of silver and gold nanoparticles using aqueous leaf extraction of Phyllanthus amarus Schum & Thonn. World Appl Sci J 2011; 13(8): 1833-40.cSarkar J, Ray S, Chattopadhyay D, Laskar A, Acharya K. Mycogenesis of gold nanoparticles using a phytopathogen Alternaria alternata. Bioprocess Biosyst Eng 2012; 35(4): 637-43.
[http://dx.doi.org/10.1007/s00449-011-0646-4] [PMID: 22009439] dCai F, Li J, Sun J, Ji Y. Biosynthesis of gold nanoparticles by biosorption using Magnetospirillum gryphiswaldense MSR-1. Chem Eng J 2011; 175: 70-5.
[http://dx.doi.org/10.1016/j.cej.2011.09.041] eHonary S, Gharaei-Fathabad E, Paji ZK, Eslamifar M. A novel biological synthesis of gold nanoparticle by Enterobacteriaceae family. Trop J Pharm Res 2012; 11(6): 887-91.

[45] Mewada A, Oza G, Pandey S, Sharon M. Extracellular synthesis of gold nanoparticles using Pseudomonas denitrificans and comprehending its stability. J Microbiol Biotech Res 2012; 2(4): 493-9.

[46] Verissimo TV, Santos NT, Silva JR, Azevedo RB, Gomes AJ, Lunardi CN. *In vitro* cytotoxicity and phototoxicity of surface-modified gold nanoparticles associated with neutral red as a potential drug delivery system in phototherapy. Mater Sci Eng C 2016; 65: 199-204.
[http://dx.doi.org/10.1016/j.msec.2016.04.030] [PMID: 27157744]

[47] Fratoddi I, Venditti I, Cametti C, Russo MV. How toxic are gold nanoparticles? The state-of-the-art. Nano Res 2015; 8(6): 1771-99.
[http://dx.doi.org/10.1007/s12274-014-0697-3]

[48] Hainfeld JF, Slatkin DN, Focella TM, Smilowitz HM. Gold nanoparticles: a new X-ray contrast agent. Br J Radiol 2006; 79(939): 248-53.
[http://dx.doi.org/10.1259/bjr/13169882] [PMID: 16498039]

[49] Kumar A, Zhang X, Liang X-J. Gold nanoparticles: emerging paradigm for targeted drug delivery system. Biotechnol Adv 2013; 31(5): 593-606.
[http://dx.doi.org/10.1016/j.biotechadv.2012.10.002] [PMID: 23111203]

[50] Young KD. The selective value of bacterial shape. Microbiol Mol Biol Rev 2006; 70(3): 660-703.
[http://dx.doi.org/10.1128/MMBR.00001-06] [PMID: 16959965]

[51] Esfandyari-Manesh M, Ghaedi Z, Asemi M, *et al.* Study of antimicrobial activity of anethole and carvone loaded PLGA nanoparticles. journal of pharmacy research 2013; 7(4): 290-5.

[52] Al-Bahrani R, Raman J, Lakshmanan H, Hassan AA, Sabaratnam V. Green synthesis of silver nanoparticles using tree oyster mushroom Pleurotus ostreatus and its inhibitory activity against pathogenic bacteria. Mater Lett 2017; 186: 21-5.
[http://dx.doi.org/10.1016/j.matlet.2016.09.069]

[53] Shankar SS, Rai A, Ahmad A, Sastry M. Rapid synthesis of Au, Ag, and bimetallic Au core-Ag shell nanoparticles using Neem (Azadirachta indica) leaf broth. J Colloid Interface Sci 2004; 275(2): 496-

502.
[http://dx.doi.org/10.1016/j.jcis.2004.03.003] [PMID: 15178278]

[54] Katas H, Moden NZ, Lim CS, *et al.* Biosynthesis and Potential Applications of Silver and Gold Nanoparticles and Their Chitosan-Based Nanocomposites in Nanomedicine. J Nanotechnol 2018; 20184290705
[http://dx.doi.org/10.1155/2018/4290705]

[55] Khandel P, Yadaw RK, Soni DK, Kanwar L, Shahi SK. Biogenesis of metal nanoparticles and their pharmacological applications: present status and application prospects. Journal of Nanostructure in Chemistry 2018; 8(3): 217-54.
[http://dx.doi.org/10.1007/s40097-018-0267-4]

# CHAPTER 11

# Novel Drug Development Strategies- A Case Study With SARS-CoV-2

**Iqbal Azad**[1,*], **Tahmeena Khan**[1], **Mohammad Irfan Azad**[2] and **Abdul Rahman Khan**[1]

[1] *Integral University, Lucknow, India*
[2] *Jamia Millia Islamia, New Delhi, India*

**Abstract:** The current epidemic of Severe Acute Respiratory Syndrome coronavirus (SARS-CoV-2) has led to a major health crisis in 2020. SARS-CoV-2 has spike protein, polyproteins, nucleoproteins, and membrane proteins with RNA polymerase, 3-chymotrypsin-like protease, papain-like protease, helicase, glycoprotein, and accessory proteins. These are probable targets to be explored for the discovery of antiviral agents, still, to date, no definite treatment or vaccine has been discovered. Virtual screening with molecular docking has its advantage to speed up the drug development procedure in an accurate manner. In this chapter, novel computational strategies for drug discovery have been elaborated. Docking tools and drug filtering rules which may efficiently assist the drug development procedure and channelize the whole process in the right direction have also been discussed. A case study with 322 natural, semi-synthetic, and synthetic derivatives of citric acid (2-hydroxy-1,2-3-propane tricarboxylic acid), in search of a potential lead molecule to combat the novel coronavirus SARS-CoV-2, has been elaborated. The derivatives were explored from the PubChem database. The obtained library of compounds was filtered through Lipinski's rules, out of which, 74 obeyed the rule and were further subjected to molecular docking investigation against the SARS-CoV-2 replicase polyprotein 1a or pp1a (ID: 6LU7), with AutoDock Vina and iGEMDOCK. Deptropine possessed the highest binding affinity, in terms of released binding energy (-7.4 kcal/mol), against the SARS-CoV-2 replicase polyprotein 1a.

**Keywords:** Citric acid, Computational strategies, Drug, Docking, Repurposing, SARS-CoV-2, Virtual screening.

## INTRODUCTION

Three coronaviruses responsible for zoonotic diseases *viz.* Severe Acute Respiratory Syndrome coronavirus (SARS-CoV), Middle East Respiratory Syndrome coronavirus (MERS-CoV), and SARS-CoV-2, have caused lethal

---

[*] **Corresponding author Iqbal Azad:** Integral University, Lucknow, India; E-mail: iqbal.azad11@gmail.com

Tahmeena Khan, Abdul Rahman Khan, Saman Raza, Iqbal Azad and Alfred J. Lawrence (Eds.)
All rights reserved-© 2021 Bentham Science Publishers

pneumonia in humans by crossing the species barrier in recent times [1]. In 2002, SARS-CoV originated in the Guangdong region of China and was transmitted through the air to adjacent regions, leading to approximately 8,098 infections and 774 deaths [2, 3]. MERS-CoV outbreak took place in 2012 in the Arabian Peninsula and became a major public health issue. MERS-CoV reached 27 countries, infecting ~2,494 people and causing 858 casualties [4]. In December 2019, in Wuhan, a novel coronavirus (SARS-CoV-2) was discovered, spreading all over the world within few months [5]. It is linked with lethal pneumonia infecting over one crore people worldwide and causing more than five lakh deaths, till 13$^{th}$ July 2020 (Table **1**). MERS-CoV was found to be originated from bats [6]; similarly, SARS-CoV and SARS-CoV-2 which are closely associated, also originated from bats. SARS-CoV-2 is a positive-sense ssRNA virus. It is a β-coronavirus like MERS-CoV and SARS-CoV. The initial viral 30 kb RNA genome is termed as an open reading frame (ORF1a/b) part and interpreted through polyproteins (pp1a and pp1ab). The remaining portion of the viral RNA genome encrypts accessory proteins as well as four important structural proteins, namely spike (S) glycoprotein, small envelope (E) protein, matrix (M) protein, and nucleocapsid (N) protein [7].

## FACTORS AFFECTING THE SPREAD OF SARS-COV-2

### Environmental Factors

SARS-CoV-2 is a positive-sense ssRNA virus that mainly causes respiratory failure [8]. In the spread of the virus, numerous factors are involved, associated with the environment and the human correlation [9], in which migration, community interactions, dispersal of the human population, agricultural development, climate transformation as well as interaction with animals find a prominent place [10]. The correlation between the viral spread with major environmental factors like humidity, ambient temperature, and wind speed, *etc.* has not been satisfactorily explored. How the virus crosses the nose, ears, eyes, and mouth, *etc.* is not well recognized, and the release of SARS-CoV-2 as droplets and aerosols have also not being investigated thoroughly [11]. Owing to the versatility and mutation of COVID-19, control and prevention have drawn serious and urgent concern [12].

Concerns have originated to establish a clear relationship between environmental factors and SARS-CoV-2 cases [13]. According to the World Health Organization (WHO) (2020), sunlight, pH variations, and high temperature may curb viral growth [14]. A study conducted in China and Italy described the association between the SARS-CoV-2 spread with several environmental factors, such as humidity, wind speed, and temperature [15, 16]. Some researchers have described the resistance of the SARS-CoV-2 at low and high temperatures and found that at 4 °C its survival is for a longer period, whereas at 70 °C the virus survived only for 5 minutes. Wang *et al.* in their study conducted in 26 areas in China with a sample size of 24,139 positive SARS-CoV-2 cases, showed that a 1 °C rise in the minimum ambient air temperature decreased the cases by 0.86% [16, 17].

Table 1. Total cases of SARS-CoV-2 in top 10 countries (https://news.google.com/covid19) till 13th July 2020.

| S. No. | Country | Confirmed Cases | Recovered Cases | Deaths |
|---|---|---|---|---|
| 1 | United States | 33,66,515 | 9,88,656 | 5,71,444 |
| 2 | Brazil | 18,66,176 | 12,13,512 | 1,37,191 |
| 3 | India | 8,78,254 | 5,53,470 | 72,151 |
| 4 | Russia | 7,33,699 | 5,04,021 | 23,174 |
| 5 | Peru | 3,26,326 | 2,42,474 | 11,439 |
| 6 | Chile | 3,15,041 | 2,83,902 | 11,870 |
| 7 | Mexico | 2,99,750 | 1,84,764 | 6,979 |
| 8 | United Kingdom | 2,90,133 | No data | 35,006 |
| 9 | South Africa | 2,76,242 | 1,34,874 | 44,830 |
| 10 | Iran | 2,59,652 | No data | 4,079 |
| 11 | **Worldwide** | **1,29,45,505** | **70,01,675** | **5,71,444** |

## Food Materials, Handlers, and Packaging

Center for Disease Control and Prevention (CDC) has reported that the spread of SARS-CoV-2 through food materials, handlers, and packages has not been recognized till yet [17]. Recently, Seymour *et al.* (2020) described the perseverance of the SARS-CoV-2 on surfaces ranging from an hour to a few days [18].

## Water and Wastewater

Adopting good personal hygiene plays a crucial role in supporting human health [17]. Bhattacharjee (2020) has reported that SARS-CoV-2 also causes intestinal infection [19]. Some recent studies have also shown that around 2-10% of the active SARS-CoV-2 cases suffer from diarrhoea [13]. In sewage, the occurrence of the SARS-CoV-2 virus has been confirmed as well as its persistence in water and wastewater resources. Its occurrence in water and wastewater resources is based on important factors like sunlight, temperature, and organic content [17].

Viruses can adsorb organic content for protection against the direct contact of sunlight. Based on the newest report of WHO, there is no clear indication that COVID-19 spreads through drinking and wastewater [18].

## Air

SARS-CoV-2 infected patients usually do not show any visible symptoms in the initial phase and may become the transmitter. Present evidence is not enough to explain the air dispersal of the virus [21]. Hence, the use of viral filters and airflow variations, and the use of face masks are suggested to avoid any risk [22]. Even after wearing masks people have exhibited signs of respiratory distress, claustrophobia, skin irritation, and uneasiness [23, 24].

## Insects

Seven coronaviruses have been recognized to be pathogenic and cause acute respiratory syndrome. The spread and progress of the SARS-CoV-2 from bats to pangolins and then to humans has been testified. Some studies have reported that there is not any evidence of the spread of SARS-CoV-2 by blood-borne arthropods like mosquitoes. The latest investigation has reported that SARS-CoV-2 can spread through insects *via* mechanical transmission [25 - 28]. The environmental factors linked with the spread of COVID-19 are presented in Fig. (1).

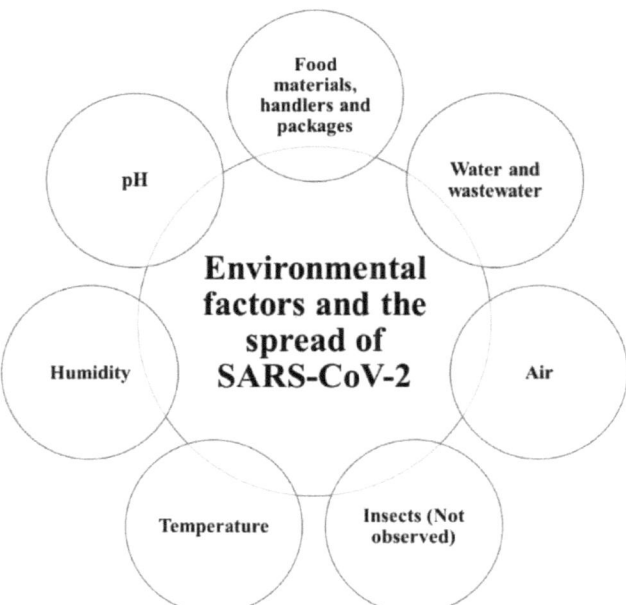

**Fig. (1).** Various environmental factors involved in the spread of SARS-CoV-2.

## Medicinal Intervention: The Scope of Virtual Screening

The process of development of new drug candidates through large progress sets and high cost has a very low success rate. The drug development procedure has been revolutionized by virtual screening [29]. Through virtual screening, the success rate of finding a new drug candidate has increased as well as the efficiency of targeted exploration, as compared to conventional drug design. Computational evaluation is widely used for the identification of drug targets. Some other important aspects explored through virtual screening are:

1. Preparation of small sets of large compound libraries of active compounds that can be experimentally confirmed.
2. Assessment of pharmacokinetic and pharmacodynamic properties, such as absorption, distribution, metabolism, excretion, and the potential for toxicity (ADMET) [30].
3. Designing new drug candidates, either based on fragmentation or by modification of functional groups.

Two broad classifications are used for virtual screening.

### *Structure-based Virtual Screening (SBVS)*

SBVS is based on the information of the target protein structure to analyse the binding energies of the drug candidate.

### *Ligand-based Virtual Screening (LBVS)*

It is based on the identification of active and inactive compounds, as well as evaluating quantitative structure-activity relation (QSAR) models [31].

Virtual screening is frequently used in the analysis of computer-aided drug design (CADD) to minimize the chances of failure, and to save time and money as the number of experimental drug candidates is significantly reduced, leading to more accurate predictions [32]. Virtual screening is done on the principle that a new lead may be developed from a single fragment, by trying different permutations and combinations *via* addition or removal of other fragments, to improve the overall activity. The major areas explored through virtual screening are presented in Fig. (**2**).

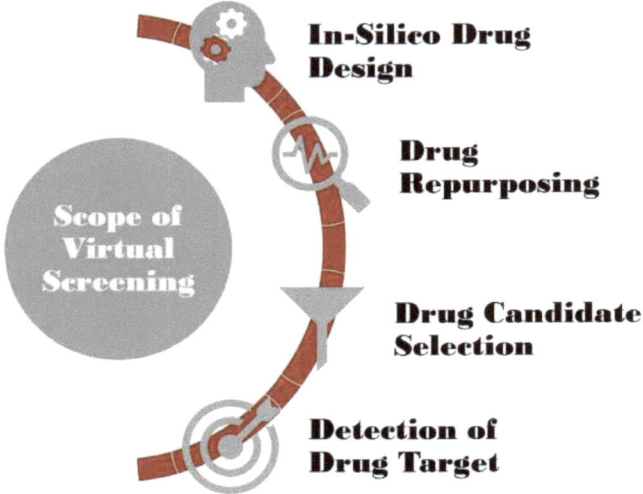

**Fig. (2).** The major scopes of virtual screening.

## *In-silico* Approaches

The protocol to find the lead compounds depends upon bioactivity information obtained from high throughput screening (HTS) operations [33]. The experimental investigation of a huge library of compounds requires a lot of time as well as expenses. *In-silico* screening significantly reduces the time and cost of drug discovery [34, 35].

## LIGAND SELECTION CRITERION AS PHARMACEUTICAL LEADS

There are many pharmacological rules, such as Lipinski's rule of five, Ghose filter, Veber Rules, MDDR-like rules, CMC-50 like rules, WDI-like rules, Bayer filter, Rule of three, and Weighted and Unweighted QED, which are applied to filter and select the most suitable hits.

## LIPINSKI'S RULE OF FIVE

Lipinski's rule of five is known as the rule of five (Ro5) or Pfizer's rule of five and was formulated by C. A. Lipinski *et al.* in 1997 [36]. This rule estimates the drug-likeness of a molecular system, based on five broadly applicable chemical as well as physical properties, to define the pharmacological or biological activity of systems that would categorize it as a probably orally active drug. The rule extensively correlates the molecular parameters of small molecules with their pharmacokinetics activity inside the human body [37]. There are five criteria proposed by Lipinski that assist in the search of a good drug candidate, *viz.,* the calculated logarithm of the octanol-water partition coefficient LogP ≤ 5, (HBD) NHOH ≤ 5, (HBA) ON ≤ 10, RB ≤ 10, and MW≤500 Da (delta).

## GHOSE FILTER

The Ghose filter, developed by AK Ghose *et al.* in 1987 [38], is used for the assessment of an ideal drug candidate. In their evaluation, they observed that the library of drugs available in the comprehensive medicinal chemistry (CMC) database (more than 80% of compounds covered) possesses a few common molecular parameters that would be used to categorize a molecular system as a potential drug candidate [39]. The molecular descriptors proposed to identify the drug-likeness of a molecular system are as follows- LogP -0.4-56, the total number of atoms 20-70, molecular weight 160-480, molar refractivity 40-130.

## VEBER'S RULES

Oral bioavailability can also be predicted by Veber's rules, proposed by Veber *et al.* in 2002. According to their findings, only molecular weight cutoff at 500 Da does not considerably differentiate between compounds with poor and good oral bioavailability [40]. According to Veber *et al.*, compounds with good oral bioavailability in rats must meet the following criteria, *viz.,* less molecular flexibility or fewer rotatable bonds (10 ≥ RB) and polar surface area (140Å$^2$ ≥ PSA) or 12 ≥ H-bond donors and acceptors with a high probability.

## MDDR-LIKE RULES

T.I. Oprea, in the year 2000, described the difference between drug and non-drug candidates and termed it as MDDR-like rule, by utilizing subsequent compound databases: MACCS-II Drug Data Report (MDDR), Physician Desk Reference, Comprehensive Medicinal Chemistry, New Chemical Entities, Current Patents Fast-alert, and the Available Chemical Directory (ACD) [41]. The ACDF and MDDRF subsets were generated by eliminating some reactive functional groups, such as acyl-halides, sulfonyl-halides, Michael acceptors, *etc.* from the ACD and MDDR databases, respectively. ACD is a non-drug and the MDDR is a drug database. He proposed the MDDR-like rule to differentiate between drug and non-drug candidates based on the number of rings, rigid bonds, and rotatable bonds [39]. For a druglike candidate, the ideal conditions to be followed are: rings ≥ 3, rigid bonds ≥ 18 and RB ≥ 6, non-druglike candidate condition are rings ≤ 2, rigid bonds ≤ 17 and RB ≤ 5.

## CMC LIKE RULES

Comprehensive Medicinal Chemistry (CMC) defined the druglike parameters of a molecular system. According to the CMC-50 rule, the ideal parameters to be followed by a drug-like candidate are logP 1.3-4.1, MW 230-390 Da, MR 70-110, and the total number of atoms 30-55 [39].

## WDI-LIKE RULES

R.D. Brown et al. defined some molecular descriptors for the search of a good drug candidate in their chapter titled 'Tools for designing diverse, drug-like, cost-effective combinatorial libraries' published in 2001 [42]. The molecular descriptors defined in WDI-like rule, which should be fulfilled by a potential lead candidate are: MW 550-12 Da, MR 120-8, RB 13-8, nHBD 12-5, nHBA 20-9, alogP 10-5.

## BAYER FILTER

Muegge et al. have established a modest pharmacophore-based filter to distinguish between druglike and non-druglike chemical entities on the fact that non-drugs are usually under-functionalized [43]. The main pharmacophore parameters of Bayer's filter are MW 200-600, logP -2-5, TPSA $\leq$ 150, Rings $\leq$ 7, RB $\leq$ 15, HBD $\leq$ 10, HBA $\leq$ 5, Carbons > 4, and Heteroatoms >1. Consequently, a molecule with less than two (under functionalization) or more than seven (over functionalization) pharmacophore specifics fails the filter.

## RULE OF THREE

M. Congreve et al., in 2003, defined some molecular descriptors for the fragment-based drug design (FBDD) that were recognized for the variety of biological targets. This molecular descriptor is known as the 'rule of three' in which a fragment of a drug candidate must have MW <300 Da, cLogP $\leq$3, HBD $\leq$3 and HBA $\leq$3. Furthermore, their findings also identified rotatable bond $\leq$3 and PSA $\leq$60 as possible convenient measures for fragment assortment [44].

## WEIGHTED AND UNWEIGHTED QED

G. R. Bickerton et al., in 2012, extensively explored the available bioactive compounds and the difficulties associated with molecular target identification [45]. Based on their findings, they developed a concept termed the quantitative estimate of drug-likeness (QED), which is directly applicable to several practical sets. Molecular descriptors suggested in the QED are MW, AlogP, PSA, HBD, HBA, RB, aromatic bonds count (ABC), and the number of structural alerts.

## DRUG REPURPOSING

Drug repurposing is an effective alternative method for exploring the activity spectrum of already known drugs. In general, the objective of drug repurposing is to discover new relationships between drugs and disease [46]. From the viewpoint of unmet therapeutic requirements, drug repurposing has emerged as a great opportunity for discovering treatment options for rare diseases [47]. Out of the 7,000 rare diseases known, more than 95% do not have any approved medicinal drugs by the US Food and Drug Administration (FDA). Fig. **3** represents some of the drugs used for repurposing for the treatment of COVID-19.

**Fig. (3).** Most recommended repurpose drugs (off label) for treatment of COVID-19.

## DRUG REPURPOSING ADVANTAGES

For efficient drug development, the scientific fraternity relies on systematically organized and information-driven drug repurposing methods, which are generally facilitated by computational techniques [48, 49].

The most common benefit of drug repurposing is that the repurposed drugs have been previously confirmed to be suitably safe and involve minimal chances to fail in efficacy trials, as the approved drugs have already passed all the clinical trials and regulatory analysis, and qualified post-marketing investigation.

## DRUG CANDIDATE SELECTION

In drug development, one of the most critical parameters is to find out the most suitable clinical applicant which makes the evaluation of ADMET, pharmacodynamic, and pharmacokinetic properties essential [42]. These evaluations offer valuable information in the initial stage of drug repurposing to minimize failure in medical trials afterward [50].

## DETECTION OF TARGETS FOR DRUGS AND THEIR MECHANISM OF ACTION

Studies have suggested that 7% of approved drugs are developed without identification of the target protein [51]. The starting point of the drug discovery procedure is to assess the interaction of a drug applicant with a target protein. Considerable efforts have been made to find the target proteins, as well as evaluate binding studies of approved drugs with G-protein coupled receptors (GPCRs) and kinases [52].

## MOLECULAR DOCKING

Almost all cellular processes involve protein-protein interaction. The three-dimensional (3-D) protein-protein interaction may provide an insight into their mechanisms and role. Molecular docking evaluation helps in the deliberation and finding of the interaction of protein-protein as well as between drug-protein, which may elucidate the possible mechanism of action of drug molecule [53]. Docking findings include binding of the ligand/drug candidate/ drug molecule to the ideal binding or active site of the target protein/DNA molecule, referred to as receptor. The interaction between the ligand and protein takes place through the following interactions:

1. Electrostatic interactions, including charge-charge, dipole-dipole and charge-dipole interactions known as coulombic forces of attraction. These forces are developed due to the presence of charged species.
2. Electrodynamic forces or van der Waals interactions, which include weak London dispersion forces and stronger dipole-dipole forces.
3. Steric forces are produced due to the overcrowding of molecules that affect chemical reactivity.
4. Solvent-related forces arise due to the interaction of solvent and protein/ligand.

Critical ligand binding confirmation, molecular mechanisms and interaction between the ligand-receptor complexes can be deliberated based on the docking outcomes. The docking outcomes can be interpreted based on the released binding energy as well as the dissociation constant ($K_d$) associated with the stability of docking complexes [54].

## TYPES OF MOLECULAR DOCKING

Molecular docking predicts the most suitable confirmation of ligand associated with specific binding energy at the binding site of the receptor protein. Based on the nature of the ligand, docking can be classified as flexible (*e.g.* ligand-protein) and rigid (*e.g.* protein-protein) (Fig. **4**).

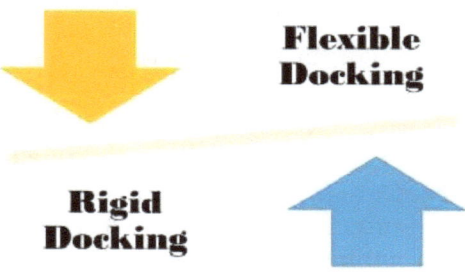

**Fig. (4).** Basic docking types.

## FLEXIBLE DOCKING

The flexible docking approach does not require prior information regarding the ligand's conformation and involves the exploration of 6+N translational, rotational and conformational modes. Flexible docking is most preferred when there is no substantial data available on the ligand conformation [37]. Four variable approaches are presently utilized for flexible docking:

a. Molecular dynamics or Monte Carlo docking of the whole molecule
b. Point combinatorial exploration
c. Ligand collection
d. Spot mapping and fragment gathering

## FLEXIBLE DOCKING: CHALLENGES AND REQUIREMENTS

Receptor protein flexibility is still a hard task to achieve; however, ligand flexibility has been achieved computationally, allowing changes in the internal geometry of the interacting molecules during the formation of a complex.

The main binding site of the target protein generally has 10-20 amino acids involved in the possible rotatable torsion. As compared to degrees of freedom (6-

12) of the druglike molecule, the number of amino acids is greater [55]. Therefore, a completely flexible receptor/ligand docking algorithm is needed as well as a higher degree of freedom.

## RIGID DOCKING: CHALLENGES AND REQUIREMENTS

Docking-based algorithms have numerous limitations, such as the requirement of the 3-D structure of the receptor protein, which reduces the interaction space of receptor protein. To overcome this problem, researchers have tried homology modelling [56]. A less time-consuming, rigid-docking approach that does not deliberate upon protein flexibility, has been used to resolve the issue [57].

If a complex has limited conformational variations, the structures of target proteins are observed as rigid forms and thus a 3D rotational, as well as translational finding, is achieved [37]. A re-positioning of the subsequent complexes might be assumed, probably using exhaustive computational calculations in which conformational flexibility might be incorporated into the protocol. Several receptor protein estimation approaches have utilized rigid-body docking practices and are applied to solve actual biological difficulties [58]. The rigid docking procedure does not allow modification in molecular conformation. Thus, to find the structural insights of the receptor protein, like its conformation, post-docking analysis is required [56 - 59]. The rigid docking procedure shows docking space. But, in some cases, adjacent-native structures cannot be created in a conventional rigid model [60, 61].

## MOLECULAR DOCKING STUDIES OF PLANT-BASED ACTIVE CONSTITUENTS IN SEARCH OF A LEAD MOLECULE TO COMBAT SARS-COV-2

### Role of Immunity

The immune system is extremely elaborate, consisting of specific cells that help combat infections and diseases by modification and regulation of malignant and foreign cells. The spleen and thymus are the main organs of the human immune system, along with lymph nodes and bone marrow. B and T are the two main types of immune cells. B cells are composed of protein, involved in the production of antibodies, and designed to identify an exact antigen, whereas, T cells are charged in nature and destroy antigens [62].

Around 80% of the COVID-19 patients have been asymptomatic so far or have displayed minor to modest symptoms, whereas 15% suffered from severe pneumonia and nearly 5% in severe cases suffered from septic shock and/or multiple organ failure and acute respiratory distress syndrome (ARDS). Fever, fatigue, and respiratory symptoms like cough, sore throat, and shortness of breath are the most common symptoms experienced by the patients. Innate and adaptive immune responses are stimulated by the infection of SARS-CoV-2. A well-functioning immune response forms the front line of resistance against the viral infection [63].

Studies have shown changes in both the innate and adaptive immune systems on the onset of the viral infection. In patients with severe COVID-19, a noticeable lowering in the levels of the total quantity of $CD4^+$, $CD8^+$, natural killer (NK) and B cells has been observed, along with a reduction in the total count of monocytes, eosinophils and basophils. Moreover, the serum levels of proinflammatory cytokines (IL-6, IL-1β, IL-2, IL-8, IL-17, G-CSF, GMCSF, IP-10, MCP-1, CCL3 and TNFα) have been found to increase remarkably, especially in serious patients. In a single-center study of a unit of 452 COVID-19 patients in Wuhan, serious patients showed a remarkable decrease in the total quantity of T cells (helper T cells and suppressor T cells), whereas an increase in the number of naive T cells was found [63]. The naive and memory T cells are very important immune components, whose stability is critical for preserving an extremely effective defence response. Naive T cells provide defence against the previously and newly unrecognized infection by a harmonized release of cytokines, while memory T cells regulate an antigen-specific immune response. Based on the comparison of the genomic sequence of SARS-CoV-2 with SARS-CoV, it is anticipated that SARS-CoV-2 can accept comparable approaches with SARS-CoV to control the host's innate immune response [64].

## PLANT-BASED RESOURCES AS NATURAL IMMUNITY BOOSTERS

The importance of medicinal plants has increased in the past few decades due to the improved productivity of novel Phyto-drugs and the increasing awareness about natural product consumption. Many plants and their constituents are enriched with varied medicinal properties, such as anti-inflammatory, antioxidant, antimicrobial, immune-modulatory, and anticancer, while some plants and their metabolites have health-improving properties [65, 66]. Numerous natural immune boosters [67, 68] such as ginger (*Zingiber officinale*), garlic (*Allium sativum*), green tea (*Camellia sinensis*), *Echinacea* (purple coneflower), black cumin (*Nigella sativa*) and citrus fruits are easily available(Fig. **5**).

## GINGER (*ZINGIBER OFFICINALE*)

Current exploration has highlighted the use of old traditional herbs to treat several illnesses through diet-based therapy. Ginger is well-known for its unique therapeutic implication. The common bioactive ingredients of ginger are α-zingiberene, α-farnesene, β-bisabolene, α-curcumin, 6-gingerol, 6-shogaol,

Paradol, Zingerones *etc* [69]. (Fig. **6**).

Ginger is used as a natural immunity booster, protecting from the effects of harmful chemicals. 50-100 mg/kg 8-gingerol has been known to block humoral and cellular immune responses *via* direct inhibition of T and B lymphocytes cells [70, 71].

**Fig. (5).** Various natural immunity boosters.

**Fig. (6).** Some active ingredients of Ginger.

## GARLIC (*ALLIUM SATIVUM L.*)

The most common bioactive ingredients of garlic are diallyl sulfide, diallyl disulfide, δ-glutamyl-S-allyl-L-cysteines, S-allyl-mercaptocysteine, S-allyl-L-cysteine sulfoxides (Fig. 7) and non-starch polysaccharides. Garlic works as a natural immune booster reduces platelet accumulation and it also acts as an effective antioxidant by scavenging free radicals [72, 73].

**Fig. (7).** Some active ingredients of Garlic.

## GREEN TEA (*CAMELLIA SINENSIS*)

Worldwide, the consumption and popularity of green tea have increased manifold due to its tremendous antioxidant, anti-inflammatory and anticancer activities. Epigallocatechin-3-gallate is a well-known active ingredient of green tea [74] which is the main antioxidant. Epigallocatechin-3-gallate controls the formation of immunoregulatory cytokines in dendrite cells and works as a suppressor of T cell activation (Fig. 8) [75, 76].

**Fig. (8).** Active ingredient of green tea.

## PURPLE CONEFLOWER (*ECHINACEA*)

*Echinacea* is infused with immunity-boosting properties. *Echinacea sp.* has several bioactive ingredients, such as betaine, sesquiterpenes, caryophyllene, polyacetylene, rosmarinic acid, glycosides, echinacoside, chicoric acid, arabinogalactan-proteins and alkyl amides (Fig. **9**) [77]. The immunomodulatory action of *Echinacea sp.* is due to alkyl amides. Arabinogalactan-protein and other bioactive ingredients of *Echinacea* act as stimulators or inhibitors of the traditional and alternate pathways of immunity. These components are also exciting targets for drug discovery and development for mucosal immune suppression and reduce the risk of upper respiratory tract infections [78], thereby activating cellular immunity and stimulating phagocytosis of neutrophils.

**Fig. (9).** Some active ingredients of *Echinacea*.

## BLACK CUMIN (*NIGELLA SATIVA*)

Black cumin seeds are traditionally used in the treatment of severe diseases. Various studies have shown their anti-inflammatory and immunomodulatory effects. Bioactive constituents of cumin also have controlling properties *via* inflammatory cells by influencing the immune system and recover helper T cells (T4) to suppress the T cell (T8) ratio [75].

## CITRUS FRUITS

Citrus fruits belong to the family *Rutaceae*. Orange, mandarin, lime, lemon, grapefruit and citron are some of the common examples of family *Rutaceae*.

Citrus fruits are mainly consumed in food, beverage and pharmaceutical industries as additives, spices, and chemoprophylactic drugs [79]. Citrus fruits are well-known robust natural immunity boosters. The main constituents of citrus fruits are citric acid, vitamin C, flavones and flavanones compounds (Fig. **10**) [80]. Due to the presence of vitamin C, they also exhibit antioxidant activity and increase white blood cell production to fight against infections. Due to the lack of the main enzyme in the biosynthetic pathway of vitamin C, it cannot be synthesized in the human body [81].

**Fig. (10).** Some active ingredients of citrus fruits.

## MOLECULAR DOCKING STUDIES WITH SOME BIOACTIVE CONSTITUENTS OF CITRUS FRUITS

In this study, 322 natural and synthetic derivatives of naturally concentrated citric acid (2-hydroxy-1,2,3-propane tricarboxylic acid) in citrus plants were explored from PubChem (https://pubchem.ncbi.nlm.nih.gov/), out of which 74 were selected using Lipinski's rules and subjected to molecular docking interactions against the COVID-19 replicase polyprotein 1a or pp1a (PDB ID: 6LU7), downloaded from Protein Data bank (http://www.rcsb.org/).

## SOFTWARES USED

The compounds were subjected to molecular docking using the MGL Tools 1.5.6 and AutoDock Vina 1.1.2 [82] to understand the drug-receptor interaction with target receptor enzymes to examine the potential binding mode with the lowest energy. Molecular docking validation was done with iGEMDOCK 2.1 [83]. For the preparation of the 2D structure of ligands, ChemDraw Professional 15.1 was used, while the 3D structure, as well as docking interactions, were visualized with the help of Biovia Discovery Studio Visualizer 2017 R2 [84].

## THE OPEN READING FRAME (ORF)

In molecular genetics and bioinformatics, an open reading frame (ORF) plays a significant step in encoding genes in genomic sequences. *In silico* analyses are used to evaluate and define the ORFs [85]. A reading frame is initiated with a specific start codon (AUG), extends over a sequence of triplets instead of amino acids, and ends with one of the three-terminal codons (UAA, UGA, UAG).

## TARGET PROTEINS
### Polyproteins (Proteases)

Covalently associated specific proteins with various functions are very common in nature and are known as polyproteins. In SARS-CoV-2, the complete proteome is composed of two long polyproteins that are encoded by a single large ORF. Polyproteins are mostly used by viruses to build their proteome [86]. Many researchers have targeted polyproteins as the inhibition of polyproteins retards the viral maturation, providing a direct mode of action to fight against a viral disease.

## SPIKE (S) PROTEIN

In general, the spike protein is composed of the ectodomain, I-transmembrane and intercellular domain region. The entry of SARS-CoV-2 into the host cell is facilitated by the transmembrane spike protein that appears in a trimeric form. Spike protein is composed of two functional subunits S1 and S2 [87]. S1 subunit regulates the binding of the host cell receptor to the virus, whereas the S2 subunit is involved in the union of the viral and cellular membranes. Various coronaviruses use separate domains inside the S1 subunit to identify the variability of receptors' surface, based on the viral species. SARS-CoV-2 interacts with the angiotensin-converting enzyme 2 (ACE2) receptor, to cross the target cell directly. Therefore, it has become the key target for vaccine development and drug design [88].

## NUCLEOCAPSID (N) PROTEIN

The nucleocapsid protein is composed of N-terminal, central linker and C-terminal. The N-terminal helps in the binding of RNA, and the C-terminal plays a significant part in dimerization [12]. The N protein causes the inhibition of cell growth in humans by the inhibition of cytokinesis. Therefore, N protein can be explored as a probable target for the development of a DNA vaccine [89].

## ENVELOPE (E) PROTEIN

The envelope protein is composed of the hydrophobic domain and cytoplasmic tail. It is the smallest transmembrane structural protein of coronavirus. The envelope protein is recognized to facilitate morphogenesis during viral assembly, the release of the virus, and is also responsible for virulence [90].

## M-PROTEIN

M protein supports intracellular homeostasis in the virus [91]. M-Protein consists of small N-terminal and large C-terminal. Coronavirus enters the host by the associated interaction with M-spike protein, M-Nucleocapsid proteins, and M-M proteins [92]. The interaction of the host takes place through the M-spike protein and the nucleocapsid-RNA complex while stabilization is related to the interaction of M-Nucleocapsid proteins. Nucleocapsid and M proteins also help in the release of the virus [93]. The structure of SARS-CoV-2 is presented in Fig. (11).

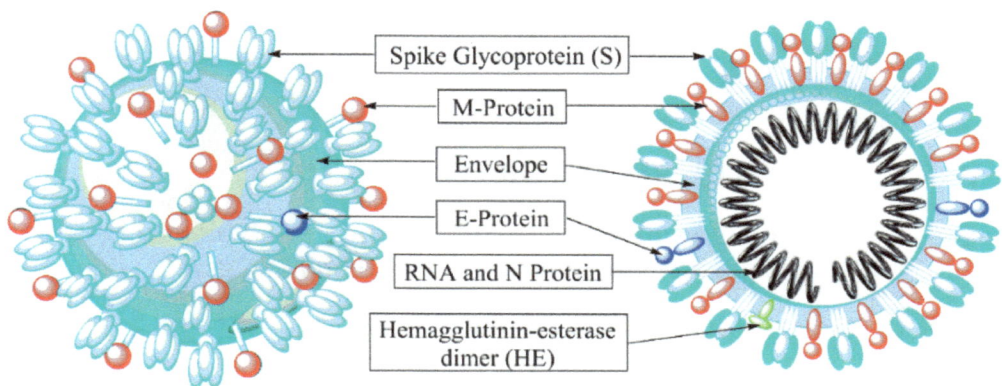

**Fig. (11).** Diagrammatic structure of novel coronavirus.

## SARS-COV HELICASE

The SARS-CoV Helicase enzyme belongs to the superfamily 1. Helicase enzymes can be used as a probable target for designing novel drugs. Intending to design the helicase inhibitors, the main issue is toxicity, leading to some side effects [94, 95]. The structures of various receptor proteins are presented in Fig. (12).

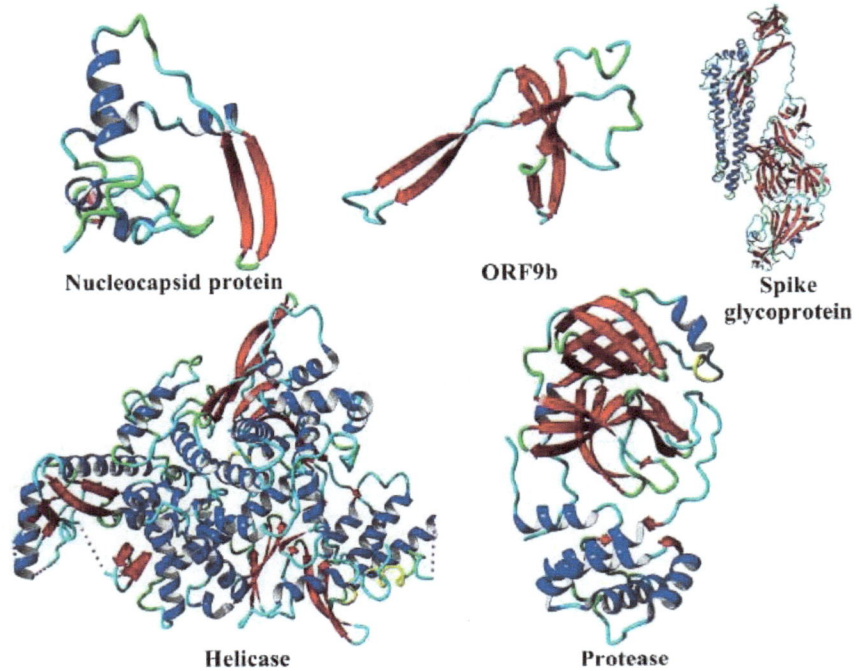

**Fig. (12).** 3D ribbon structure of various receptor proteins of SARS-CoV-2.

## PREPARATION OF THE RECEPTOR FOR DOCKING

The X-ray crystal structures of the COVID-19 replicase polyprotein 1a or pp1a (ID: 6LU7) were obtained from the Protein Data Bank in PDB format.

## PREPARATION OF LIGANDS FOR DOCKING

The X-ray crystal structures of the ligand molecules (natural and synthetic) were downloaded from the PubChem in SDF format. Biovia Discovery Studio was used to visualize the structures and was used to convert SDF format into PDB as well as in Mol format. The structures of the top five ligands and the reference drug chloroquine are shown as (Fig. **13**).

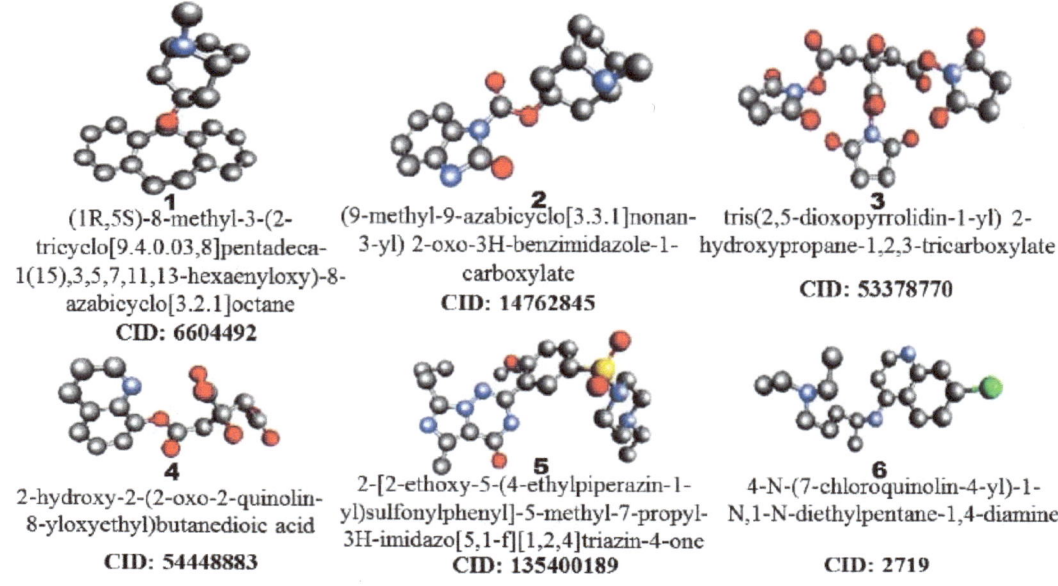

**Fig. (13).** Top five active natural lead derivatives (1-5) and control drug (6, Chloroquine).

## AUTODOCK VINA

Auto Dock Vina is a scoring function with improved accuracy and speed for molecular docking as well as virtual screening studies. During the optimization process, Vina follows a sophisticated gradient optimization protocol [82]. Vina does not require a grid parameter as well as a docking parameter file, it simply runs with PDBQT files of ligand and target protein, which is easily prepared by MGL tools. In this study, grid space centre coordinates were set at -26.00×13.00×59.00Å along, x, y, z axes, and grid space sizes are 35.00×35.00×35.00Å along, x, y, z axes.

## IGEMDOCK

iGEMDOCK is an automated graphical program used for the combined study of docking, screening, and post-screening investigation [83, 96]. Also, iGEMDOCK automatically identifies the main interior cavity and pocket of the receptor protein and proposes a possible pocket site [84]. The input file of iGEMDOCK is PDB/MOL and is generated as a PDB file. The optimized file of the ligand and receptor protein was directly uploaded, and docking variables were with a genetic algorithm (GA) for the docking analysis. Very slow docking was performed with the docking parameters, population size: 800, generations: 80, and number of solutions: 10. The best docking pose, having the least energy, was automatically obtained after docking. The iGEMDOCK scoring function is represented as equation **1**.

$$Fitness = vdW + H\ bond + Elec \tag{1}$$

where, vdW = van der Waals energy, H bond = hydrogen bonding energy and Elect = electrostatic energy

## RESULTS AND DISCUSSION

During the preliminary evaluation, a total of 322 natural and synthetic derivatives of citric acid were generated as a library. The obtained library was subjected to pharmacological filtration on Lipinski's parameters. 74 derivatives obeyed the rule of five and were further subjected to virtual screening *via* AutoDock Vina with the help of VEGA ZZ v3.2.0. Chloroquine was selected as the control drug, due to its potential binding ability with the viral proteins. The top 10 ligands, along with their binding energies upon binding with the receptor protein polyprotein 1a, are presented in Table **2**.

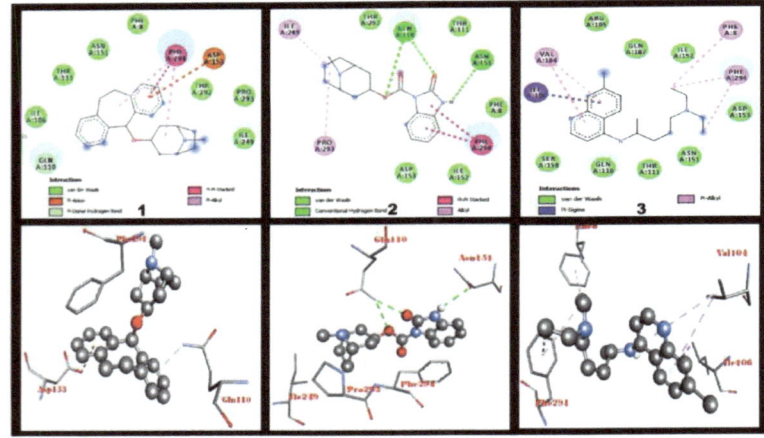

**Fig. (14).** 2D and 3D docking interaction poses of top two natural lead derivatives (1-2) and control drug (3, Chloroquine).

Out of the 74 screened ligands, six exhibited potential binding with the target protein. The control drug exhibited binding energy *viz.* -5.6 kcal/mol with the polyprotein 1a and the least binding energy was released by deptropine citrate, with a binding affinity of -7.4 kcal/mol, followed by benzimidazole-1-carboxylate (-7.3 kcal/mol) and 2,5-dioxopyrrolidin (-7.2 kcal/mol) (Table **2**). To recognize the best docking pose, the docked complex was viewed in BDS visualizer, along with the most interacting amino acid residues in the active site which were PHE8,

VAL104, ARG105, GLN107, ASN151, ASP153, and THR292, respectively. All the selected ligands also exhibited hydrogen bonding interactions, while the control drug bonded through van der Waals interaction with ARG105, GLN107, ASN151, ILE152, ASP153, and SER158, respectively, in the active site. The docking poses of the two best ligands and control drugs are presented in Fig. (**14**).

Table 2. Docking interactions of the top five natural lead derivatives (1-5) and control drug (6, Chloroquine) with polyprotein 1a or pp1a generated using iGEMDOCK and AutoDock Vina.

| S. No. | CID of Ligands | AutoDock Vina | iGEMDOCK | Docking interactions | | |
|---|---|---|---|---|---|---|
| | | BE [a] | TE (kcal/mol)[b] | vdW[c] | H Bond[d] | π-π, π-Other |
| 1 | 6604492 | -7.4 | -89.83 | PHE8, ILE106, THR111, ASN151, ILE249, THR292, PRO293 | GLN110 | Asp153, phe294 |
| 2 | 14762845 | -7.3 | -89.05 | PHE8, ILE152, ASP153, THR111, THR292 | GLN110, ASN151 | ILE249, PRO293, PHE294 |
| 3 | 53378770 | -7.2 | -88.24 | PHE8, VAL104, ARG105, GLN107, ASN151, THR292 | GLN110, PHE294 | ILE106, ILE249, PRO293 |
| 4 | 54448883 | -7.1 | -88.22 | THR25, THR26, LEU27, MET49, HIS164, GLN189 | ASN142, GLY143, CYS145, GLU166 | HIS41, MET165 |
| 5 | 135400189 | -7.1 | -83.80 | PHE8, VAL104, ARG105, GLN107, ASN151, ASP153, THR292 | GLN110, PHE294 | ILE106, ILE249, PRO293 |
| 6 | 2719 | -5.6 | -74.61 | ARG105, GLN107, ASN151, ILE152, ASP153, SER158 | - | PHE8, VAL104, ILE106, PHE294 |

[a]Binding Energy (kcal/mol)

[b]Total Energy (kcal/mol)

[c]van der Waal forces

[d]Hydrogen Bond

## CONCLUDING REMARKS

Virtual screening, including drug repurposing, is a useful and relatively less explored protocol to boost pharmaceutical research. Molecular docking is another tool that aids drug target and receptor protein identification. The chapter has highlighted some important virtual screening procedures in search of lead identification for SARS-CoV-2, which can also be explored for other novel diseases. Pharmacologists may discover the current databases to assimilate their content into a modified repurposing pipeline, whereas the experimentalists can take advantage of the protocol recorded while starting novel drug repurposing premises. In the future, an online platform with user-friendly handling should be developed for drug repurposing, to decrease operator time, multi-step investigation, and human involvement. This program should have three characteristic approaches for successful lead identification: 1) Automated comparable reverse docking database, 2) Small molecule library database covering drugs and non-drugs, 3) Receptor protein database having a wide range of protein structures from the protein data bank.

## CONSENT FOR PUBLICATION

Not Applicable.

## CONFLICT OF INTEREST

The author confirms that this chapter contents have no conflict of interest.

## ACKNOWLEDGEMENT

Declared none.

## REFERENCES

[1] Gorbalenya AE, Baker SC, Baric RS, *et al*. Coronaviridae Study Group of the International Committee on Taxonomy of Viruses. The species Severe acute respiratory syndrome-related coronavirus: classifying 2019-nCoV and naming it SARS-CoV-2. Nat Microbiol 2020; 5(4): 536-44.
[http://dx.doi.org/10.1038/s41564-020-0695-z] [PMID: 32123347]

[2] Walls AC, Park YJ, Tortorici MA, Wall A, McGuire AT, Veesler D. Structure, Function, and Antigenicity of the SARS-CoV-2 Spike Glycoprotein. Cell 2020; 181(2): 281-292.e6.
[http://dx.doi.org/10.1016/j.cell.2020.02.058] [PMID: 32155444]

[3] Wu D, Wu T, Liu Q, Yang Z. The SARS-CoV-2 outbreak: What we know. Int J Infect Dis 2020; 94: 44-8.
[http://dx.doi.org/10.1016/j.ijid.2020.03.004] [PMID: 32171952]

[4] Omrani AS, Al-Tawfiq JA, Memish ZA. Middle East respiratory syndrome coronavirus (MERS-CoV): animal to human interaction. Pathog Glob Health 2015; 109(8): 354-62.
[http://dx.doi.org/10.1080/20477724.2015.1122852] [PMID: 26924345]

[5]     Astuti I, Ysrafil. Severe Acute Respiratory Syndrome Coronavirus 2 (SARS-CoV-2): An overview of viral structure and host response. Diabetes Metab Syndr 2020; 14(4): 407-12.
[http://dx.doi.org/10.1016/j.dsx.2020.04.020] [PMID: 32335367]

[6]     Ujike M, Taguchi F. Incorporation of spike and membrane glycoproteins into coronavirus virions. Viruses 2015; 7(4): 1700-25.
[http://dx.doi.org/10.3390/v7041700] [PMID: 25855243]

[7]     Brown TA. Understanding a Genome Sequence, Genomes. 2nd ed., Garland Science 2002.

[8]     World Health Organization. 2020.https://www.who.int/dg/speeches/detail/who-director-genera--s-opening-remarks-at-the-media-briefing-on-covid-19---16-march-2020

[9]     Swerdlow DL, Finelli L. Preparation for Possible Sustained Transmission of 2019 Novel Coronavirus: Lessons From Previous Epidemics. JAMA 2020; 323(12): 1129-30.
[http://dx.doi.org/10.1001/jama.2020.1960] [PMID: 32207807]

[10]    Dakin HD, Dunham EK. The Disinfection of Drinking Water. BMJ 1917; 1: 682-4.
[http://dx.doi.org/10.1136/bmj.1.2943.682] [PMID: 20768599]

[11]    Thienemann F, Pinto F, Grobbee DE, et al. World Heart Federation Briefing on Prevention: Coronavirus Disease 2019 (COVID-19) in Low-Income Countries. Glob Heart 2020; 15(1): 31.
[http://dx.doi.org/10.5334/gh.778] [PMID: 32489804]

[12]    World Health Organization. https://www.who.int/2018.

[13]    Wang M, Jiang A, Gong L, et al. Temperature Significant Change COVID-19 Transmission in 429 Cities. medRxiv 2020.
[http://dx.doi.org/10.1101/2020.02.22.20025791]

[14]    Pica N, Bouvier NM. Environmental factors affecting the transmission of respiratory viruses. Curr Opin Virol 2012; 2(1): 90-5.
[http://dx.doi.org/10.1016/j.coviro.2011.12.003] [PMID: 22440971]

[15]    Peci A, Winter AL, Li Y, et al. Effects of Absolute Humidity, Relative Humidity, Temperature, and Wind Speed on Influenza Activity in Toronto, Ontario, Canada. Appl Environ Microbiol 2019; 85(6): e02426-18.
[http://dx.doi.org/10.1128/AEM.02426-18] [PMID: 30610079]

[16]    Bhattacharjee S. Statistical Investigation of Relationship between Spread of Coronavirus Disease (COVID-19) and Environmental Factors Based on Study of Four Mostly Affected Places of China and Five Mostly Affected Places of Italy. 2020.

[17]    Eslami H, Jalili M. The role of environmental factors to transmission of SARS-CoV-2 (COVID-19). AMB Express 2020; 10(1): 92.
[http://dx.doi.org/10.1186/s13568-020-01028-0] [PMID: 32415548]

[18]    Seymour N, Yavelak M, Christian C, Chapman B. Seymour N, Yavelak M, Christian C, Chapman B, D. M. COVID-19 FAQ for Foodservice: Receiving and Food Packaging. 2020.

[19]    US Food and Drug Administration. Food Safety and the Coronavirus Disease 2019.https://www.fda.gov/food/food-safety-during-emergencies/food-safet--and-coronavirus-disease-2019-covid-19

[20]    Xiao F, Tang M, Zheng X, Liu Y, Li X, Shan H. Evidence for Gastrointestinal Infection of SARS-CoV-2. Gastroenterology 2020; 158(6): 1831-1833.e3.
[http://dx.doi.org/10.1053/j.gastro.2020.02.055] [PMID: 32142773]

[21]    WHO. Updated WHO Advice for International Traffic in Relation to the Outbreak of the Novel Coronavirus 2019-NCoV. 2020. 2020, 1–4.

[22]    Li JO, Lam DSC, Chen Y, Ting DSW. Novel Coronavirus disease 2019 (COVID-19): The importance of recognising possible early ocular manifestation and using protective eyewear. Br J Ophthalmol

2020; 104(3): 297-8.
[http://dx.doi.org/10.1136/bjophthalmol-2020-315994] [PMID: 32086236]

[23] Malhotra N, Gupta N, Ish S, Ish P. COVID-19 in intensive care. Some necessary steps for health care workers. Monaldi Arch Chest Dis 2020; 90(1): 161-2.
[http://dx.doi.org/10.4081/monaldi.2020.1284] [PMID: 32210421]

[24] Barratt R, Shaban RZ, Gilbert GL. Clinician perceptions of respiratory infection risk; a rationale for research into mask use in routine practice. Infect Dis Health 2019; 24(3): 169-76.
[http://dx.doi.org/10.1016/j.idh.2019.01.003] [PMID: 30799181]

[25] Cramer A, Tian E, Yu SH, et al. Disposable N95 Masks Pass Qualitative Fit-Test But Have Decreased Filtration Efficiency after Cobalt-60 Gamma Irradiation. Cold Spring Harbor Laboratory Press 2020.

[26] Yu A-Y, Tu R, Shao X, Pan A, Zhou K, Huang J. A comprehensive Chinese experience against SARS-CoV-2 in ophthalmology. Eye Vis (Lond) 2020; 7: 19.
[http://dx.doi.org/10.1186/s40662-020-00187-2] [PMID: 32289038]

[27] Oh MD. Environmental Contamination and Viral Shedding in MERS Patients. Clin Infect Dis 2016; 62(12): 1615.
[http://dx.doi.org/10.1093/cid/ciw178] [PMID: 27025831]

[28] Shankar A, Saini D, Roy S, et al. Cancer Care Delivery Challenges Amidst Coronavirus Disease - 19 (COVID-19) Outbreak: Specific Precautions for Cancer Patients and Cancer Care Providers to Prevent Spread. Asian Pac J Cancer Prev 2020; 21(3): 569-73.
[http://dx.doi.org/10.31557/APJCP.2020.21.3.569] [PMID: 32212779]

[29] Hughes J P, Rees S S, Kalindjian S B, Philpott K L. Principles of Early Drug Discovery
[http://dx.doi.org/10.1111/j.1476-5381.2010.01127.x]

[30] Kruger A, Gonçalves Maltarollo V, Wrenger C, Kronenberger T. ADME Profiling in Drug Discovery and a New Path Paved on Silica. Drug Discovery and Development - New Advances, Vishwanath Gaitonde, Partha Karmakar and Ashit Trivedi, IntechOpen 2020.
[http://dx.doi.org/10.5772/intechopen.86174]

[31] Lionta E, Spyrou G, Vassilatis DK, Cournia Z. Structure-based virtual screening for drug discovery: principles, applications and recent advances. Curr Top Med Chem 2014; 14(16): 1923-38.
[http://dx.doi.org/10.2174/1568026614666140929124445] [PMID: 25262799]

[32] Maia EHB, Assis LC, de Oliveira TA, da Silva AM, Taranto AG. Structure-Based Virtual Screening: From Classical to Artificial Intelligence. Front Chem 2020; 8: 343.
[http://dx.doi.org/10.3389/fchem.2020.00343] [PMID: 32411671]

[33] de Souza Neto LR, Moreira-Filho JT, Neves BJ, et al. In silico Strategies to Support Fragment-to-Lead Optimization in Drug Discovery. Front Chem 2020; 8: 93.
[http://dx.doi.org/10.3389/fchem.2020.00093] [PMID: 32133344]

[34] Macalino SJY, Billones JB, Organo VG, Carrillo MCO. In Silico Strategies in Tuberculosis Drug Discovery. Molecules 2020; 25(3): 665.
[http://dx.doi.org/10.3390/molecules25030665] [PMID: 32033144]

[35] Sharma A, Jacob A, Tandon M, Kumar D. Orphan drug: Development trends and strategies. J Pharm Bioallied Sci 2010; 2(4): 290-9.
[http://dx.doi.org/10.4103/0975-7406.72128] [PMID: 21180460]

[36] Lipinski CA, Lombardo F, Dominy BW, Feeney PJ. Experimental and computational approaches to estimate solubility and permeability in drug discovery and development settings. Adv Drug Deliv Rev 2001; 46(1-3): 3-26.
[http://dx.doi.org/10.1016/S0169-409X(00)00129-0] [PMID: 11259830]

[37] Khan T, Lawrence AJ, Azad I, Raza S, Joshi S, Khan AR. Computational Drug Designing and Prediction Of Important Parameters Using in silico Methods- A Review. Curr Comput Aided Drug Des 2019; 15(5): 384-97.

[http://dx.doi.org/10.2174/1573399815666190326120006] [PMID: 30914032]

[38] Ghose AK, Viswanadhan VN, Wendoloski JJ. A knowledge-based approach in designing combinatorial or medicinal chemistry libraries for drug discovery. 1. A qualitative and quantitative characterization of known drug databases. J Comb Chem 1999; 1(1): 55-68.
[http://dx.doi.org/10.1021/cc9800071] [PMID: 10746014]

[39] Shaik NA, Hakeem KR, Banaganapalli B, Elango R. 2019.

[40] Veber DF, Johnson SR, Cheng HY, Smith BR, Ward KW, Kopple KD. Molecular properties that influence the oral bioavailability of drug candidates. J Med Chem 2002; 45(12): 2615-23.
[http://dx.doi.org/10.1021/jm020017n] [PMID: 12036371]

[41] Oprea TI. Property distribution of drug-related chemical databases. J Comput Aided Mol Des 2000; 14(3): 251-64.
[http://dx.doi.org/10.1023/A:1008130001697] [PMID: 10756480]

[42] Brown RD, Hassan M, Waldman M. Combinatorial library design for diversity, cost efficiency, and drug-like character. J Mol Graph Model 2000; 18(4-5): 427-437, 537.
[http://dx.doi.org/10.1016/S1093-3263(00)00072-3] [PMID: 11143560]

[43] Muegge I, Heald SL, Brittelli D. Simple selection criteria for drug-like chemical matter. J Med Chem 2001; 44(12): 1841-6.
[http://dx.doi.org/10.1021/jm015507e] [PMID: 11384230]

[44] Congreve M, Carr R, Murray C, Jhoti H. A 'rule of three' for fragment-based lead discovery? Drug Discov Today 2003; 8(19): 876-7.
[http://dx.doi.org/10.1016/S1359-6446(03)02831-9] [PMID: 14554012]

[45] Bickerton GR, Paolini GV, Besnard J, Muresan S, Hopkins AL. Quantifying the chemical beauty of drugs. Nat Chem 2012; 4(2): 90-8.
[http://dx.doi.org/10.1038/nchem.1243] [PMID: 22270643]

[46] Hodos RA, Kidd BA, Shameer K, Readhead BP, Dudley JT. In silico methods for drug repurposing and pharmacology. Wiley Interdiscip Rev Syst Biol Med 2016; 8(3): 186-210.
[http://dx.doi.org/10.1002/wsbm.1337] [PMID: 27080087]

[47] Rodriguez-Monguio R, Spargo T, Seoane-Vazquez E. Ethical imperatives of timely access to orphan drugs: is possible to reconcile economic incentives and patients' health needs? Orphanet J Rare Dis 2017; 12(1): 1.
[http://dx.doi.org/10.1186/s13023-016-0551-7] [PMID: 28057032]

[48] Mucke HAM. Drug Repositioning in the Mirror of Patenting: Surveying and Mining Uncharted Territory. Front Pharmacol 2017; 8: 927.
[http://dx.doi.org/10.3389/fphar.2017.00927] [PMID: 29326592]

[49] Andrade EL, Bento AF, Cavalli J, *et al.* Non-clinical studies in the process of new drug development - Part II: Good laboratory practice, metabolism, pharmacokinetics, safety and dose translation to clinical studies. Braz J Med Biol Res 2016; 49(12)e5646
[http://dx.doi.org/10.1590/1414-431x20165646] [PMID: 27982281]

[50] Fogel DB. Factors associated with clinical trials that fail and opportunities for improving the likelihood of success: A review. Contemp Clin Trials Commun 2018; 11: 156-64.
[http://dx.doi.org/10.1016/j.conctc.2018.08.001] [PMID: 30112460]

[51] Schenone M, Dančík V, Wagner BK, Clemons PA. Target identification and mechanism of action in chemical biology and drug discovery. Nat Chem Biol 2013; 9(4): 232-40.
[http://dx.doi.org/10.1038/nchembio.1199] [PMID: 23508189]

[52] Sriram K, Insel PAG. G Protein-Coupled Receptors as Targets for Approved Drugs: How Many Targets and How Many Drugs? Mol Pharmacol 2018; 93(4): 251-8.
[http://dx.doi.org/10.1124/mol.117.111062] [PMID: 29298813]

[53] Santos R, Ursu O, Gaulton A, *et al.* A comprehensive map of molecular drug targets. Nat Rev Drug Discov 2017; 16(1): 19-34.
[http://dx.doi.org/10.1038/nrd.2016.230] [PMID: 27910877]

[54] Ofran Y, Rost B. Analysing six types of protein-protein interfaces. J Mol Biol 2003; 325(2): 377-87.
[http://dx.doi.org/10.1016/S0022-2836(02)01223-8] [PMID: 12488102]

[55] Meza Menchaca T, Juárez-Portilla C, Zepeda C. 2020.

[56] Salmaso V, Moro S. Bridging Molecular Docking to Molecular Dynamics in Exploring Ligand-Protein Recognition Process: An Overview. Front Pharmacol 2018; 9: 923.
[http://dx.doi.org/10.3389/fphar.2018.00923] [PMID: 30186166]

[57] Mosca R, Pons C, Fernández-Recio J, Aloy P. Pushing structural information into the yeast interactome by high-throughput protein docking experiments. PLOS Comput Biol 2009; 5(8)e1000490
[http://dx.doi.org/10.1371/journal.pcbi.1000490] [PMID: 19714207]

[58] Feixas F, Lindert S, Sinko W, McCammon JA. Exploring the role of receptor flexibility in structure-based drug discovery. Biophys Chem 2014; 186: 31-45.
[http://dx.doi.org/10.1016/j.bpc.2013.10.007] [PMID: 24332165]

[59] Dar AM, Mir S. Molecular Docking: Approaches, Types, Applications and Basic Challenges. J Anal Bioanal Tech 2017; 8: 1-3.
[http://dx.doi.org/10.4172/2155-9872.1000356]

[60] Sethi A, Joshi K, Sasikala K, Alvala M. 2020.

[61] Forli S, Huey R, Pique ME, Sanner MF, Goodsell DS, Olson AJ. Computational protein-ligand docking and virtual drug screening with the AutoDock suite. Nat Protoc 2016; 11(5): 905-19.
[http://dx.doi.org/10.1038/nprot.2016.051] [PMID: 27077332]

[62] Zimmermann KA. https://www.livescience.com/26579-immune-system.html

[63] Newton AH, Cardani A, Braciale TJ. The host immune response in respiratory virus infection: balancing virus clearance and immunopathology. Semin Immunopathol 2016; 38(4): 471-82.
[http://dx.doi.org/10.1007/s00281-016-0558-0] [PMID: 26965109]

[64] Rouse BT, Sehrawat S. Immunity and immunopathology to viruses: what decides the outcome? Nat Rev Immunol 2010; 10(7): 514-26.
[http://dx.doi.org/10.1038/nri2802] [PMID: 20577268]

[65] Sentinel A. Clin Chim Acta 2020; 508: 122-9. Sun, D. wei; Zhang, D.; Tian, R. hui; Li, Y.; Wang, Y. shi; Cao, J.; Tang, Y.; Zhang, N.; Zan, T.; Gao, L.; *et al.* The Underlying Changes and Predicting Role of Peripheral Blood Inflammatory Cells in Severe COVID-19 Patients:
[http://dx.doi.org/10.1016/j.cca.2020.05.027] [PMID: 32417210]

[66] Chaplin DD. Overview of the immune response. J Allergy Clin Immunol 2010; 125(2) (Suppl. 2): S3-S23.
[http://dx.doi.org/10.1016/j.jaci.2009.12.980] [PMID: 20176265]

[67] Amit Koparde A, Chandrashekar Doijad R, Shripal Magdum C. Natural Products in Drug Discovery. 2019.
[http://dx.doi.org/10.5772/intechopen.82860]

[68] Atanasov AG, Waltenberger B, Pferschy-Wenzig EM, *et al.* Discovery and resupply of pharmacologically active plant-derived natural products: A review. Biotechnol Adv 2015; 33(8): 1582-614.
[http://dx.doi.org/10.1016/j.biotechadv.2015.08.001] [PMID: 26281720]

[69] Funk JL, Frye JB, Oyarzo JN, Timmermann BN. Comparative effects of two gingerol-containing Zingiber officinale extracts on experimental rheumatoid arthritis. J Nat Prod 2009; 72(3): 403-7.
[http://dx.doi.org/10.1021/np8006183] [PMID: 19216559]

[70] Lu J, Guan S, Shen X, *et al.* Immunosuppressive activity of 8-gingerol on immune responses in mice. Molecules 2011; 16(3): 2636-45.
[http://dx.doi.org/10.3390/molecules16032636] [PMID: 21441866]

[71] Semwal RB, Semwal DK, Combrinck S, Viljoen AM. Gingerols and shogaols: Important nutraceutical principles from ginger. Phytochemistry 2015; 117: 554-68.
[http://dx.doi.org/10.1016/j.phytochem.2015.07.012] [PMID: 26228533]

[72] Bayan L, Koulivand PH, Gorji A. Garlic: a review of potential therapeutic effects. Avicenna J Phytomed 2014; 4(1): 1-14.
[PMID: 25050296]

[73] Liang D, Qin Y, Zhao W, *et al.* S-allylmercaptocysteine effectively inhibits the proliferation of colorectal cancer cells under *in vitro* and *in vivo* conditions. Cancer Lett 2011; 310(1): 69-76.
[http://dx.doi.org/10.1016/j.canlet.2011.06.019] [PMID: 21794975]

[74] Chacko SM, Thambi PT, Kuttan R, Nishigaki I. Beneficial effects of green tea: a literature review. Chin Med 2010; 5: 13.
[http://dx.doi.org/10.1186/1749-8546-5-13] [PMID: 20370896]

[75] Yan Z, Zhong Y, Duan Y, Chen Q, Li F. Antioxidant mechanism of tea polyphenols and its impact on health benefits. Anim Nutr 2020; 6(2): 115-23.
[http://dx.doi.org/10.1016/j.aninu.2020.01.001] [PMID: 32542190]

[76] Min SY, Yan M, Kim SB, *et al.* Green Tea Epigallocatechin-3-Gallate Suppresses Autoimmune Arthritis Through Indoleamine-2,3-Dioxygenase Expressing Dendritic Cells and the Nuclear Factor, Erythroid 2-Like 2 Antioxidant Pathway. J Inflamm (Lond) 2015; 12: 53.
[http://dx.doi.org/10.1186/s12950-015-0097-9] [PMID: 26379475]

[77] Hudson J, Vimalanathan S. Echinacea-A Source of Potent Antivirals for Respiratory Virus Infections. Pharmaceuticals 2011; 4: 1019-31.
[http://dx.doi.org/10.3390/ph4071019]

[78] Zhai Z, Liu Y, Wu L, *et al.* Enhancement of innate and adaptive immune functions by multiple Echinacea species. J Med Food 2007; 10(3): 423-34.
[http://dx.doi.org/10.1089/jmf.2006.257] [PMID: 17887935]

[79] Lv X, Zhao S, Ning Z, *et al.* Citrus fruits as a treasure trove of active natural metabolites that potentially provide benefits for human health. Chem Cent J 2015; 9: 68.
[http://dx.doi.org/10.1186/s13065-015-0145-9] [PMID: 26705419]

[80] Turner T, Burri B. Potential Nutritional Benefits of Current Citrus Consumption. Agriculture 2013; 3: 170-87.
[http://dx.doi.org/10.3390/agriculture3010170]

[81] Mahmoud AM, Hernández Bautista RJ, Sandhu MA, Hussein OE. Beneficial Effects of Citrus Flavonoids on Cardiovascular and Metabolic Health. Oxid Med Cell Longev 2019; 20195484138
[http://dx.doi.org/10.1155/2019/5484138] [PMID: 30962863]

[82] Trott O, Olson AJ. AutoDock Vina: improving the speed and accuracy of docking with a new scoring function, efficient optimization, and multithreading. J Comput Chem 2010; 31(2): 455-61.
[PMID: 19499576]

[83] Hsu KC, Chen YF, Lin SR, Yang JM. iGEMDOCK: a graphical environment of enhancing GEMDOCK using pharmacological interactions and post-screening analysis. BMC Bioinformatics 2011; 12 (Suppl. 1): S33.
[http://dx.doi.org/10.1186/1471-2105-12-S1-S33] [PMID: 21342564]

[84] Azad I, Nasibullah M, Khan T, Hassan F, Akhter Y. Exploring the novel heterocyclic derivatives as lead molecules for design and development of potent anticancer agents. J Mol Graph Model 2018; 81: 211-28.
[http://dx.doi.org/10.1016/j.jmgm.2018.02.013] [PMID: 29609141]

[85] Sieber P, Platzer M, Schuster S. The Definition of Open Reading Frame Revisited. Trends Genet 2018; 34(3): 167-70.
[http://dx.doi.org/10.1016/j.tig.2017.12.009] [PMID: 29366605]

[86] Davis DA, Soule EE, Davidoff KS, Daniels SI, Naiman NE, Yarchoan R. Activity of human immunodeficiency virus type 1 protease inhibitors against the initial autocleavage in Gag-Pol polyprotein processing. Antimicrob Agents Chemother 2012; 56(7): 3620-8.
[http://dx.doi.org/10.1128/AAC.00055-12] [PMID: 22508308]

[87] Belouzard S, Millet JK, Licitra BN, Whittaker GR. Mechanisms of coronavirus cell entry mediated by the viral spike protein. Viruses 2012; 4(6): 1011-33.
[http://dx.doi.org/10.3390/v4061011] [PMID: 22816037]

[88] Du L, He Y, Zhou Y, Liu S, Zheng BJ, Jiang S. The spike protein of SARS-CoV--a target for vaccine and therapeutic development. Nat Rev Microbiol 2009; 7(3): 226-36.
[http://dx.doi.org/10.1038/nrmicro2090] [PMID: 19198616]

[89] McBride R, van Zyl M, Fielding BC. The coronavirus nucleocapsid is a multifunctional protein. Viruses 2014; 6(8): 2991-3018.
[http://dx.doi.org/10.3390/v6082991] [PMID: 25105276]

[90] Satarker S, Nampoothiri M. Structural Proteins in Severe Acute Respiratory Syndrome Coronavirus-2. Arch Med Res 2020; 51(6): 482-91.
[http://dx.doi.org/10.1016/j.arcmed.2020.05.012] [PMID: 32493627]

[91] Schoeman D, Fielding BC. Coronavirus envelope protein: current knowledge. Virol J 2019; 16(1): 69.
[http://dx.doi.org/10.1186/s12985-019-1182-0] [PMID: 31133031]

[92] de Haan CAM, Rottier PJM. Molecular interactions in the assembly of coronaviruses. Adv Virus Res 2005; 64: 165-230.
[http://dx.doi.org/10.1016/S0065-3527(05)64006-7] [PMID: 16139595]

[93] Prajapat M, Sarma P, Shekhar N, et al. Drug targets for corona virus: A systematic review. Indian J Pharmacol 2020; 52(1): 56-65.
[http://dx.doi.org/10.4103/ijp.IJP_115_20] [PMID: 32201449]

[94] Arndt AL, Larson BJ, Hogue BG. A conserved domain in the coronavirus membrane protein tail is important for virus assembly. J Virol 2010; 84(21): 11418-28.
[http://dx.doi.org/10.1128/JVI.01131-10] [PMID: 20719948]

[95] Datta A, Brosh RM Jr. New Insights Into DNA Helicases as Druggable Targets for Cancer Therapy. Front Mol Biosci 2018; 5: 59.
[http://dx.doi.org/10.3389/fmolb.2018.00059] [PMID: 29998112]

[96] Shadrick WR, Ndjomou J, Kolli R, Mukherjee S, Hanson AM, Frick DN. Discovering new medicines targeting helicases: challenges and recent progress. J Biomol Screen 2013; 18(7): 761-81.
[http://dx.doi.org/10.1177/1087057113482586] [PMID: 23536547]

# SUBJECT INDEX

## A

Acid(s) 20, 23, 62, 65, 70, 117, 146 173, 185, 189, 190, 191 192, 238, 253, 254, 259
   Arachidonic 23
   bile 20
   carboxylic 70
   chicoric 253
   Chloroauric 223
   citric 238, 254, 259
   fatty 20
   formic 185, 189, 190, 191, 192
   lactic 117
   Lewis 115
   monomethyl arsonic 65
   mycophenolic 173
   nucleic 62
   oxalinic 146
   rosmarinic 253
Activity 3, 13, 18, 27, 40, 60, 62, 66, 85, 90, 142, 154, 198, 202, 203, 207, 229, 230, 232, 242, 243, 252
   anticancer 252
   antimicrobial 154, 229, 230, 232
   biological 60, 62, 243
   bronchodilator 40
   catalytic 90
   enzymatic 66
   mining 85
   synergistic 40
Acute respiratory distress syndrome (ARDS) 249
Advanced oxidation processes (AOPs) 141, 151, 152
Agents 61, 66, 146, 170, 220, 229, 238
   antiviral 238
   chelating 61, 66, 170
   prophylactic 146
   therapeutic 220, 229
Agricultural industry 109
Air pollutants 3, 4, 6, 10, 13, 18, 23, 24, 26, 33
   harmful 13
Air quality index (AQI) 6, 9, 19
Airway 38, 44
   remodelling process 38
   wall mechanics 44
Allergic disorders 10
*Allium sativum* 250, 252
Antibiotic 148, 150, 151, 154, 155, 158
   adjuvants 154, 155, 158
   contamination 148, 150, 151
Antibiotic pollution 148, 149, 150, 167
   effects of 148, 149, 150
   increasing 167
Antibiotic-resistant bacterium 150
Antibiotics 141, 142, 144, 145, 146, 147, 148, 149, 150, 151, 154, 155, 156, 157, 158, 159, 160, 167, 191
   beta-lactam 155
   hybrid 160
   prophylactic 150
Antibodies 42, 230, 231, 249
   therapeutic 42
Anti-inflammatory mediators 37
Antioxidant activity 254
Anti-pseudomonal beta-lactams 150
Aromatic bonds count (ABC) 245
Arsenical dermatitis 65
Arsenic 65, 73
   decontamination efficiency 73
   induced cytotoxicity 65
Arsenicosis 66
Arsenic poisoning 64, 66, 68, 69, 70, 71
   treatment of 66, 68
Artificial distillation systems 102
Aryl hydrocarbon hydroxylase (AHH) 27
Asthma 1, 9, 10, 13, 18, 20, 32, 33, 35, 36, 38, 43, 48
Atherosclerosis 20, 22, 23
Atomic 61, 73, 220
   absorption spectrophotometer (AAS) 61, 73
   force microscopy 220

*Subject Index*

# B

Bacteria 20, 36, 68, 85, 117, 142, 143, 144, 145, 146, 149, 153, 154, 155, 166, 167
 gut 146
 oral plaque 117
 pathogenic 68
 reduced antibiotic-resistant 153
 resistant 149
Bacterial 144, 145, 146
  cell wall synthesis 144
  diseases 146
  protein synthesis 145
Biosynthesis 62, 142, 143, 159, 160, 221, 222
  bacterial cell wall 142, 143, 159
  catecholamine 62
  fatty acid 160
Blood 22, 100
  cholesterol 22
  related diseases 100
Bronchiolitis 1
Bronchitis 8, 10, 24
  chronic 24
Bronchodilator effect, prolonged 39
Bronchodilators 39, 41
  and combinations of bronchodilators 39
Bronchopneumonia 20

# C

Cancer 10, 11, 20, 58, 64, 65, 66, 167, 208, 230, 231
 breast 167, 231
 prostate 167
Capillary gel electrophoresis (CGE) 118
Carcinogenesis 62, 65, 66
Cardiovascular 1 8, 20, 23, 64, 230
  diseases 1, 8, 20, 64, 230
  homeostasis 23
Cellulose 89, 99, 101, 102, 169
 acetate (CA) 89, 99, 101, 102, 169
 acetate membranes 101, 169
 triacetate (CTA) 101
Cerebrovascular diseases 20

Chelation therapy 61, 66, 68, 70
Chromatographic techniques 168, 183, 188, 193
Chromatography 118, 187
 chelation-ion 118
 ion-exchange 118
 ion-pair 118
 micellar electrokinetic 118
Colourimetric sensors signal 120
Commercial water treatment industry 87
Comprehensive medicinal chemistry (CMC) 244, 245
Computational 42, 43, 45
 fluid dynamics 45
 learning paradigms 43
 lung modelling 43
 medicinal simulation 43
 modelling simulation 42
Computer-aided drug design (CADD) 199, 242
Contaminants 62, 99, 108, 177, 232
 heavy metal 62
 inorganic 108
Contamination 64, 93, 101, 108, 167
 bacterial 108
 natural 64
Copolymer, polyether-based 90
Coronary arteries 22
 heart's 22
Coronary 11, 22
 arteriosclerosis 11
 artery disease 22
Coronaviruses 238, 241
Corrosion 97, 105
Corticosteroids, systemic 42
Coughing 8, 10, 20, 24, 27, 28
Cystic fibrosis 117, 134, 211
Cytochrome oxidase 117
Cytokines 38, 250, 252
 immunoregulatory 252
 proinflammatory 250
Cytokinesis 255

## D

Dental caries 116
Desalination 104, 105, 110
   methods 110
   processes 104, 105, 110
Devices, sensory 115
Diabetes mellitus 230
Diagnosis of asthma and COPD 48
Diarrhoea 66, 71, 220, 223, 240
Disease progression 42, 43
Disorders, renal 41
Dispersive liquid-liquid micro-extraction (DLLME) 168, 174, 175, 188
Dispersive solid-phase extraction (DSPE) 182, 183
DLLME technique 174
DNA 26, 62, 116, 118, 145, 198, 231, 255
   and nuclear proteins 62
   gyrases 145
   synthesis 145
   transcription 145
   vaccine 255
DNA damage 7, 62
   interactions cause 62
Docking 249, 258
   analysis 258
   based algorithms 249
Drug 25, 28, 246
   metabolizing enzymes 25, 28
   repurposing methods 246
Dyspnea 8, 45
Dysregulation, immunologic 26

## E

Eco-hazardous wastes 198, 199
Eczema 26
Efflux 156
   mechanism 156
   pumps 156
Electrodialysis 88, 104, 108
Electrodynamic forces 247
Electron capture detectors (ECD) 168, 185

Electronic energy transfer (EETs) 23, 24, 124
Electrophoretic mobility 118
Electrostatic interactions 115, 121, 122, 131, 231, 247
Endothelial dysfunction 23
Energy 13, 25, 28, 67, 97, 102, 104, 106, 107, 109, 110, 229, 258, 259
   consumption 106
   electrostatic 259
   hydrogen bonding 259
   metabolism 28
   production 97
   renewable 13, 110
   solar 102, 110
Environmental 4, 66
   protection agency (EPA) 66
   tobacco smoking (ETS) 4
Environment and public health organization 68
Enzymatic reactions 69
Enzymes 18, 20, 21, 25, 28, 37, 65, 66, 69, 98, 128, 144, 155, 159, 255, 256
   aminoglycoside-modifying 159
   angiotensin-converting 255
   antioxidant 37
   antiprotease 37
   beta-lactamase 155
   helicase 256
   metabolic 65
   monooxygenase 20
   transpeptidase 144
Estrogen receptors (ERs) 25, 26
European pharmaceutical review (EPR) 167
Evaporative light scattering detector (ELSD) 186
Extraction, microwave-assisted solvent 182
Extraction technique 174, 178
   solid phase 178
   ultrasonic 178

## F

Filtration processes 96, 98, 99, 101
Flame 168, 185

ionisation detector (FID) 168, 185
  thermionic detectors (FTD) 185
Food and drug administration (FDA) 63, 69, 246
Fragment-based drug design (FBDD) 245

# G

Gas chromatography (GC) 166, 167, 168, 171, 179, 181, 183, 184, 185, 186, 188, 191, 193
Gaseous contaminants 5
Gases 5, 8, 19, 23, 24, 33, 89, 186, 258
  harmful 8
  industrial-suitable 89
  irritating 24
  toxic 19
Gastrointestinal disorders 66, 220
Gold nanoparticles synthesis 230
G-protein coupled receptors (GPCRs) 247
Greenhouse gas effect 5
Green tea 250, 252
  active ingredient of 252
Growth 3, 7, 38, 71, 143, 144, 146, 153, 157, 169, 170, 220
  bacterial 38, 144, 169, 170
  breast 71
  cancerous 220

# H

Haemoglobin formation 62
Hashimoto disease 115
Hazards 33, 36, 117
  induced environmental 117
  occupational 33, 36
Headaches 8, 10, 220
Health 1, 2, 8, 9, 10, 12, 116, 117, 149, 166, 167, 206, 208, 211, 212
  atmospheric 212
  environmental 206
Heart 4, 9, 10, 18, 20, 22, 23
  attacks 10, 22
  diseases 4, 9, 18, 23

failure 20
Hemodialysis 88, 100
Hemoprotein 20
High-performance liquid chromatography (HPLC) 173, 179, 183, 186, 190
High-resolution mass spectrometry (HRMS) 186
High throughput screening (HTS) 243
Hormones, steroid 20
Human 6, 107, 215
  immunodeficiency virus (HIV) 207, 215
  respiratory system 6
Hypercholesterolemia 22
Hyperkeratosis 65

# I

Imaging techniques 43
Immune response 250
  innate 250
Immune system 65, 71, 249, 253
  compromised 65
Immunohistochemical staining 26
Immunological diseases 8
Impact on respiratory health 3, 5, 7, 9, 11, 13
Infections 27, 28, 146, 149, 150, 153, 155, 232, 239, 240, 249, 250, 254
  drug-resistant bacterial 149
  intestinal 240
  malaria 28
  pseudomonas 150
  urinary tract 146
Inflammation 8, 23, 24, 34, 36, 38, 41, 45
  respiratory 8
Inflammatory 26, 230
  diseases 230
  skin disease 26
Inhaled corticosteroids (ICs) 38, 39, 40, 41
Inhibition 145
  of bacterial protein synthesis 145
  of nucleic acid synthesis 145
Insecticides 84, 86, 198, 199, 200, 204, 205, 213
  agricultural 199

pyrethroid ether 213
synthetic 199
International agency for research on cancer 11
Ion 61, 118, 119, 134
   chromatography 118, 134
   exchange chromatography (IEC) 118
   selective electrodes (ISEs) 61, 119
Iron coated 73
   charcoal treatment 73
   coarse sand treatment 73
Ischemia-reperfusion injury (IRI) 24

## J

Japan international cooperation agency (JICA) 67

## K

*Klebsiella pneumonia* 229
Kreb's cycle 116

## L

LC-MS method 187
Lethal genetic disease 117
Ligand-based virtual screening (LBVS) 242
Lipinski's 238, 243, 254, 259
   parameters 259
   rules 238, 243, 254
Liquid chromatography (LC) 166, 167, 168, 173, 179, 181, 186, 188, 190, 192, 193
   high-performance 179
Liquid-liquid extraction techniques 182
Lung 1, 4, 9, 8, 11, 13, 18, 43
   cancer 1, 4, 9, 11, 18
   disease 8, 9, 13, 43

## M

Machine learning methods 45
Macrolides erythromycin 183

Mass spectroscopy (MS) 166, 167, 173, 183, 185, 186, 188, 193
Mechanical vapor compression (MVC) 104, 107
Mechanisms 23, 32, 36, 38, 39, 43, 62, 115, 118, 155, 158, 159, 181, 186, 247
   airflow 32
   biomedical 36
   of cardiotoxicity of air pollutants 23
   toxicological 62
Membrane-protected system 179
Membranes technology 82
Mental disorder 117
Meso-porous Membranes 91
Metal 59, 65, 66, 70, 89
   organic framework (MOF) 89
   toxicity 59, 65, 66, 70
Methods 89
   pyrolysis 89
   sol-gel 89
Methylate arsenic trioxide 65
Microbes 27, 141, 148, 149, 150, 151, 154, 156
   antibiotic-resistant 149, 154
   drug-resistant 148
Microporous silica membrane 96
Microwave-assisted solvent extraction (MASE) 182, 183
Middle east respiratory syndrome 238
Modelling 42, 46, 49, 249
   computational 42, 46, 49
   mathematical 49
   mechanism-based 49
   tried homology 249
Molecular docking studies 249, 254
Myocardial infarction (MI) 11, 20, 22

## N

National clean air programme (NCAP) 1, 12
Natural killer (NK) 250
Neutrophil elastase 37
*Nigella sativa* 250, 253
Nitrogen oxides (NOx) 1, 5, 8, 19, 23, 26

## Subject Index

Non-nucleoside reverse transcriptase inhibitors (NNRTIs) 207, 215
Nucleic acid synthesis 145
Nucleophilicity 131
Nucleoproteins 238
Nutritional malfunctioning 36

## O

Optical 120, 133
  chemosensors for anions 120
  sensing by molecular assemblies 133
Ordinary differential equations (ODE) 45
ORFs, single large 255
Organ systems, cardiovascular 18
Osmosis 104
Osteoporosis 116
Oxidative stress 23, 24, 26, 36, 37, 62, 65
  systemic 23

## P

Paper Industry 98
Paracetamol 173, 180, 192
Peptidoglycan precursors 145
Peroxy-acyl nitrates (PANs) 19, 89, 99
Pesticides 4, 84, 86, 98, 199
Phagocytosis 253
Pharmacokinetics activity 244
Physicochemical methods, traditional 148
Plants 2, 8, 13, 20, 61, 67, 84, 93, 146, 148, 149, 173, 181, 221, 222, 223, 229, 230, 250
  food-producing 149
  industrial processing 93
  mediated biological process 221
  medicinal 250
  nuclear power 2
  thermal power 84
  wastewater treatment 148, 173, 181
PLE technique 183
Pollutants 1, 2, 3, 5, 7, 8, 21, 24, 27, 29, 33, 34, 36, 84, 85, 86, 209
  environmental 21
  gaseous 34
  hazardous 209
  transportation-related 8
Pollution 1, 5, 6, 11, 18, 32, 63, 67, 85, 86, 141, 150, 160
  agricultural 63
  oil 86
  traffic-related 11
Pressurized liquid extraction (PLE) 168, 182, 188, 192
Principal component analysis (PCA) 45
Process 68, 84, 85, 86, 87, 88, 90, 91, 92, 93, 94, 97, 99, 101, 102, 104, 105, 108, 152, 174
  agricultural 84, 85
  centrifugation 174
  filtering 99
  metallurgical 97
  traditional 68
Processing 93, 94
  heavy chemical 94
  juice 94
  osmosis 93
Production 24, 38, 62, 63, 65, 98, 99, 116, 117, 215, 221, 226, 249
  cytokine 38
  glass fibre 98
  industrial 63
Prostate biopsy 150
Protease 37, 238, 255
  antiprotease 42
  antiprotease imbalance 37
Protein(s) 20, 26, 28, 98, 100, 142, 143, 144, 239, 247, 249, 254, 255, 256, 259, 261
  data bank 254, 257, 261
  expression 26
  flexibility 249
  nucleocapsid 255
  small proline-rich 26
  stress-related 28
  viral 259
PubChem database 238
Public health 64, 68
  crisis 64
  Organization 68

Pulmonary oedema 8

# R

Radiation 36, 124, 183
  microwave 183
  solar 36
Radioactive 58, 86
  metals 58
  waste 86
Raw sewage disposal 86
Reactive oxygen species (ROS) 24, 25, 26, 36, 37, 62
Receptor 25, 26, 115, 120, 121, 124, 125, 129, 131, 134, 200, 247, 248, 249, 255, 256, 257, 258
  anion-selective 115
  estrogen 26
  protein 248, 249, 256, 257, 258
  ryanodine 200
  synthetic 121
Recycling 83, 96, 97, 98, 111
  industrial 96
  promoting wastewater 111
Regulation, immune 208
Release 37, 115, 116, 198, 199, 207, 209, 213, 216, 231, 239, 256
  degraded chemical 209
  inflammatory mediator 37
Residue oil fly ash (ROFA) 37
Resonance 224, 232
  distinctive surface plasmon 224
  surface plasmon 232
Respiratory 8, 9, 10, 11, 18, 19, 20, 33, 241
  diseases 8, 10
  distress 19, 241
  infections 9, 11, 18
  toxicants 33
  tract infections 20
Reverse osmosis (RO) 93, 99, 101, 104, 108
Rheological properties 119, 120
Rheumatoid arthritis 22
RNA polymerase 238

# S

SARS-CoV 238, 239, 250, 256
  helicase enzyme 256
SARS-CoV-2 238
  novel coronavirus 238
  replicase polyprotein 238
Scanning electron microscopy 220, 222
  field emission 220
Severe acute respiratory syndrome 238
Sewage 63, 85, 96, 99, 147, 149, 187, 189, 190, 193, 240
  municipal 96
  plants 99
  sludge 149, 189, 190
  system 63
Silver nanoparticle formation 226, 229
Simulated medicinal modeling 42
Single inhaler triple therapy 41
Skin 8, 10, 20, 26, 27, 42, 58, 64, 65, 220, 241
  barrier defects 26
  bruising 42
  diseases 20
  eruptions 220
  irritation 241
  lesions 64
  sclerosis 58
Smoke 2, 4, 5, 24, 27, 33, 35, 47
  cigarette 4, 24
  induced clinical phenotype 47
  tobacco 2, 5
Smoking by-products 24
Soil sediments 84
Solar heating 105
Solid-phase 168, 170, 173, 176, 177, 178, 179, 180, 181, 183, 188, 189, 190, 192
  extraction (SPE) 168, 170, 173, 176, 177, 178, 183, 188, 189, 190, 192
  micro-extraction (SPME) 168, 179, 180, 181, 188
Sources 3, 63
  of air pollution 3
  of arsenic 63
Spike protein, transmembrane 255

## Subject Index

Spiral-wound membranes 94, 95
Stability 82, 96, 97, 100, 169, 170, 171, 248, 250
   mechanical 82, 96, 101
   thermal 90, 96, 100
Sterilization 98
Steroidal hormones 166
Steroidogenesis 166
Stir bar sorptive extraction (SBSE) 168, 181, 182, 188
Stomach diseases 220
Streptomycin 142, 146
Structure-based virtual screening (SBVS) 242
Sulphur chemiluminescence detectors (SCD) 185
Supercritical fluid extraction (SFE) 182
Surface 224, 229, 232
   enhanced resonance Raman spectroscopy 229
   plasmon resonance (SPR) 224, 232

## T

Target proteins polyproteins 255
Techniques 28, 46, 61, 68, 110, 118, 120, 148, 160, 166, 168, 169, 171, 174, 176, 179, 182, 183, 189, 220, 223, 246, 146
   analytical 166, 183
   computational 246
   developed physicochemical 148
   economical 160
   gel electrophoresis 28
   machine learning 46
   membrane desalination 110
   metal determination 61
   sand filtration 68
Technologies 13, 66, 67, 68, 76, 82, 87, 90, 91, 103, 109, 221
   chemical engineering 103
   environmental 90
   evaporation 87
   green 109
   membrane separation 82
   promoting solar energy 13

tried-and-tested 109
Thermal 104, 106, 185
   conductivity detector (TCD) 185
   vapor compression (TVC) 104, 106
Tobacco 4, 33, 35
   burning 4
   consumption 35
Toxicity 42, 59, 60, 61, 62, 63, 65, 198, 199, 214, 217, 242
   chemical 59
   corticosteroid 42
   heavy metal-induced 62
   radiological 59
Toxic metals 61, 70, 71
Toxicological profile 213, 216, 217
Traditional water resources 86
Transmission electron microscopy 220
Tubular membrane filtration processes 95

## U

Ultraviolet-visible spectroscopy 222
UV-visible 119, 220, 223, 228
   absorption spectrum 228
   spectrophotometer 119
   spectroscopy 220, 223

## V

Vacuum UV (VUV) 152
Vanadium support film 91
Vancomycin-resistant Enterococci 149
Vascular disease 22
Veber's rules 244
Volatile organic compounds (VOCs) 19
Vomiting 8, 66, 70, 71

## W

Waals 199, 231, 247, 259, 260
   energy 259
   forces 199, 231, 260
   interactions 247, 260

Waste 64, 84, 97, 96, 107, 151, 166, 167, 198, 199, 206
  disposal 64, 97
  domestic 166
  ecotoxic 198
  industrial 84
  metabolic 206
  metabolized 206
  pharmaceutical 167
Wastewater 82, 86, 87, 92, 94, 96, 98, 101, 102, 148, 160, 186, 187, 189, 190, 192, 240, 241
  filtering 94
  management 92
  recycled 102
  resources 240
  treatment 94, 98, 99, 101, 160
Water 67, 68, 76, 83, 84, 86, 87, 90, 94, 95, 96, 97, 102, 103, 106, 110, 147
  arsenic-free 67, 76
  contaminated 94
  filtration 90
  polluted 95
  pollution 84, 86, 147
  purification of 68, 97, 110
  quality of 68, 84, 87
  recycled 96
  saline 102, 103, 106
  salty 83
  treated 87
Water treatment 83, 86, 102
  by desalination 102
  plant 86
  strategies 83
Whey protein concentrations 95
World health organization (WHO) 1, 2, 5, 12, 58, 64, 68, 69, 152, 239, 241

# X

Xenobiotic(s) 18, 20, 25, 26, 27, 28, 167
  harmful 167
  metabolism 25
  metabolizing enzymes 28

X-ray 61, 119, 257
  crystal structures 257
  diffractometer 119
  fluorescence (XRF) 61

# Z

*Zingiber officinale* 250

www.ingramcontent.com/pod-product-compliance
Lightning Source LLC
Chambersburg PA
CBHW051143220526
45473CB00003B/645